哥特建筑与雕塑装饰艺术

第3卷

曹峻川　甄影博　编译
[英] 詹姆斯 · 凯拉韦 · 科林　绘

江苏凤凰科学技术出版社

前言

公元第二个千年开端不久，在诺曼王朝即将上台的前夕，英国的宗教建筑逐渐转变为后来在诺曼王朝中的独特样式，即我们现在所定义的“盎格鲁－诺曼”风格。这种风格最初由“忏悔者”爱德华介绍到英国，或者说是由卡纽特大帝，然后通过他们应用于自己领地中大量的教堂建设中。可以说，在诺曼王朝统治下迅速发展的教堂建筑，实际上在王朝来临之前就已经建立了相当完善的体系。盎格鲁－诺曼建筑的建筑师们竭尽所能，使其风格更加完美，从现存的一些建筑便可见一斑。尽管它们发展得如此之好，但建筑如同陆地上其他事物一样不能长久稳定。如我们所见，一种建筑形式或样式，一旦到达其成熟期就会或多或少被其他的建筑形式和样式所取代，这是一种潜在的规则。但是全然不顾这种潜在规则的影响，盎格鲁－诺曼王朝保留了大量古代传统建筑来宣称其永久的建立。低矮、笨重的比例，沉重且自承重的墙体，矩形的叠内拱，方形的柱顶板及柱础，以及严格说来肤浅的装饰——所有这些似乎都在诉说着一种从罗马退化到古罗马式的“宏伟”样式，而非真正从自身土壤中发展起来的伟大样式——中世纪的建筑准备用古代建筑的标准来衡量自身的力量。同时，在盎格鲁－诺曼时期，基督教迅速发展，大量的教堂建筑欠缺，仅有一些巴西利卡式的建筑——它们在不久后就由于异教徒的起源而被移除，尽管巴西利卡自身不是异教的。因此，在回顾盎格鲁－诺曼建筑的最终结束和及盎格鲁－哥特建筑建筑样式完全建立之间这段时期时，人们更多沉浸在过去的建筑样式中而不是追寻一种卓越的替代者。经历这段关于建筑样式的挣扎之后，新建筑样式的基本元素开始与旧的建筑特征相融合，哥特建筑逐渐获得了一种特定的形式，它更加明亮、深邃宽敞及高贵，相比于早期的英国样式，随即便展现出其优越性。

《哥特建筑手册》的作者曾评述到：“这种样式如此优美，自身十分完美，或许可以想象在任何的历史时期或者地方能有与其并驾齐驱的建筑样式，或者建造其的匠人和说服各个教堂的智者没能做到这一点，未来的后代也不能看待能与其媲美的建筑。”

亨利三世（1207 年—1272 年）统治后期，哥特建筑在细部及组合方式上出现了一些新奇的方式。被所在墙体清晰分隔并由连续的披水石及披水饰结合在一起的尖顶窗，由大尺度并被竖框分隔成多个窗扇的窗户所取代；竖框的引入使得以丰富的几何形体布满窗户的花格窗饰也随之产生。线脚上粗的凸起与深的凹槽之间的交替让步于更加丰富和优美曲折的新组合方式；小柱子不再分散布置或者被捆绑成束，反而是更加坚实地连在一起；卷叶形花饰作为人们最喜爱的哥特装饰，更多地从自然树木及植物中吸取元素；不同于从一簇向上伸展的茎中伸出的波浪状三叶饰，几片叶子更趋向于一种环绕的形式，然后包围它们所附着的物体。更加丰富和更具差异性的装饰也从少数哥特建筑中往外蔓延，赋予哥特式建筑更加精巧的层次性。

如此哥特建筑逐渐从早期的英国样式进入到盛装哥特样式——也是最为人称赞的样式——与爱德华时代一起形成的完美的盎格鲁 - 哥特艺术。随着这种样式的发展，一些特征的差异性更加显著，同时，相对于早期对于几何形式精确性的追求，现在的哥特艺术更加倾向于优美的波浪状流动线条。

出现在早期英国哥特样式中的空间上垂直性的趋势与盎格鲁 - 诺曼时期罗马式的水平延展的样式产生强烈对比。在盛装哥特样式中，主要的结构线条形成一种锥形边缘，而非垂直或水平的。为了在这个基本规则下实现这一系列变化，盎格鲁 - 哥特样式第三个清晰的时期的特征便是由垂直线条以及与其正交的同等重要的线条所界定的。这个最新的华美样式，因其线条的突出地位而被命名为垂直哥特式，逐渐取代了盛装哥特样式。就像作为一个更加成熟的样式，盛装哥特由于其精美及和谐的丰富性逐渐替代早期英国的哥特样式一样。作为一种新的样式，垂直哥特式建筑暂时保留了部分之前样式的特征，并与其自身特殊的特征相结合：作为垂直哥特式的第一个时期，也想要达到盛装哥特式的壮美，但用太高的赞美来评价是很困难的。然而，随着都铎式建筑中平坦拱的出现，更加丰富多样和细致的嵌板及其它装饰也随之诞生，这也清晰表明了这个时代在建筑品味上的衰退。而建筑历史中的一次后退往往是致命的。

因此，中世纪的教堂建筑尽管在衰退，但整体还是很壮美，由扇形花格饰布满的拱顶频繁出现在最后几个伟大的作品中——之后，建筑史中长时间的衰退时代就来临了。

曹峻川　甄影博

目录

第1章

桂饰

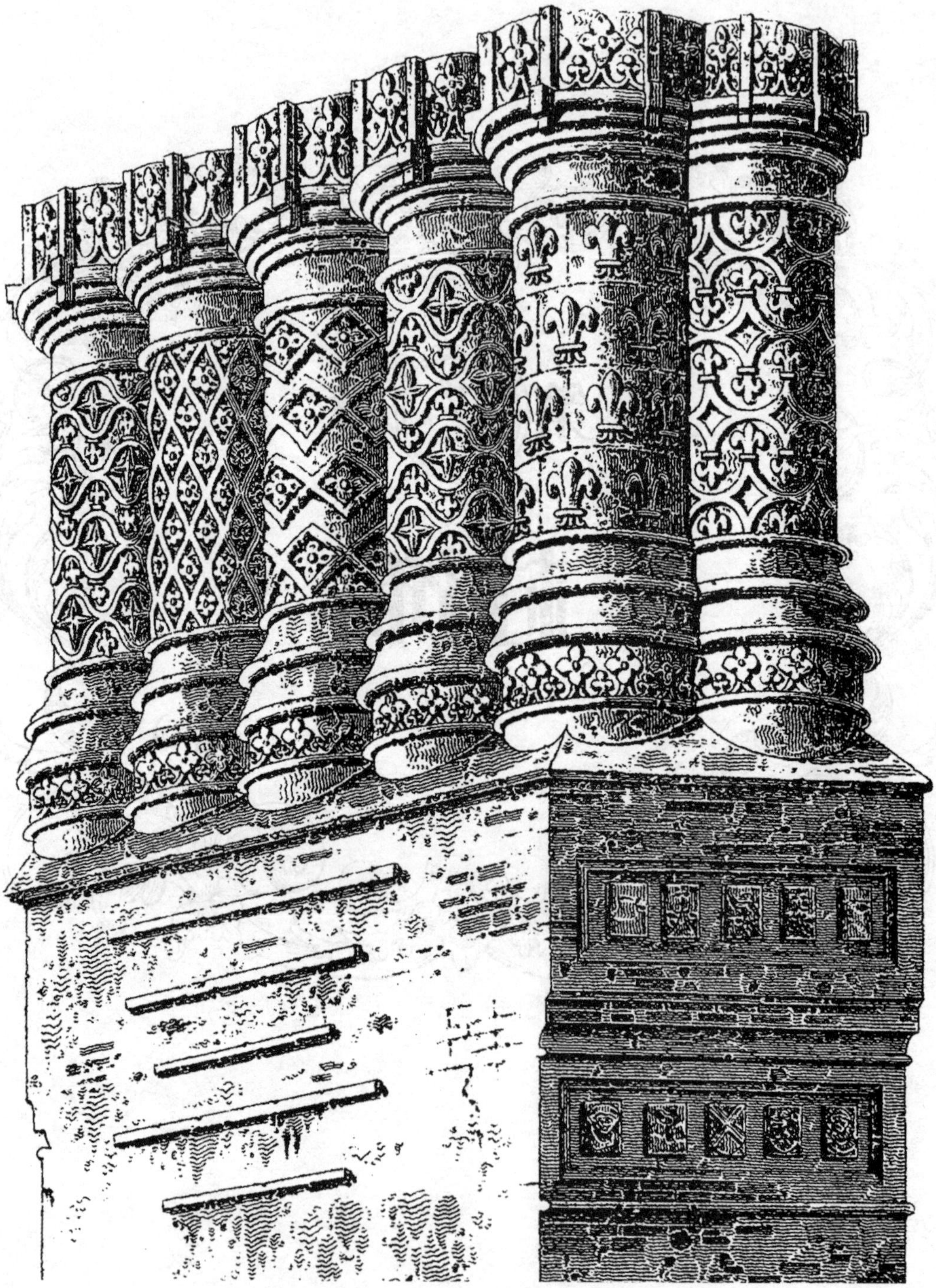

诺福克巴沙姆礼堂烟囱筒身

a. 温彻斯特大教堂北耳堂　b. 约克郡惠特比修道院　c. 伦敦白塔

诺坦普顿郡圣彼得教堂

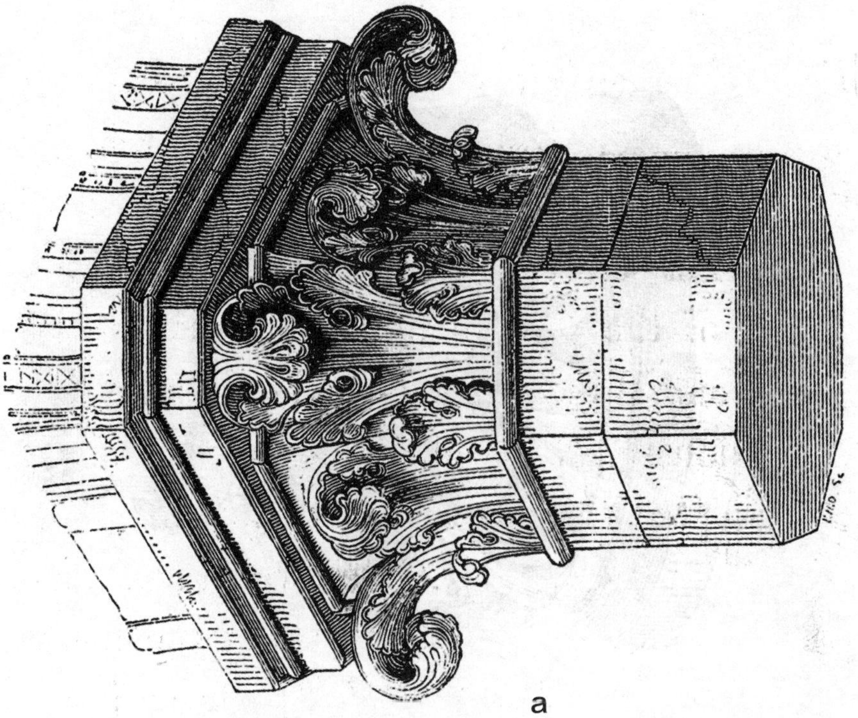

a

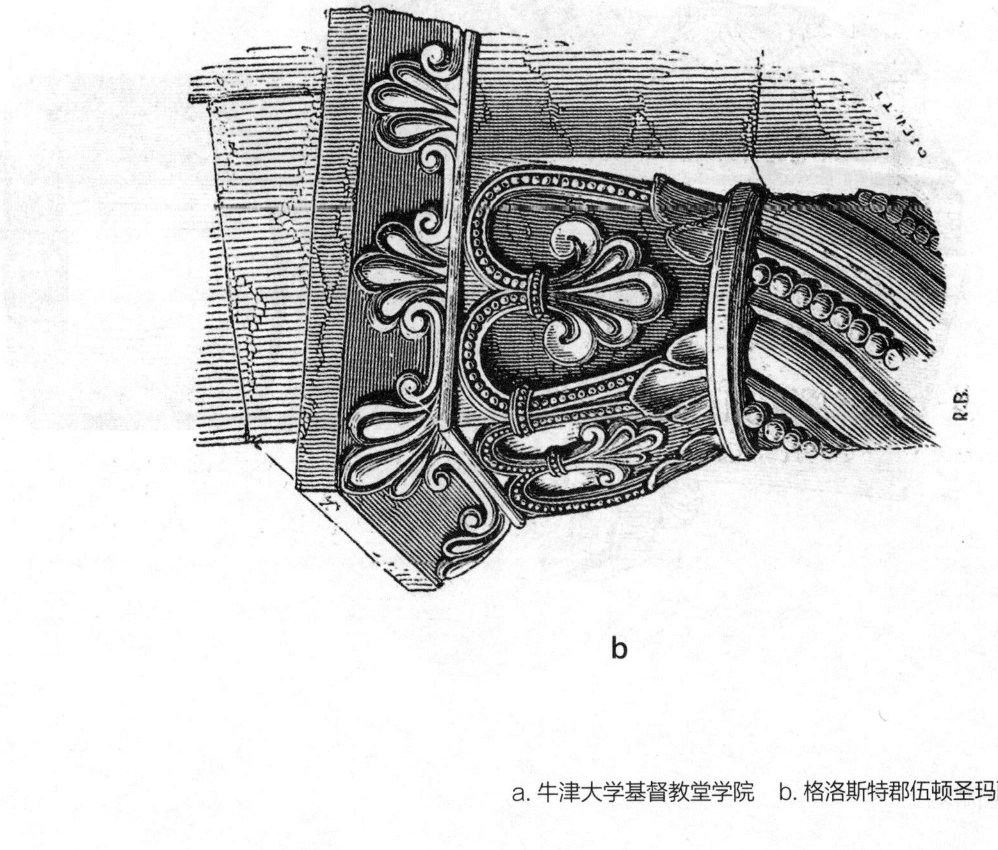

b

a. 牛津大学基督教堂学院　b. 格洛斯特郡伍顿圣玛丽教堂

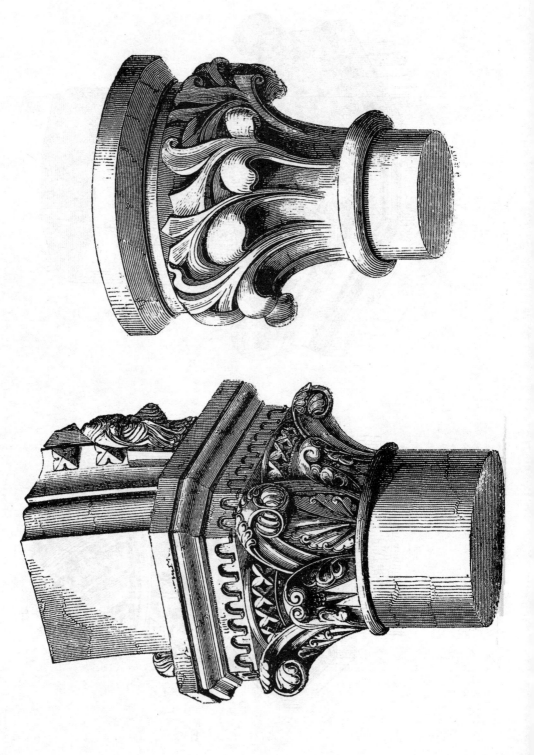

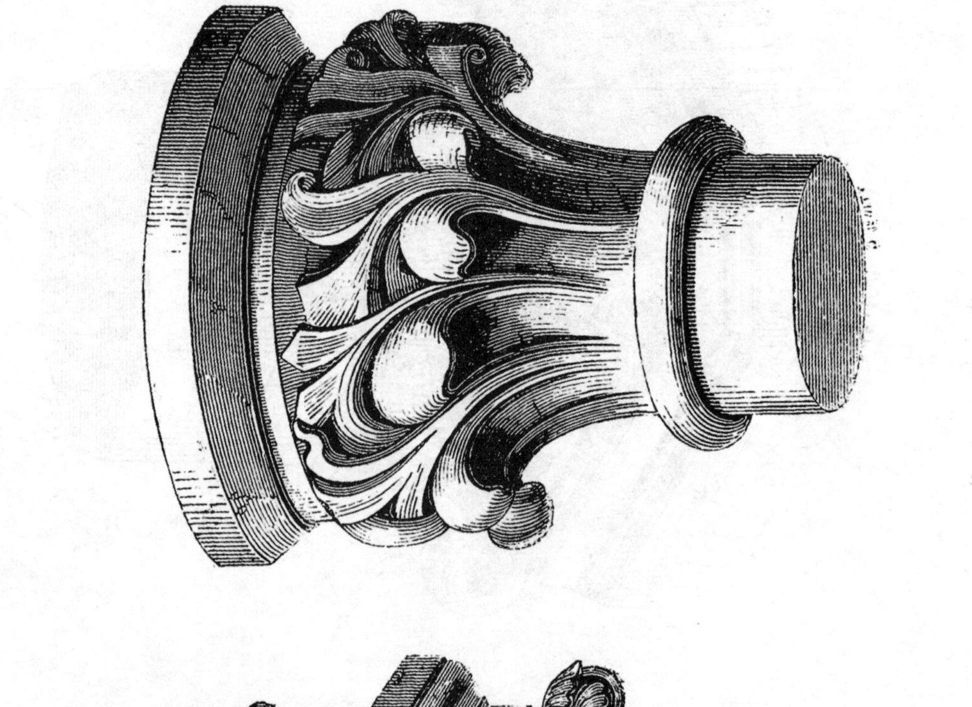

哥特式石制装饰

a

b

a. 林肯教堂长老会　b. 林肯教堂北耳堂

a

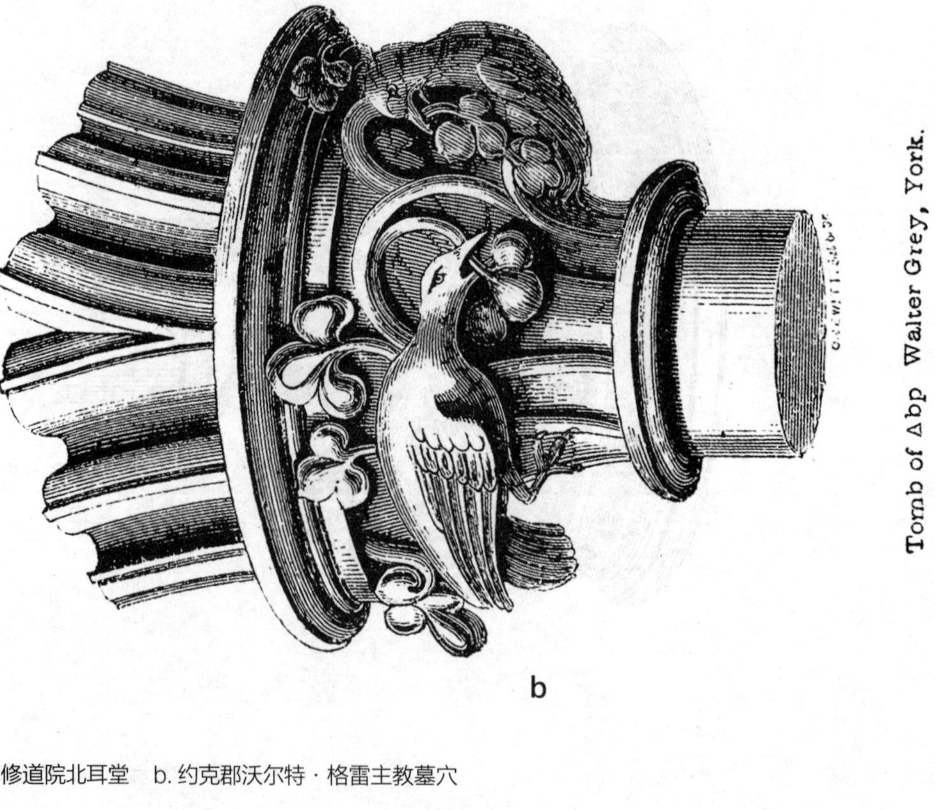

b

a. 罗姆塞修道院北耳堂　b. 约克郡沃尔特 · 格雷主教墓穴

林肯教堂长老会

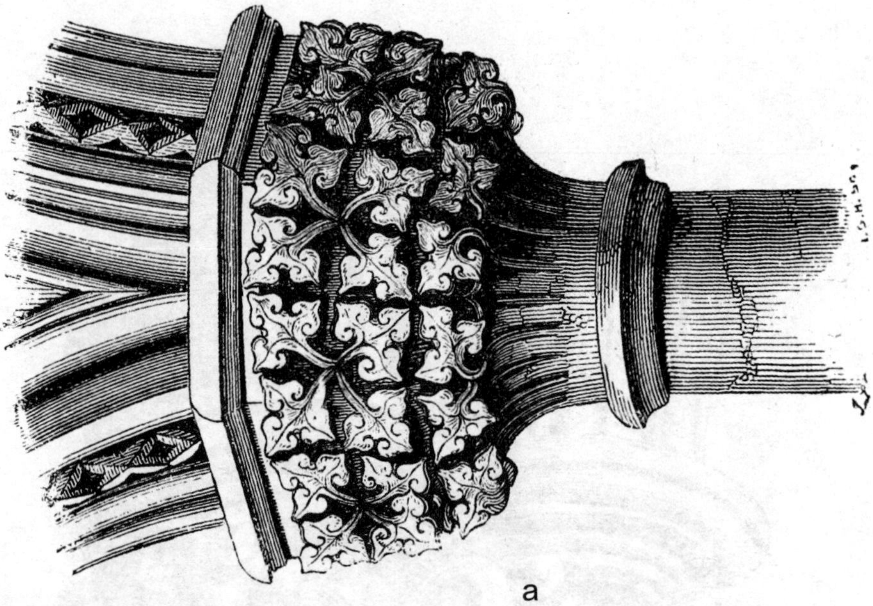

a

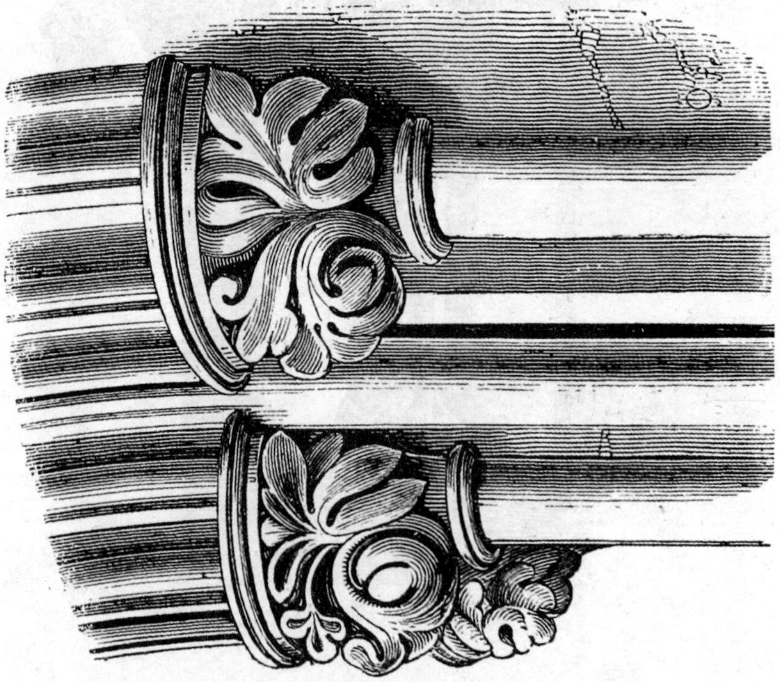

b

a. 贝弗利大教堂耳室　b. 牛津郡多尔切斯特修道院

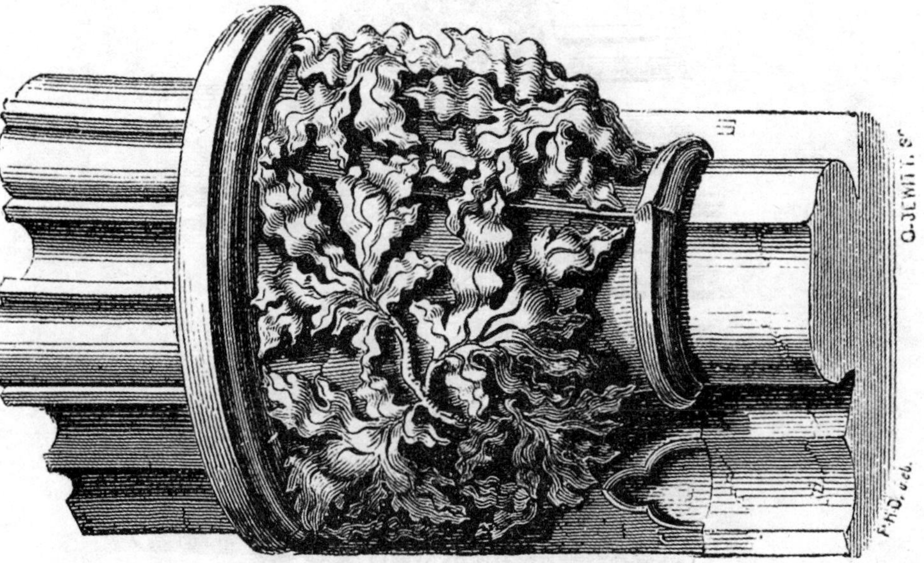

a

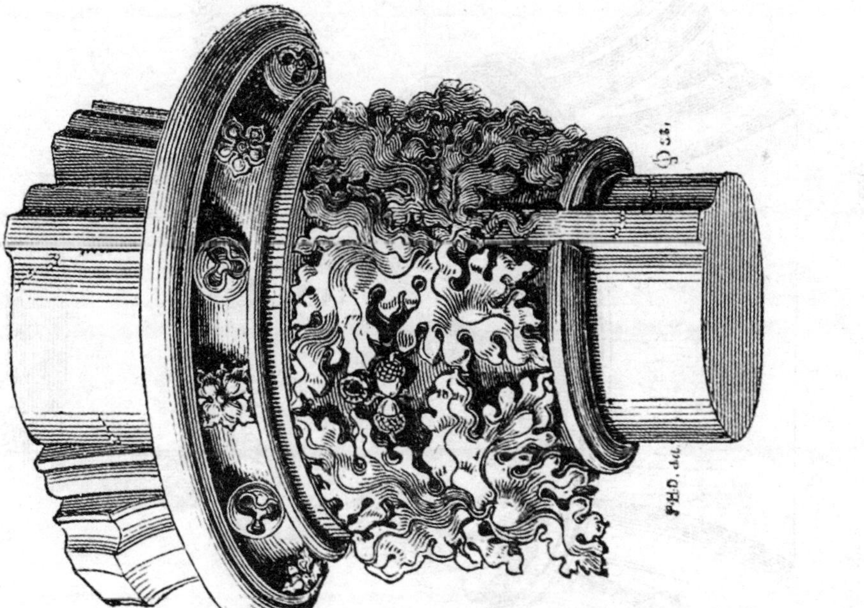

b

a. 索斯韦尔大教堂　b. 林肯大教堂

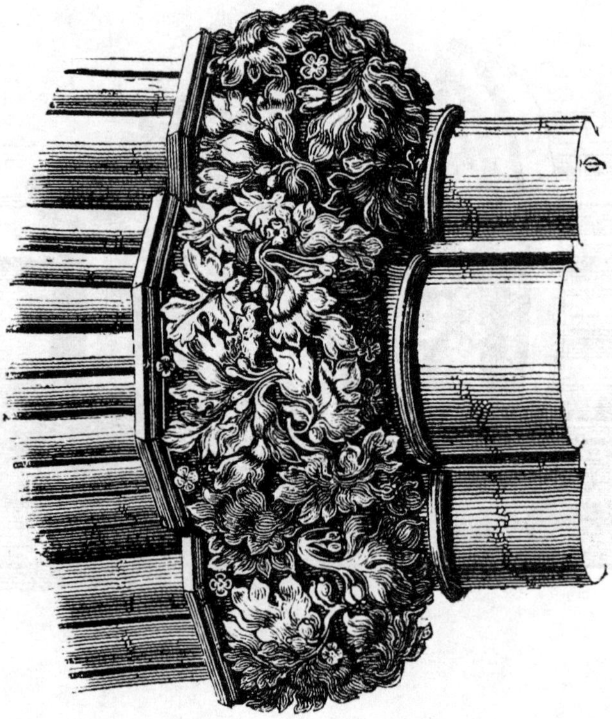

a

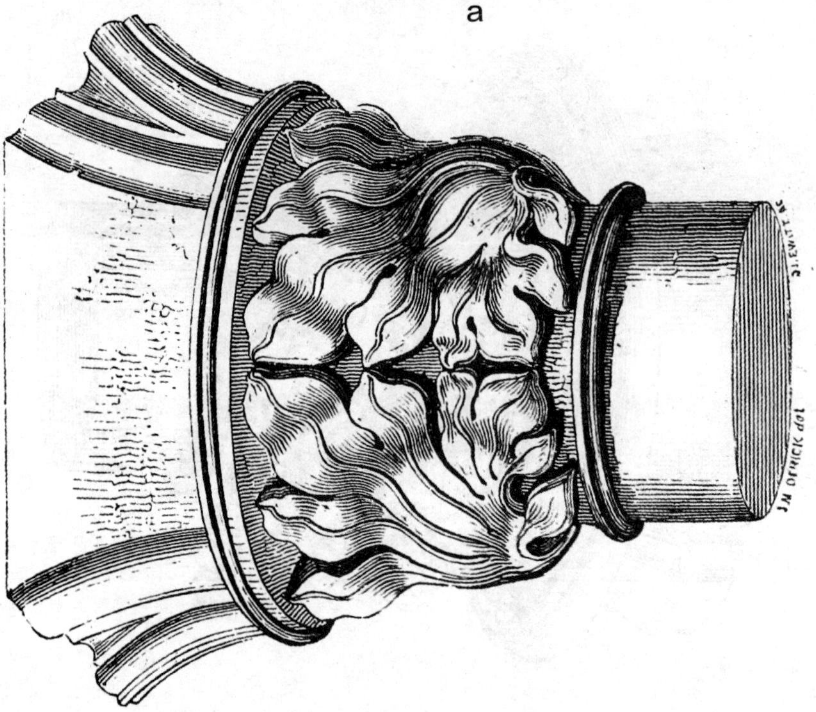

b

a. 约克大教堂　b. 贝弗利大教堂

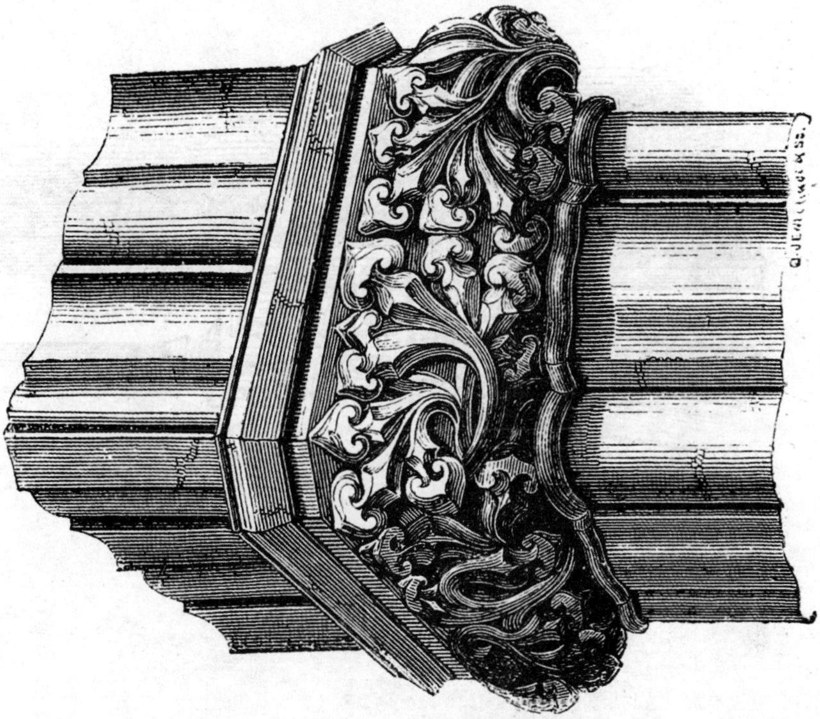

a

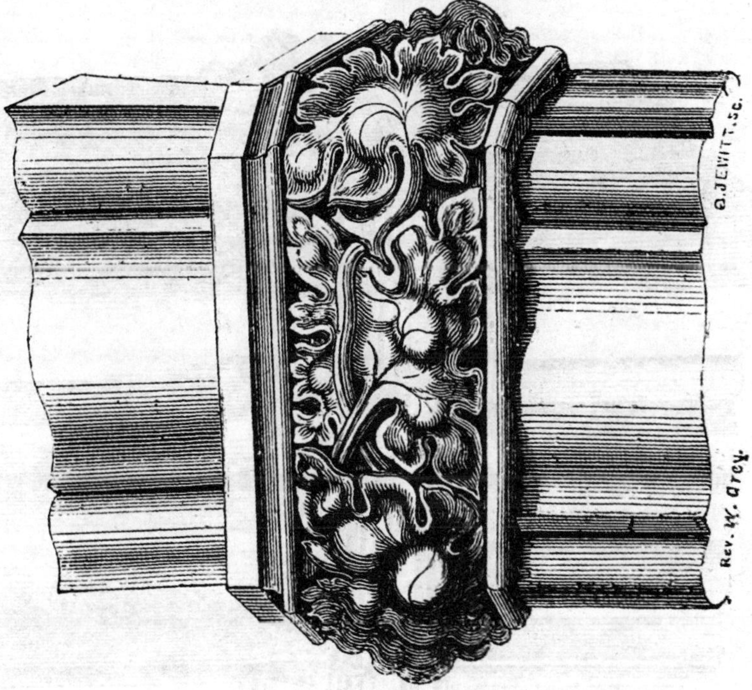

b

a. 德文郡斯托克因泰因海德　b. 德文郡肯顿教堂

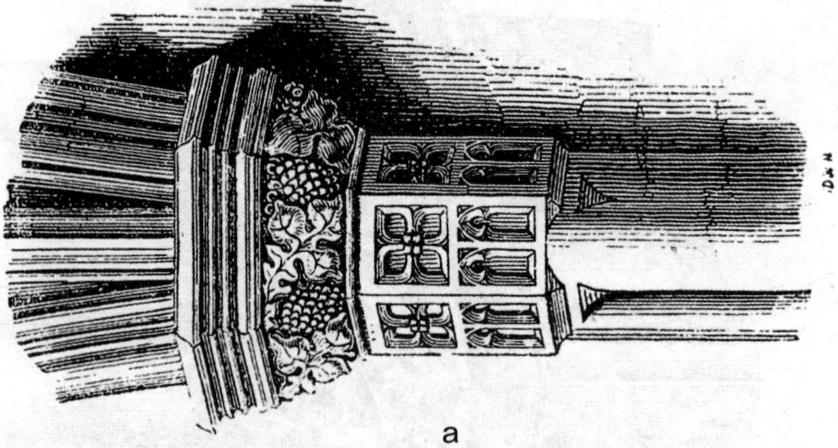

a

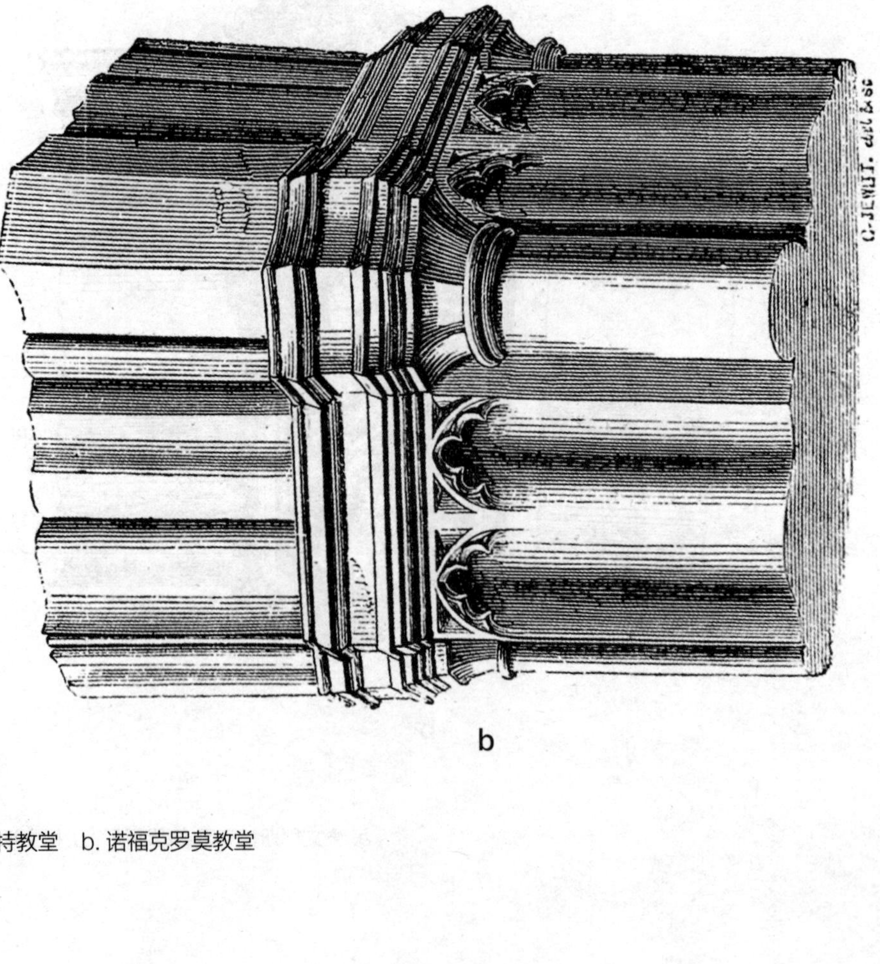

b

a. 温彻斯特教堂　b. 诺福克罗莫教堂

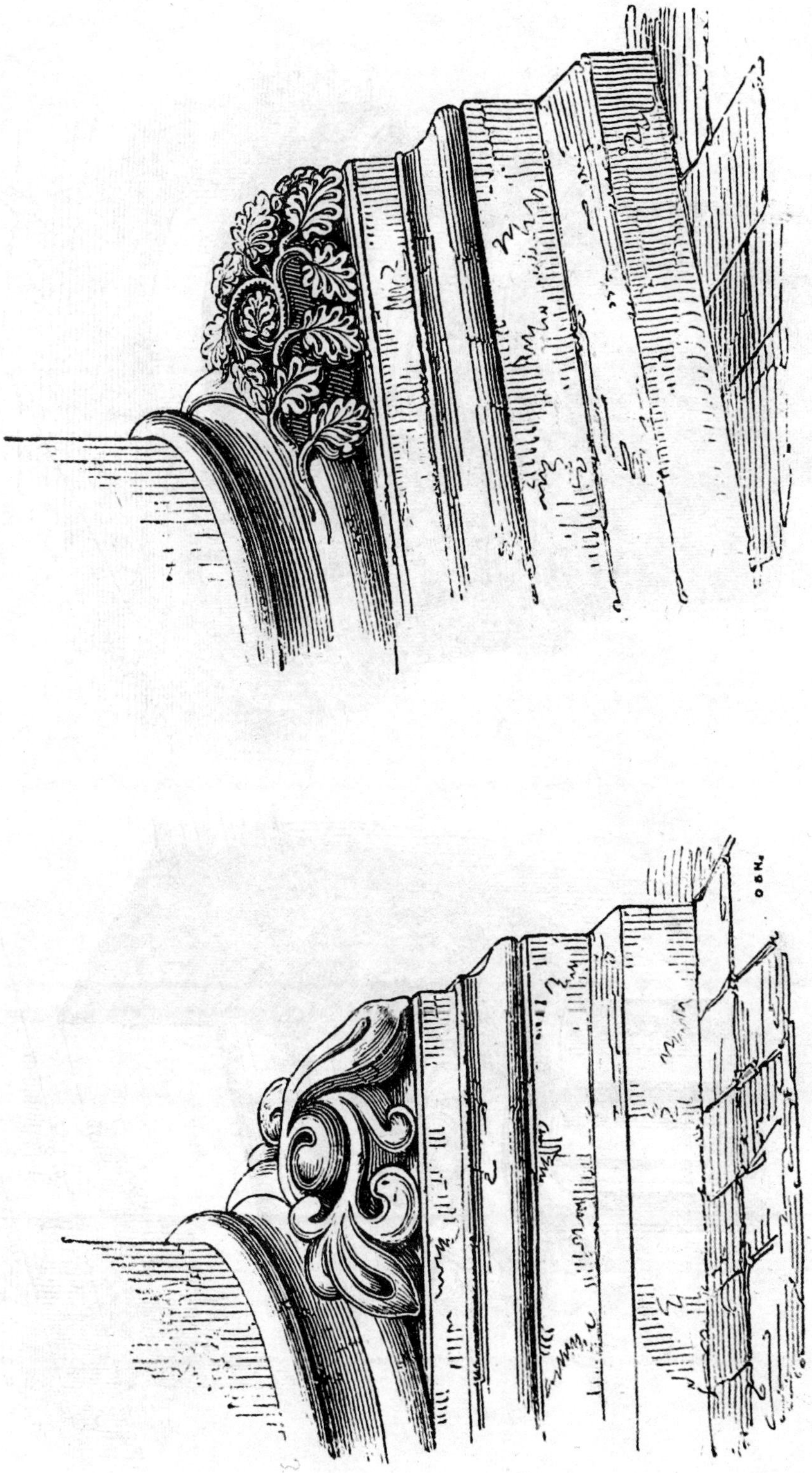

温彻斯特圣十字教堂

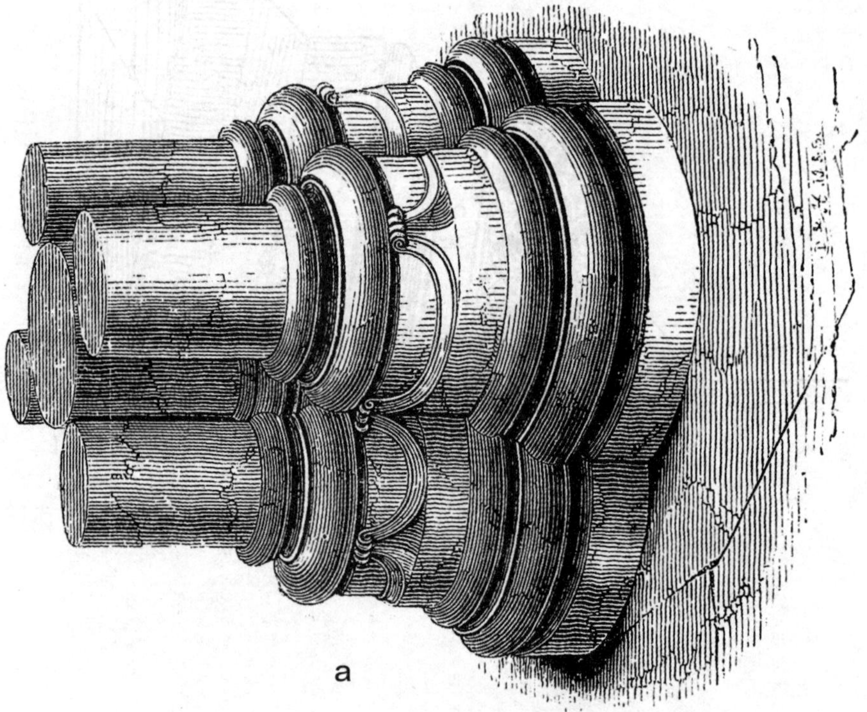

a

b

a. 坎特伯雷大教堂　b. 林肯教堂牧师会礼堂

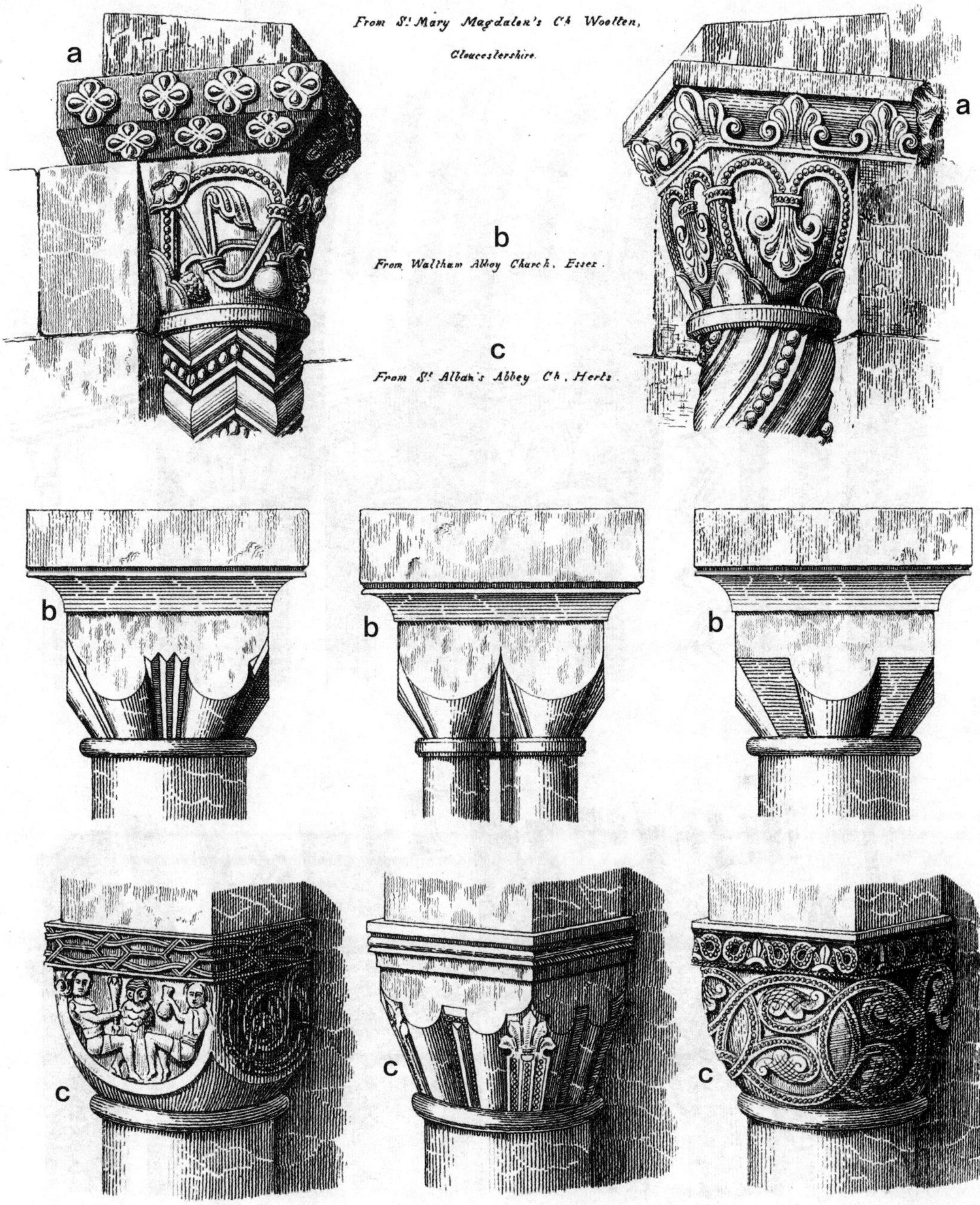

a. 来自洛斯特郡伍顿圣玛利亚玛达肋纳教堂　b. 来自萨塞克斯郡沃尔瑟姆修道院教堂　c. 来自赫特福德郡圣阿尔班修道院教堂

Capitals and details from the S. door of Sempringham Ch

瑟布 汉教堂南门柱头及细部

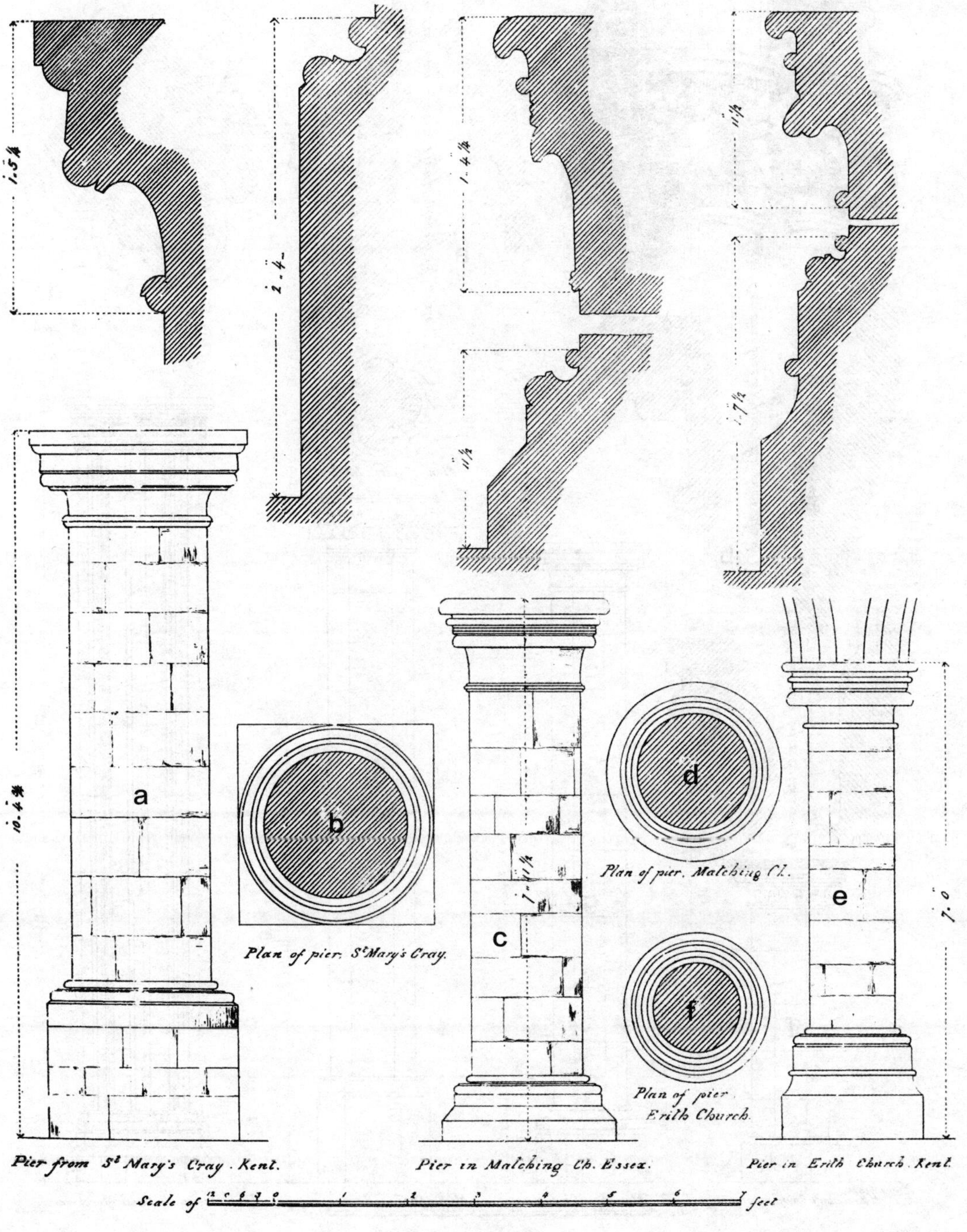

柱头及柱础剖面：a. 来自肯特郡圣玛丽克雷的柱子　b. 圣玛丽克雷柱子的平面　c. 艾塞克斯马尔欣格教堂内柱子　d. 马尔欣格教堂柱子的平面　e. 肯特郡埃里思教堂柱子　f. 埃里思教堂柱子平面

第1章 柱饰

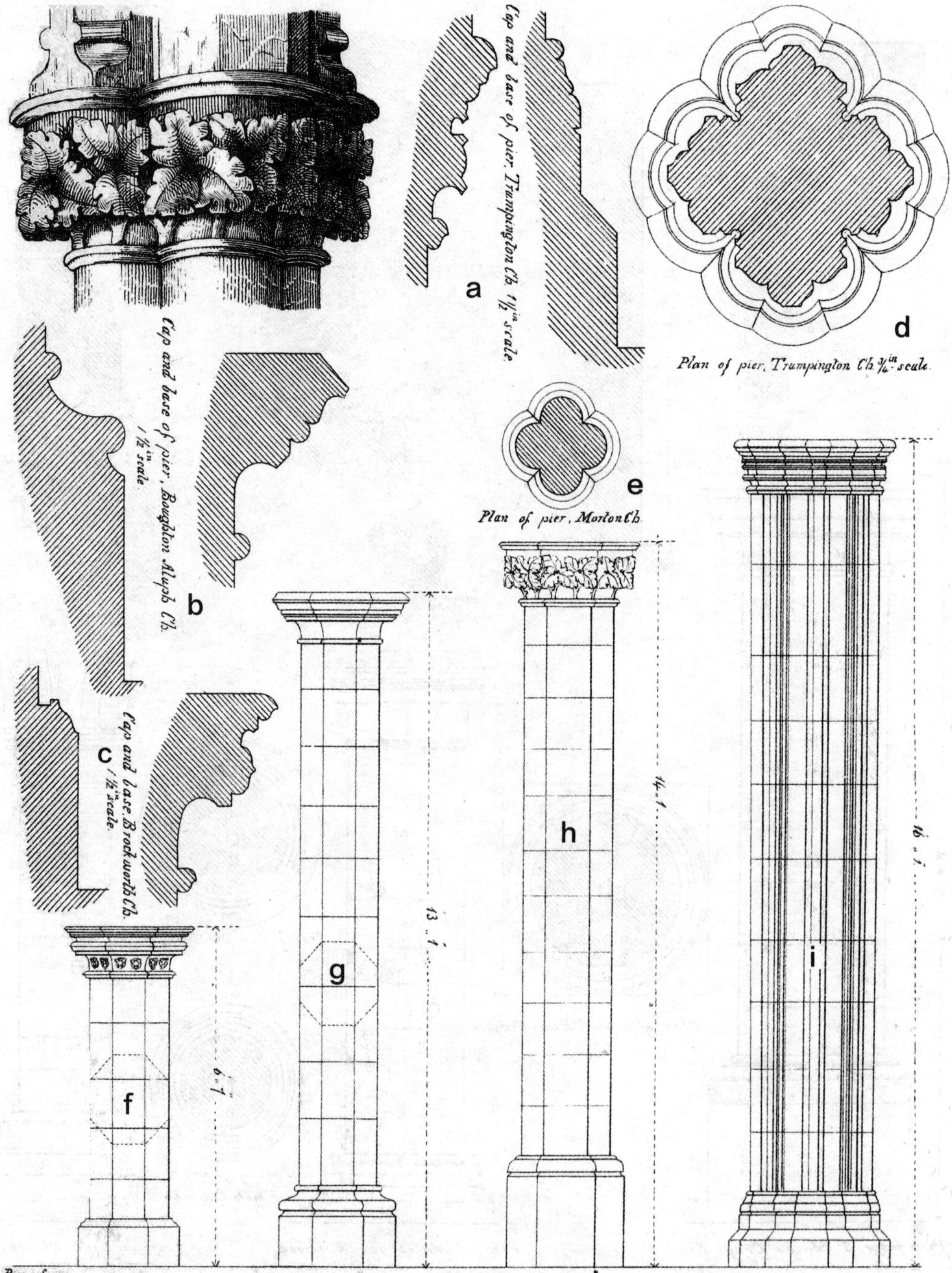

a. 特兰平顿教堂柱子柱头及柱础　b. 鲍顿阿鲁教堂柱子柱头及柱础　c. 布罗克沃思教堂柱头及柱础　d. 特兰平顿教堂柱子平面　e. 默顿教堂柱子平面　f. 格洛斯特郡布罗克沃思教堂柱子　g. 肯特郡鲍顿阿鲁教堂　h. 林肯郡默顿教堂　i. 剑桥郡特兰平顿教堂

第1章 柱饰

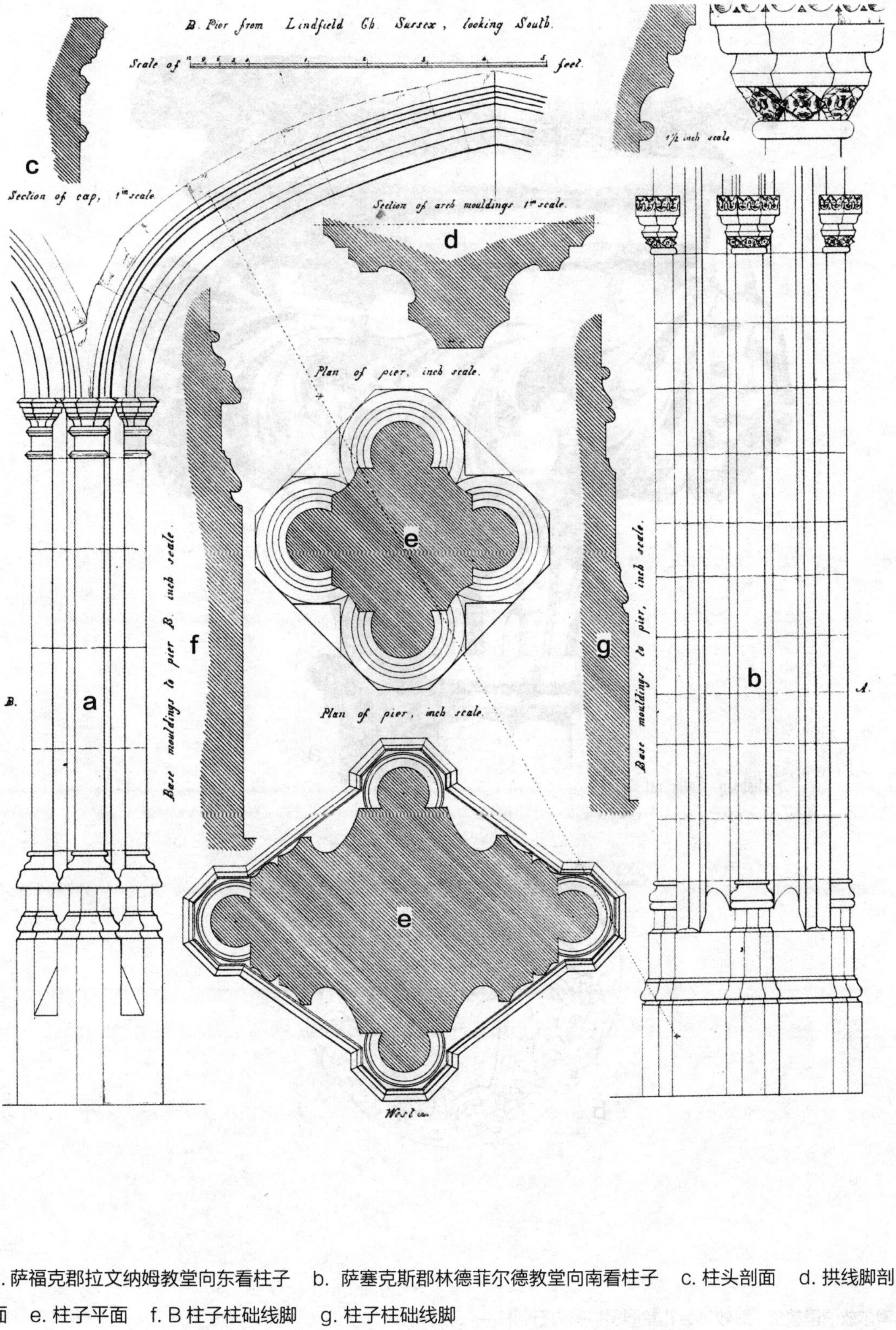

a. 萨福克郡拉文纳姆教堂向东看柱子　b. 萨塞克斯郡林德菲尔德教堂向南看柱子　c. 柱头剖面　d. 拱线脚剖面　e. 柱子平面　f. B柱子柱础线脚　g. 柱子柱础线脚

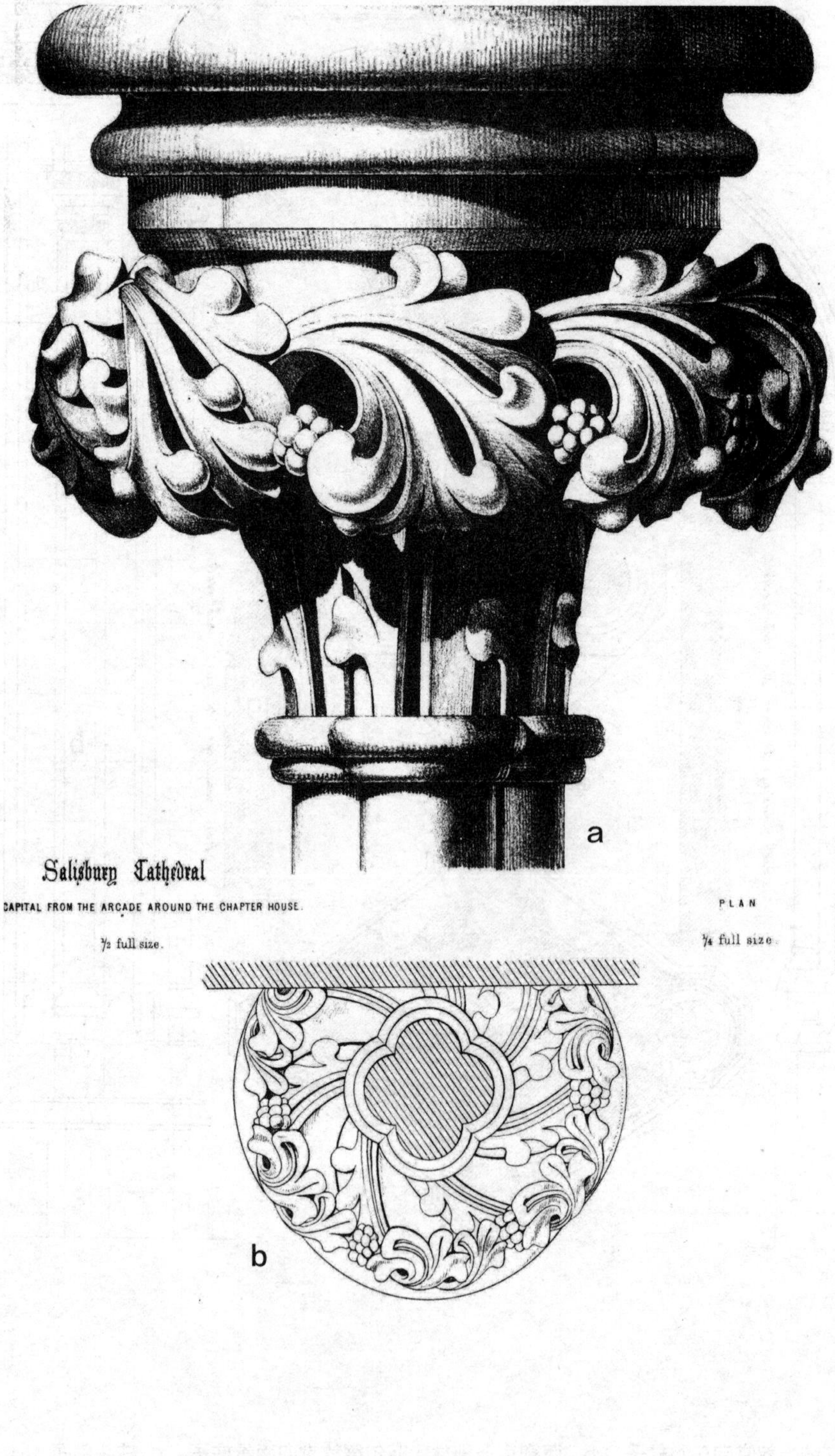

索尔兹伯里教堂：a. 牧师会礼堂周围拱廊内石制柱头　b. 平面

a

TION OF THE ARCADE SHEWING
POSITION OF THE ORNAMENTS

b

Salisbury Cathedral.

STONE ORNAMENTS FROM THE ARCADE AROUND THE CHAPTER HOUSE.

½ full size.

索尔兹伯里教堂：a. 牧师会礼堂周围拱廊内石制装饰　b. 展示装饰位置的拱廊剖面

威尔斯教堂：a. 通向牧师会礼堂地下室通道内的石制柱头　b. 柱础轮廓

威尔斯教堂：a. 通向牧师会礼堂地下室通道内石制柱头　b. 柱身平面以及柱础立面

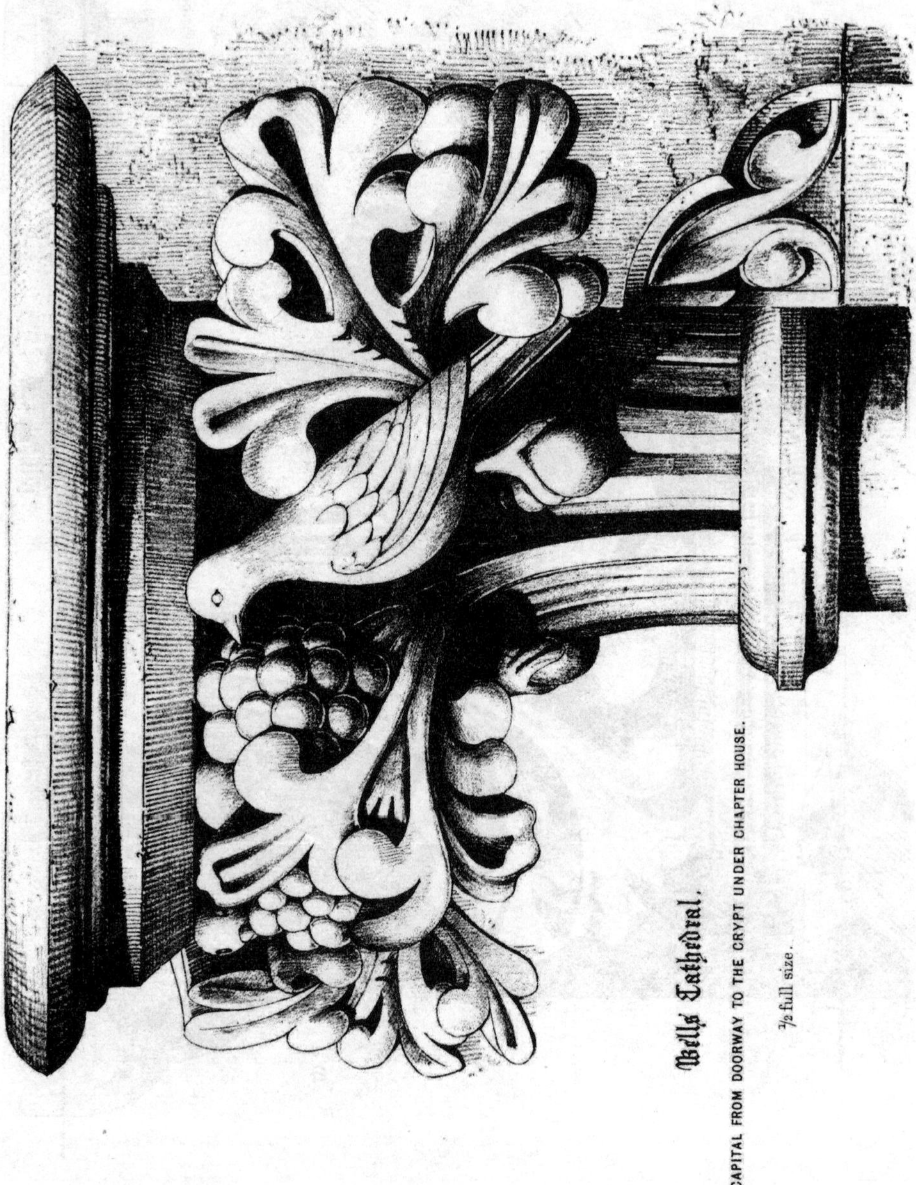

Wells Cathedral.

STONE CAPITAL FROM DOORWAY TO THE CRYPT UNDER CHAPTER HOUSE.

½ full size.

威尔斯教堂通向牧师会礼堂地下室大门上石制柱头

第2章

门

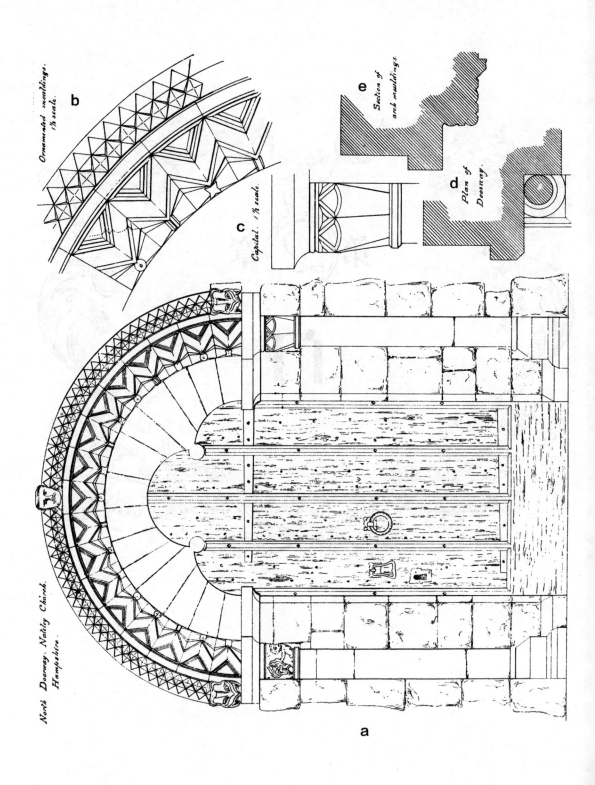

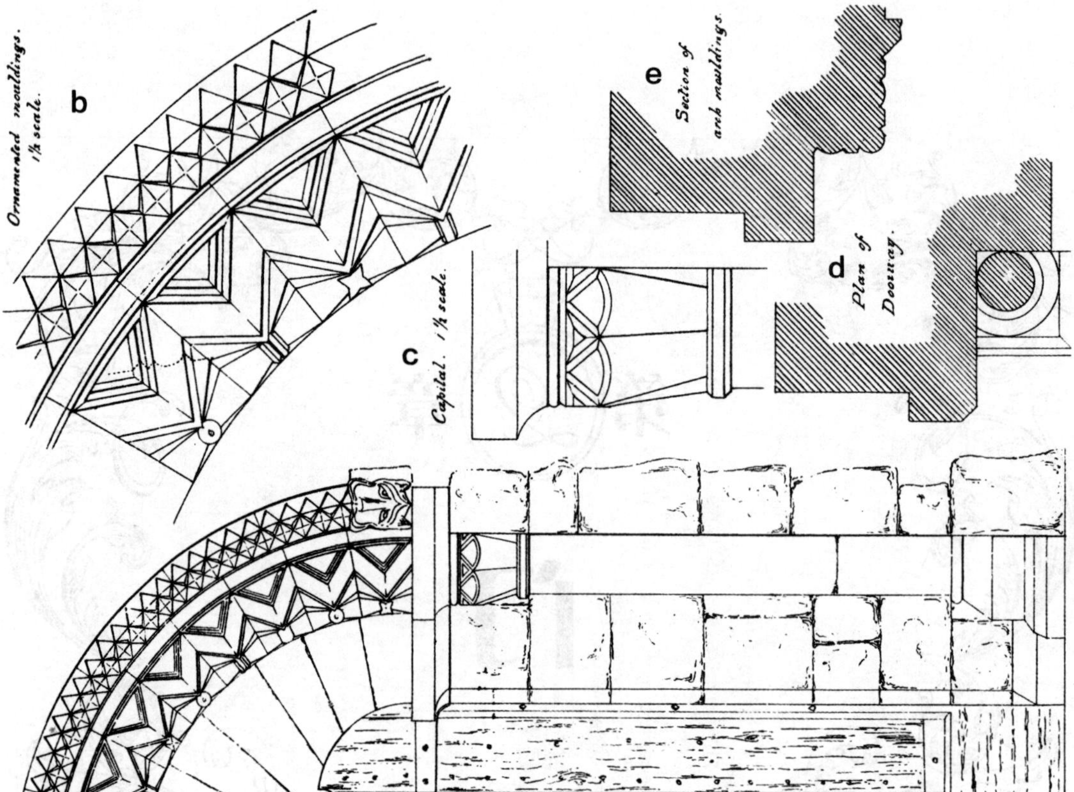

a. 汉普君纳特雷教堂北门　b. 装饰性线脚　c. 柱头　d. 大门平面　e. 拱线脚剖面

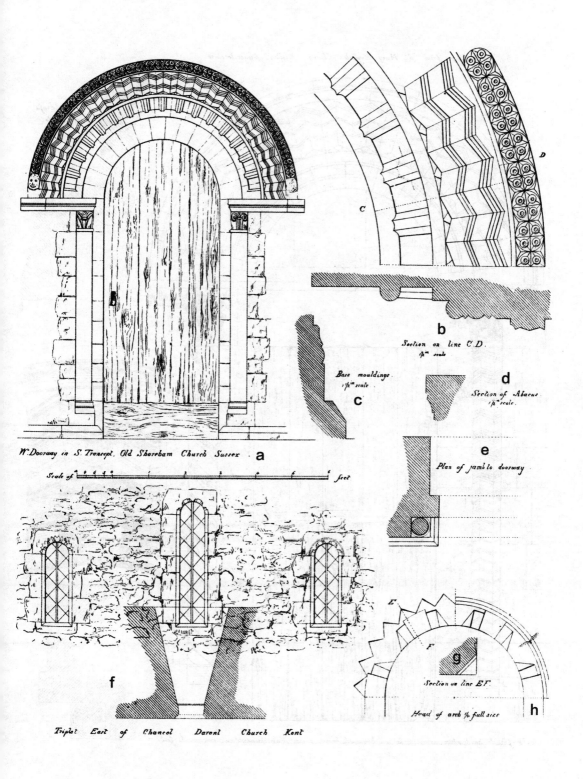

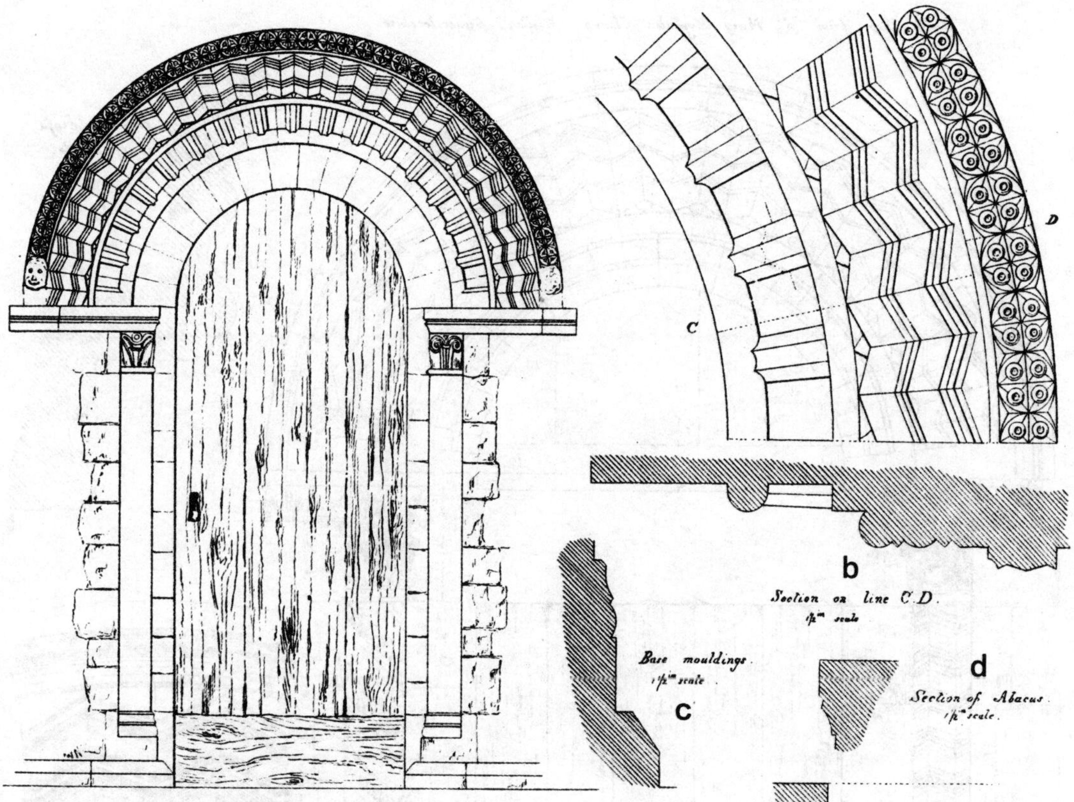

a. 萨塞克斯郡老肖勒姆教堂南耳堂西门 b. C–D 处的剖面 c. 柱础线脚 d. 柱顶板剖面 e. 大门侧壁平面 f. 肯特郡达伦特教堂东侧三个窗户 g. E–F 处的剖面 h. 拱顶部

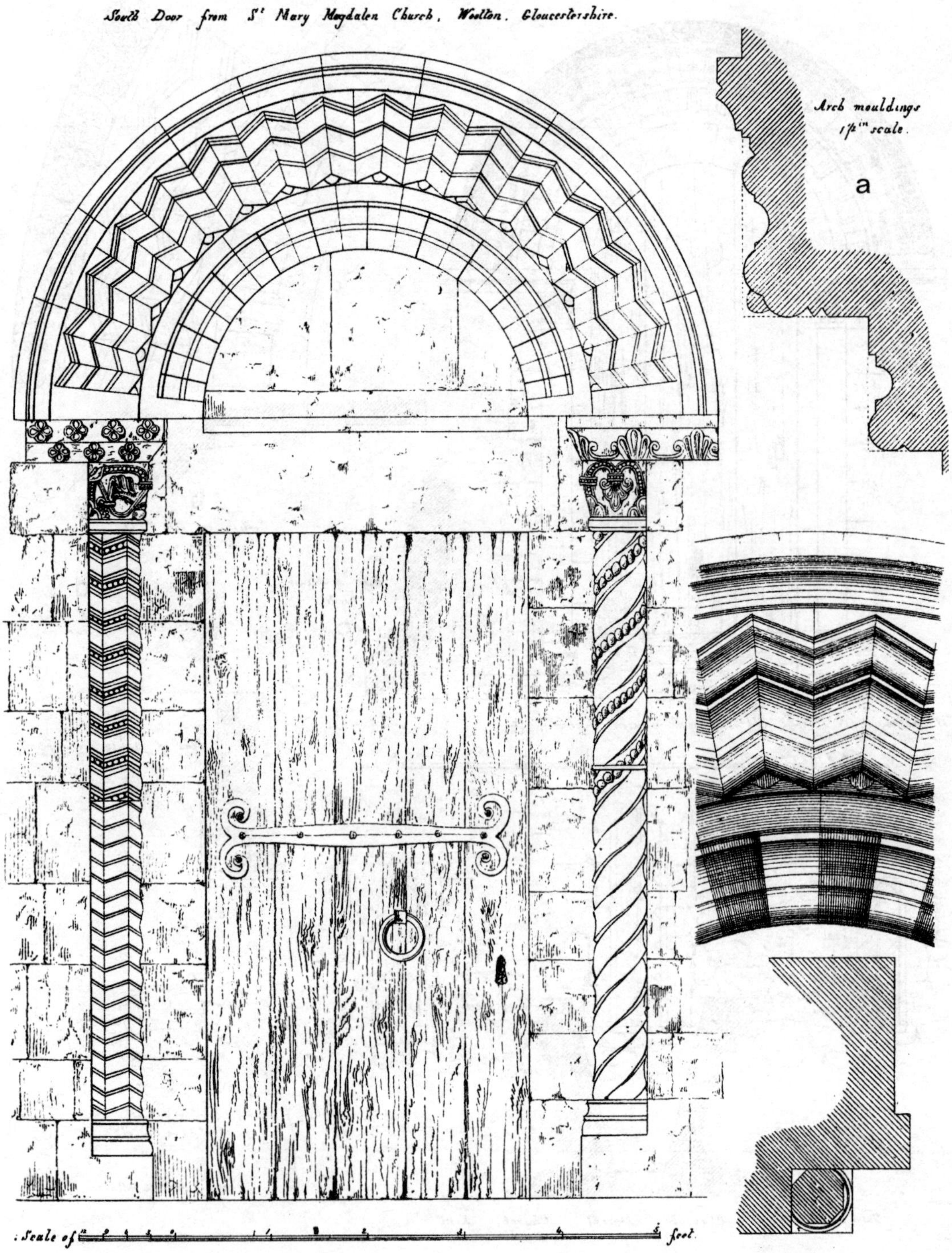

格洛斯特郡伍顿圣玛利亚玛达肋纳教堂南门：a. 拱线脚

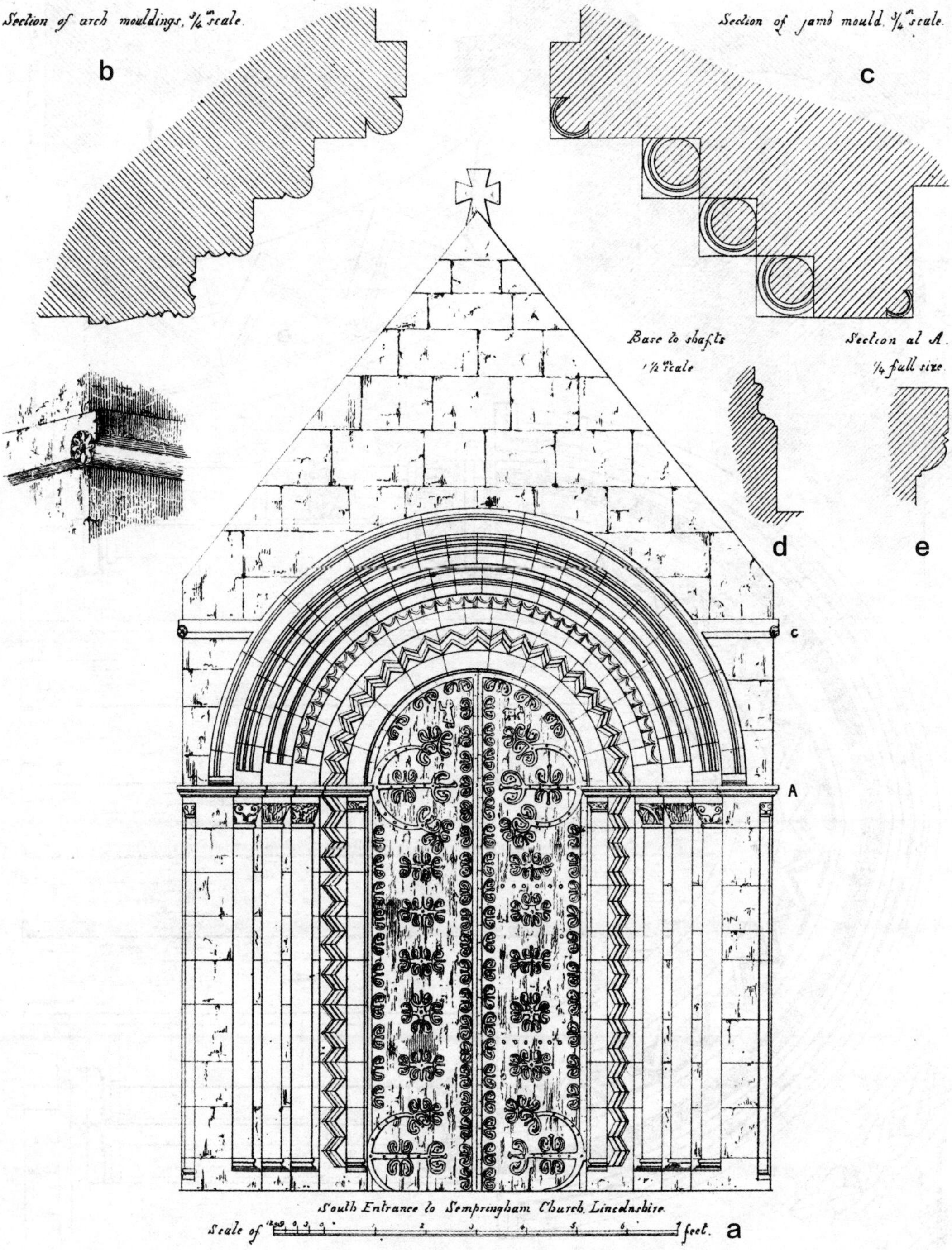

a. 林肯郡瑟布 汉教堂南侧入口 b. 拱线脚剖面 c. 侧壁剖面 d. 柱础 e.A 处的剖面

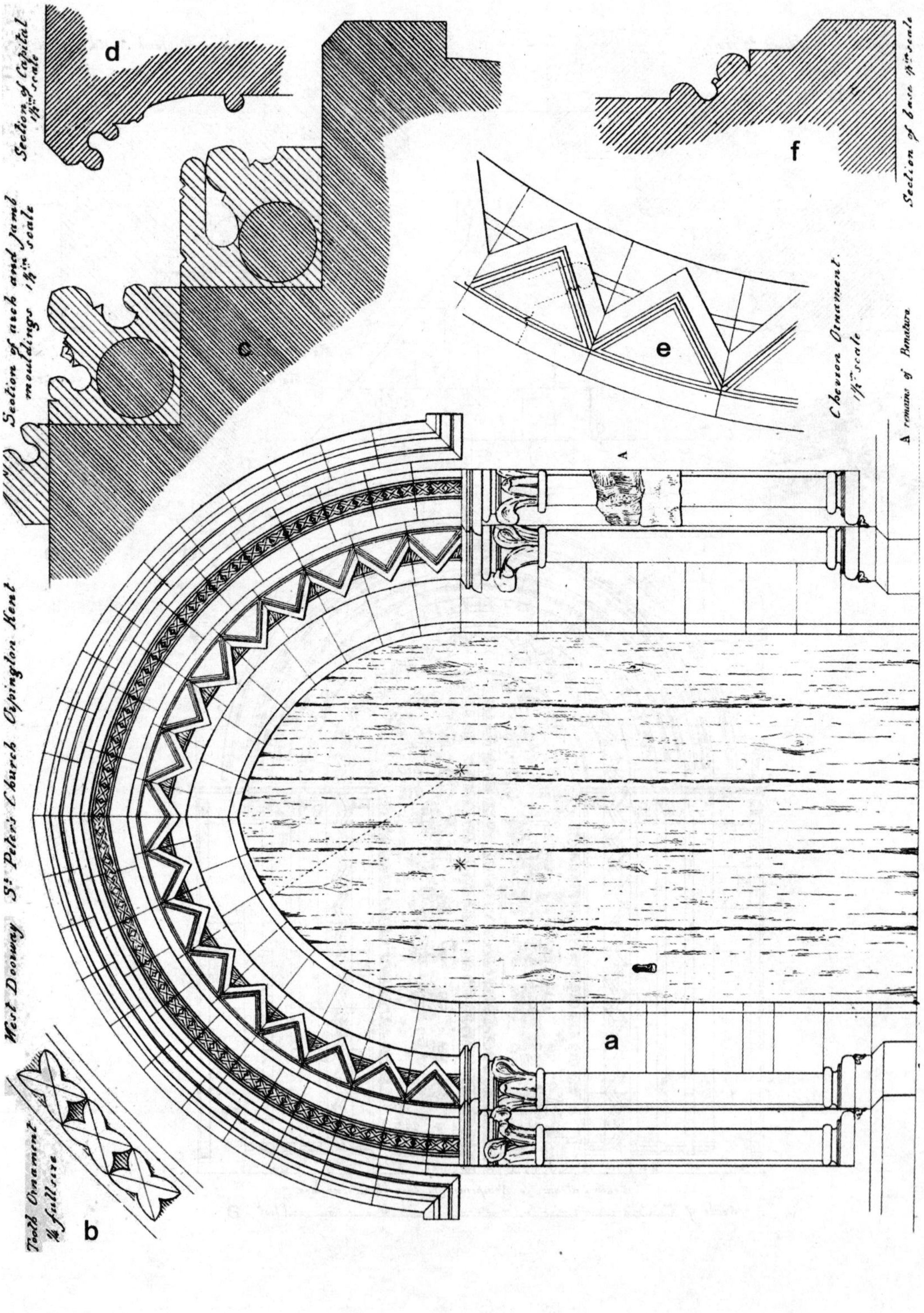

a. 肯特郡奥平顿圣彼得教堂西门　b. 齿形装饰　c. 拱线脚及侧壁线脚剖面　d. 柱头剖面　e. 山形纹样装饰　f. 柱础剖面

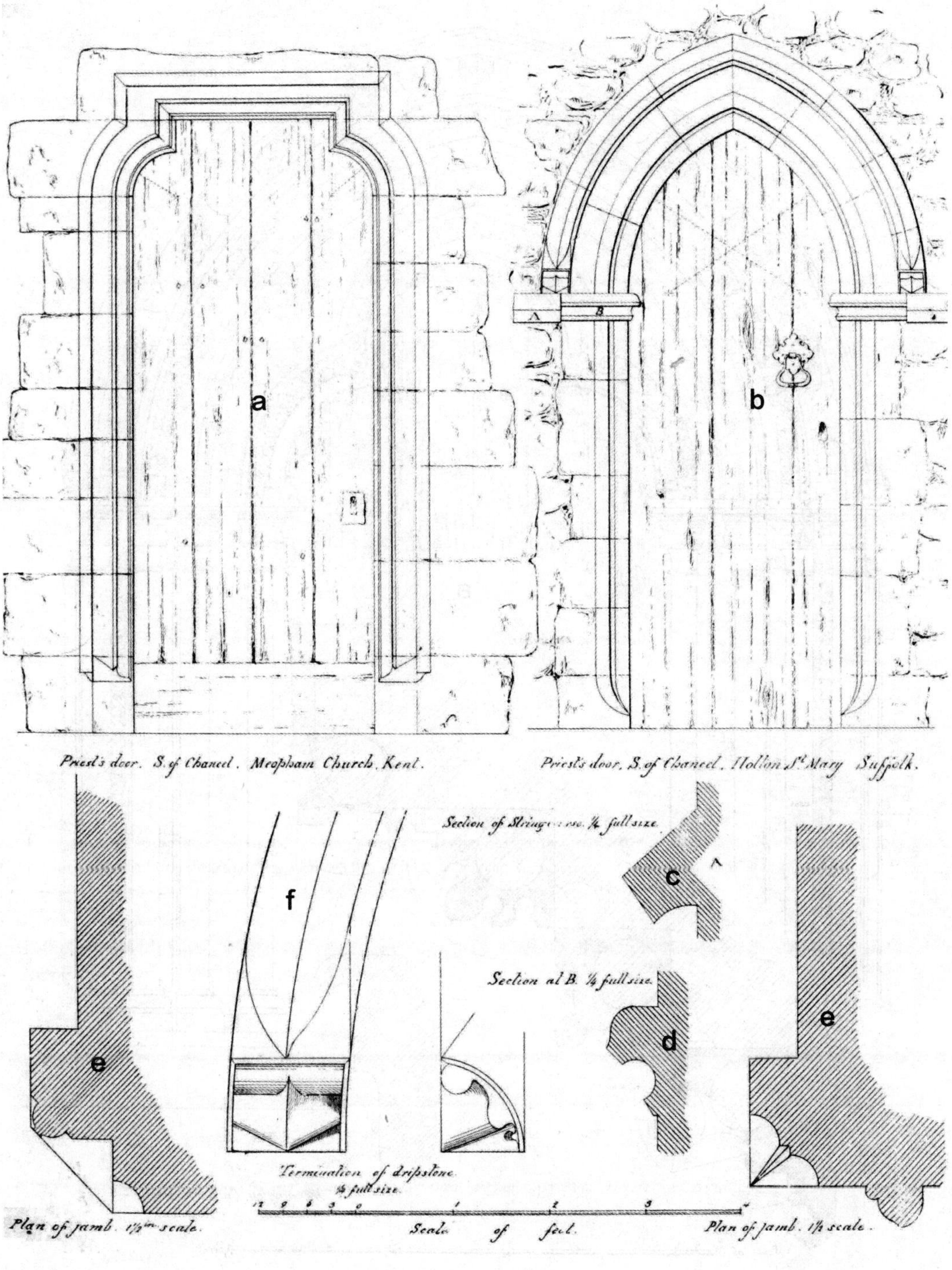

a. 肯特郡梅珀姆教堂圣坛南侧牧师之门　b. 萨福克郡霍尔顿圣玛丽教堂圣坛南侧牧师之门　c. 束带层剖面　d.B处的剖面　e. 侧壁平面　f. 披水石端部

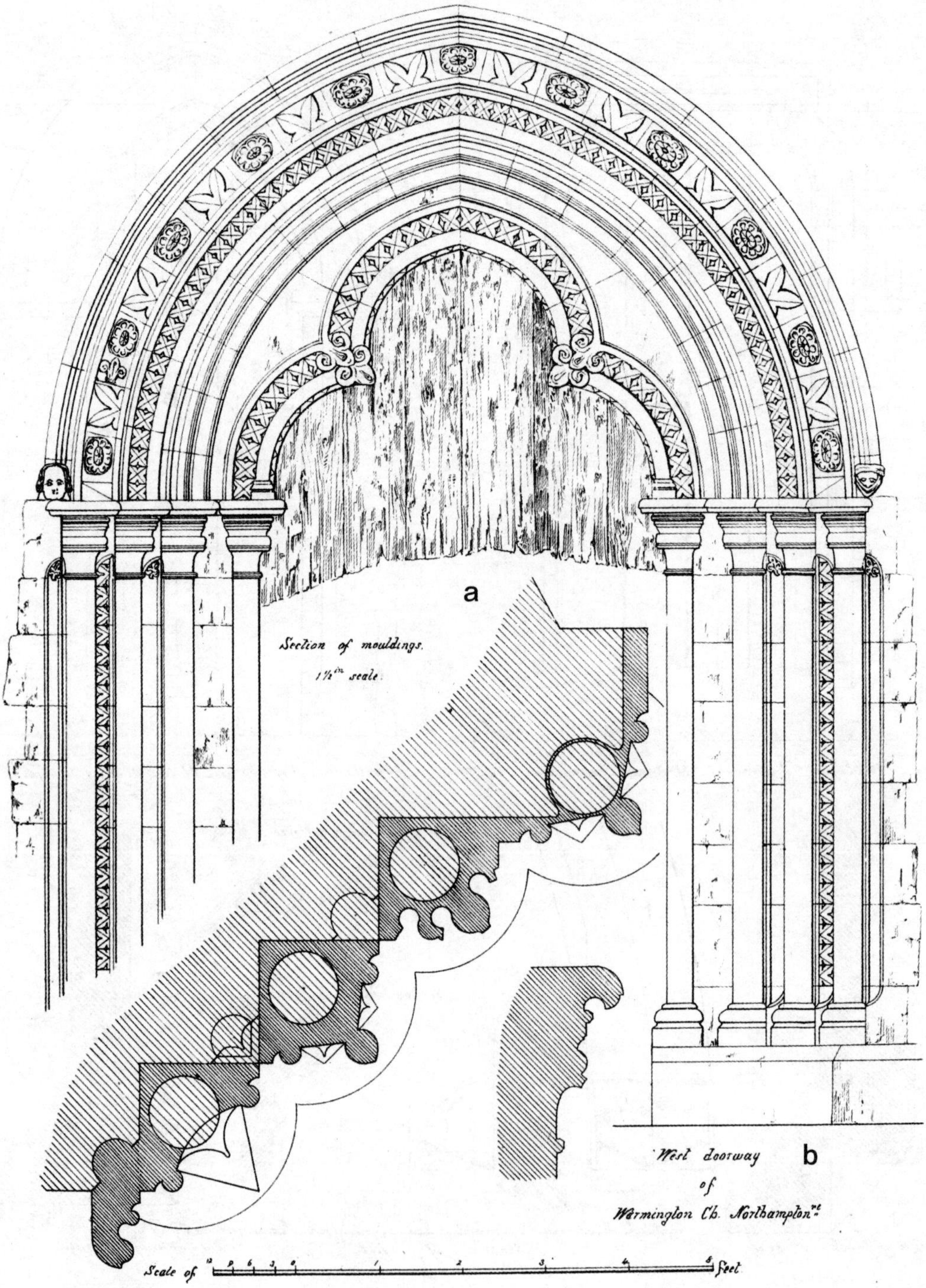

a. 北安普敦郡沃明顿教堂西门　b. 线脚剖面

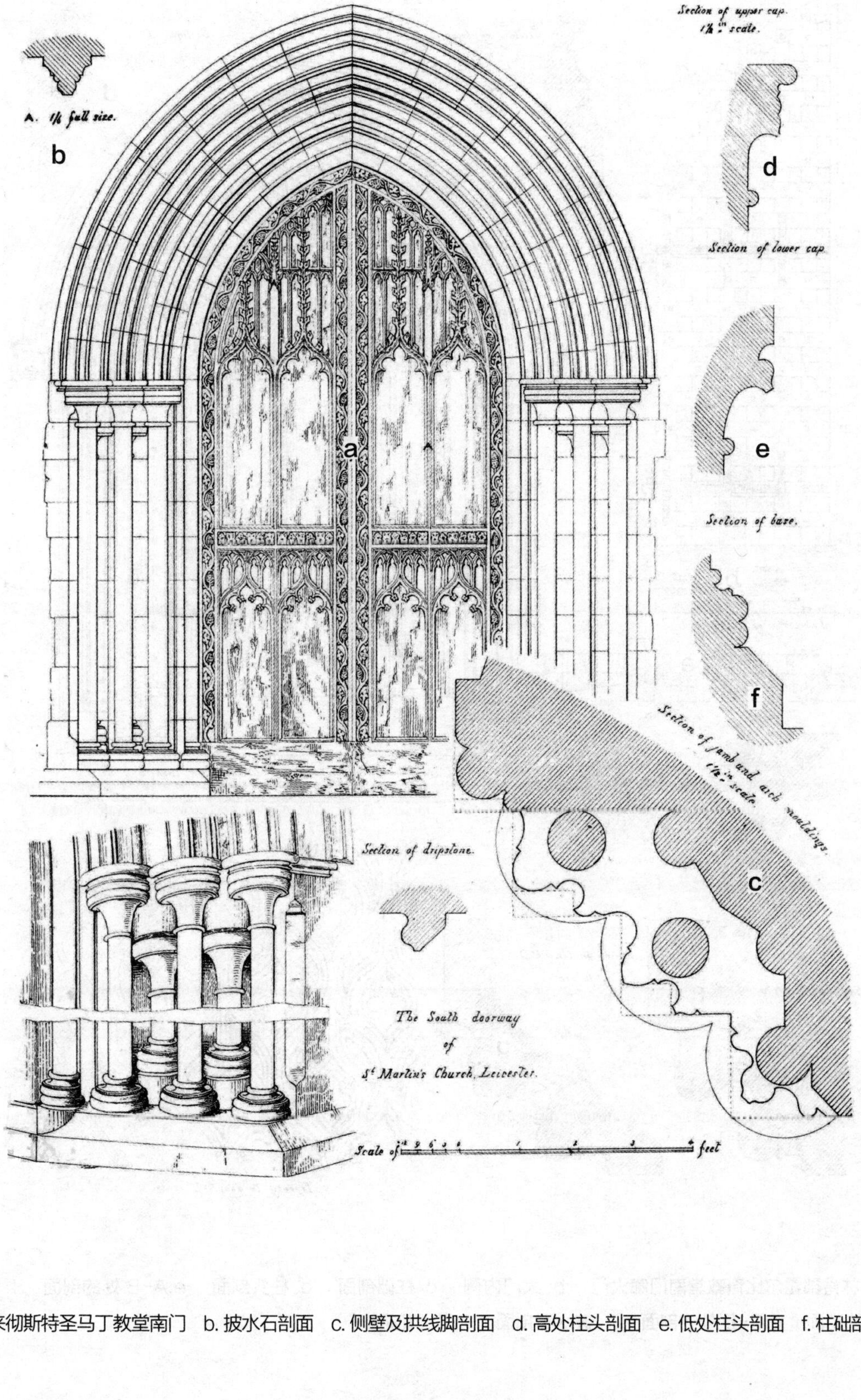

a. 来彻斯特圣马丁教堂南门　b. 披水石剖面　c. 侧壁及拱线脚剖面　d. 高处柱头剖面　e. 低处柱头剖面　f. 柱础剖面

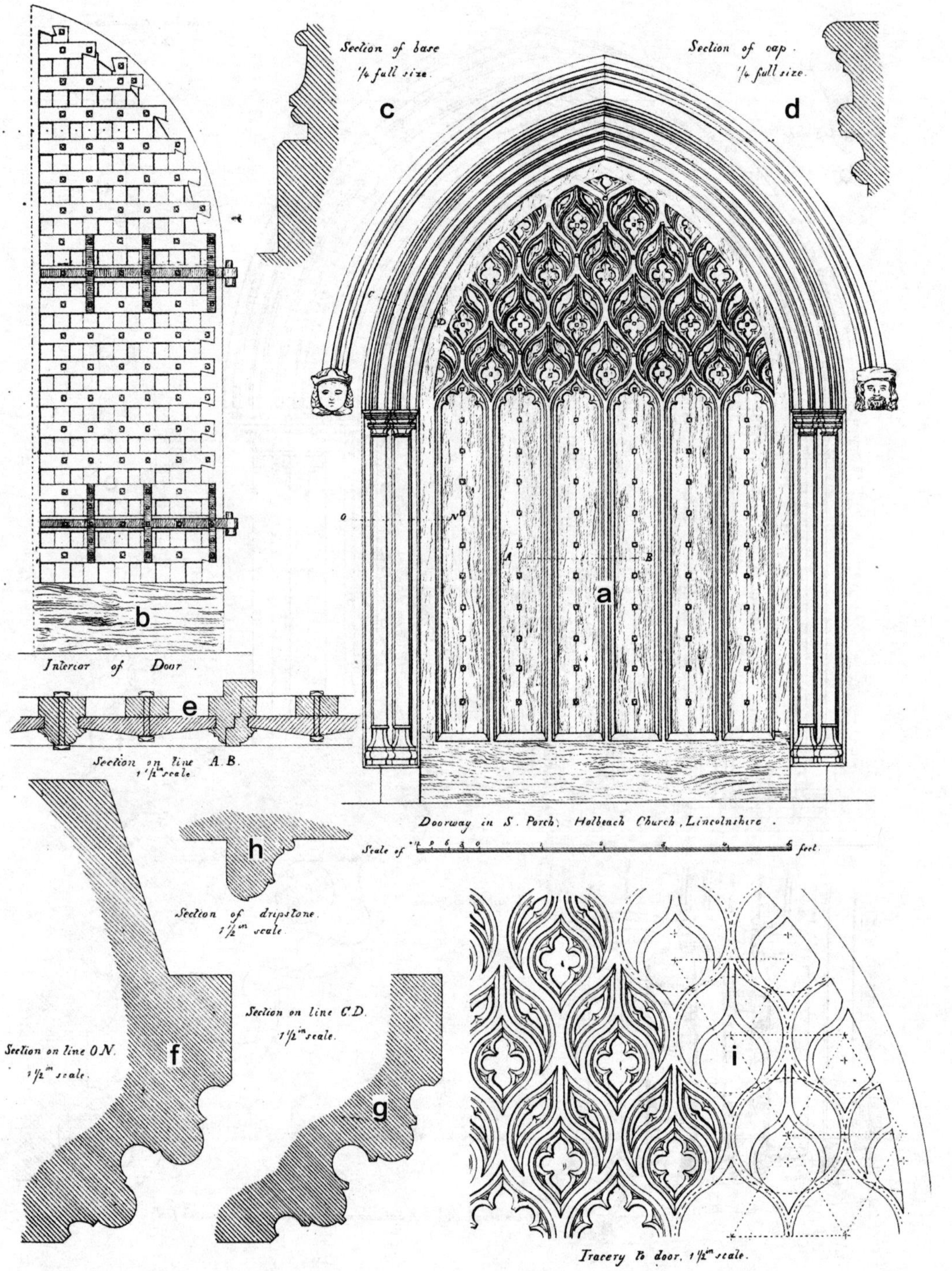

a. 林肯郡霍尔比奇教堂南门廊大门　b. 大门内侧　c. 柱础剖面　d. 柱头剖面　e.A–B 处的剖面　f. O–N 处的剖面　g.C–D 处的剖面　h. 披水石剖面　i. 门上花格饰

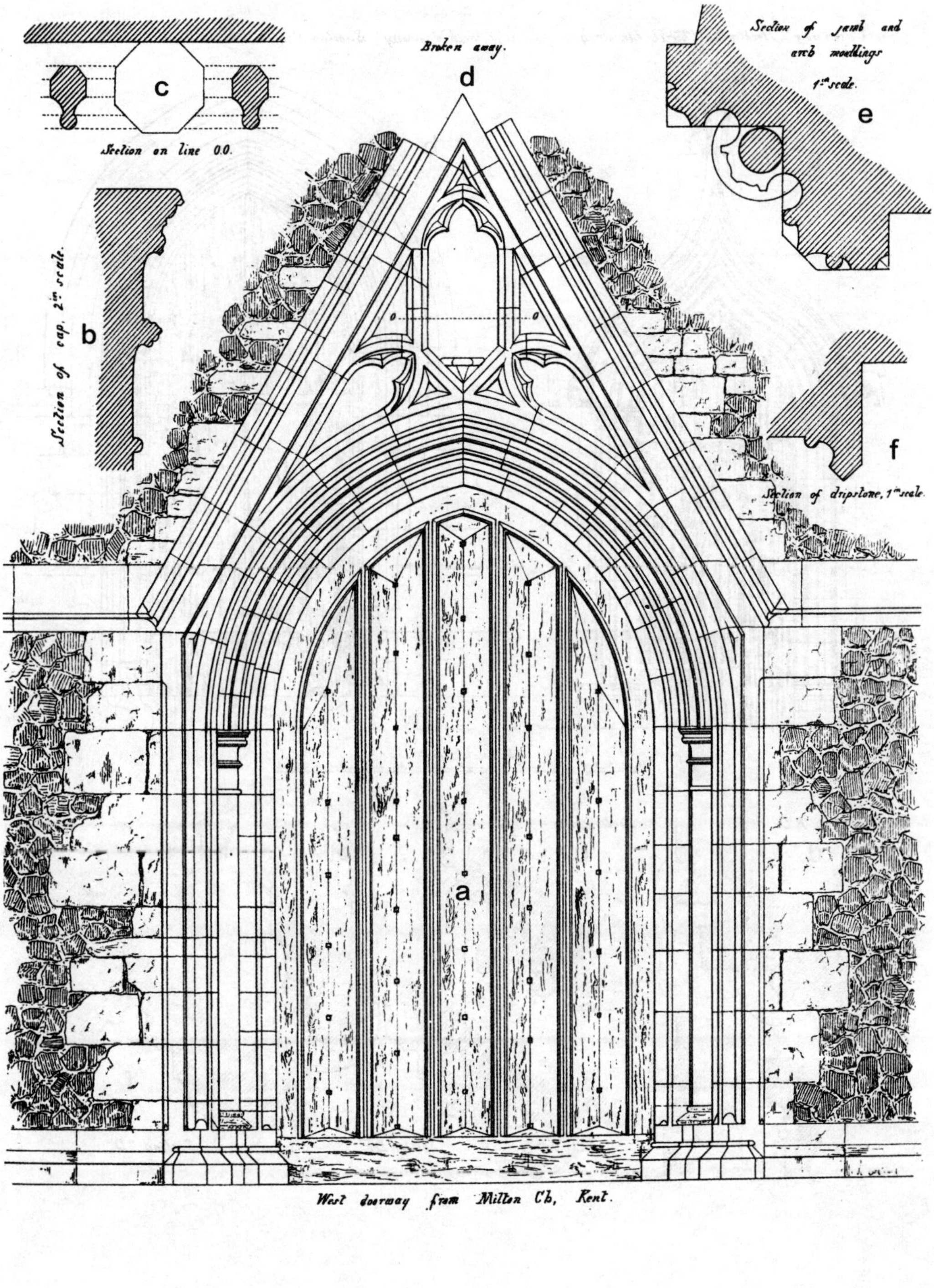

a. 肯特郡米尔顿教堂西门　b. 柱头剖面　c.O–O 处的剖面　d. 损坏　e. 侧壁及拱线脚剖面　f. 披水石剖面

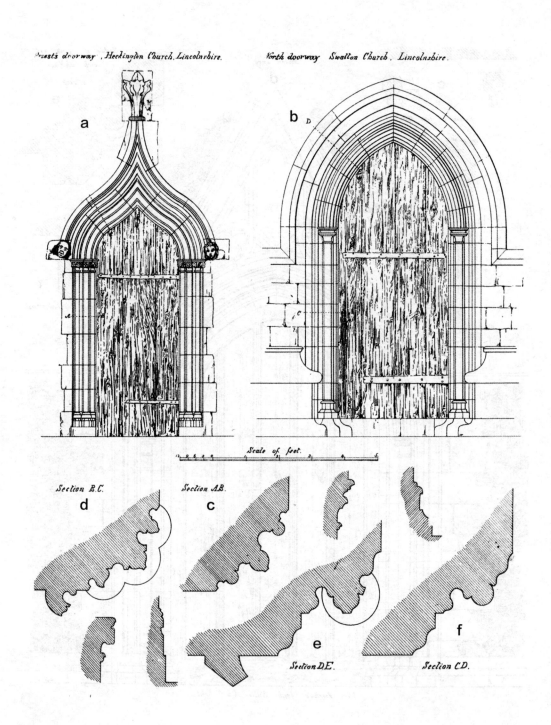

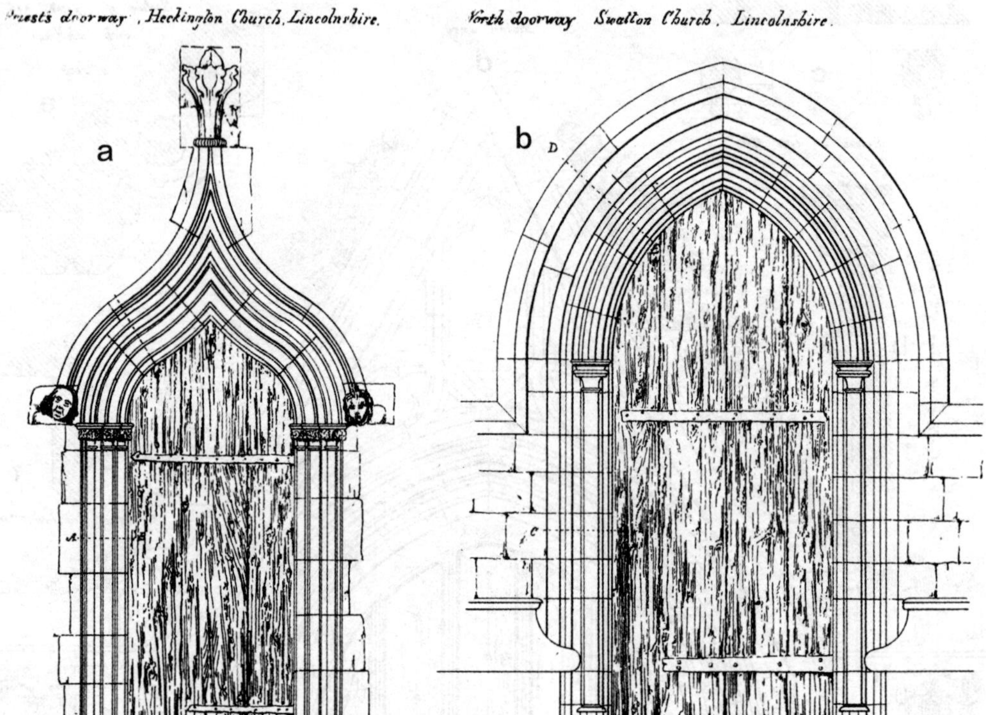

a. 林肯郡赫金顿教堂牧师之门　b. 林肯郡斯瓦顿教堂北门　c. A–B处的剖面　d.B–C处的剖面　e.D–E处的剖面　f.C–D 处的剖面

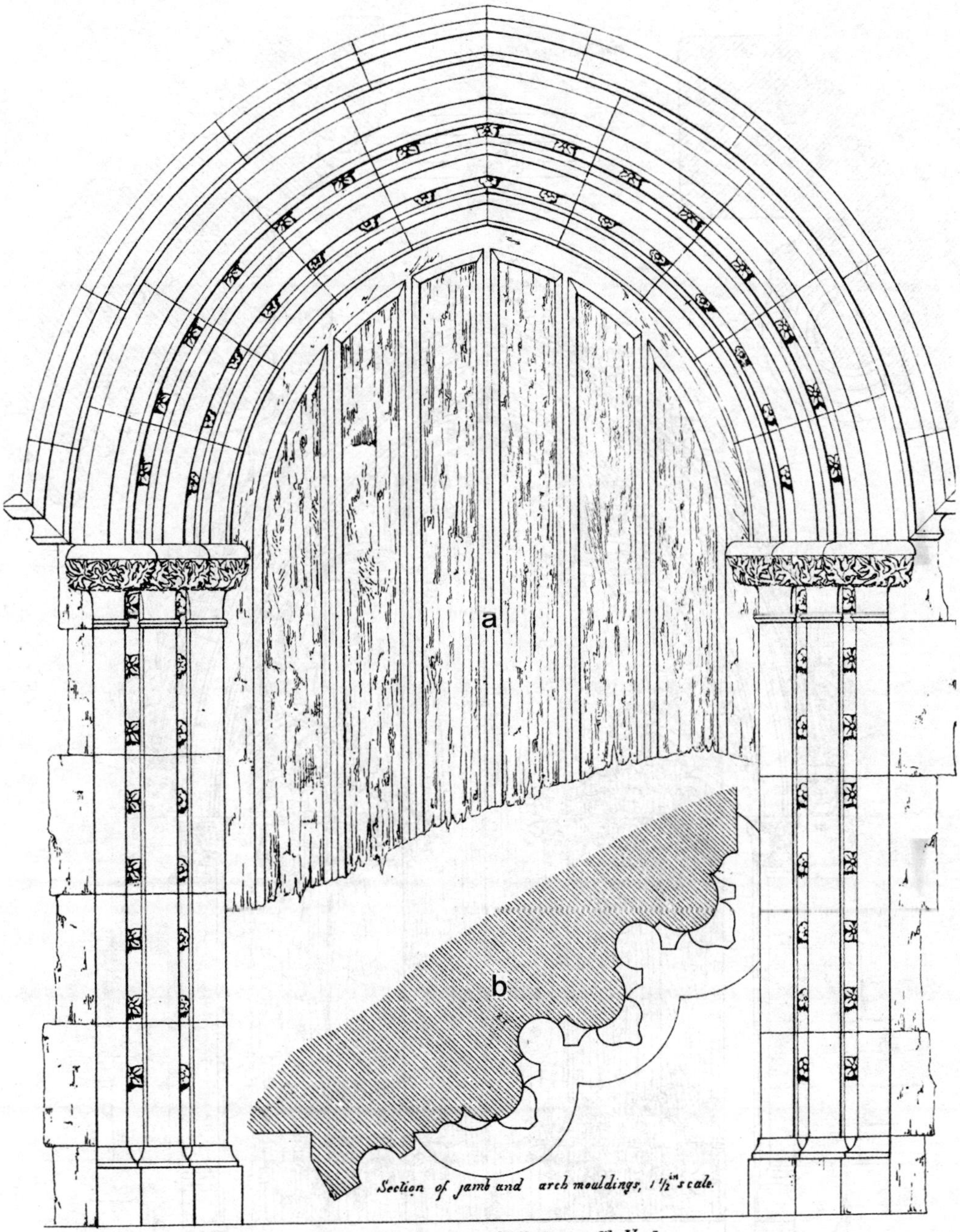

a. 赫特福德郡北米姆斯教堂西门　b. 侧壁及拱线脚剖面

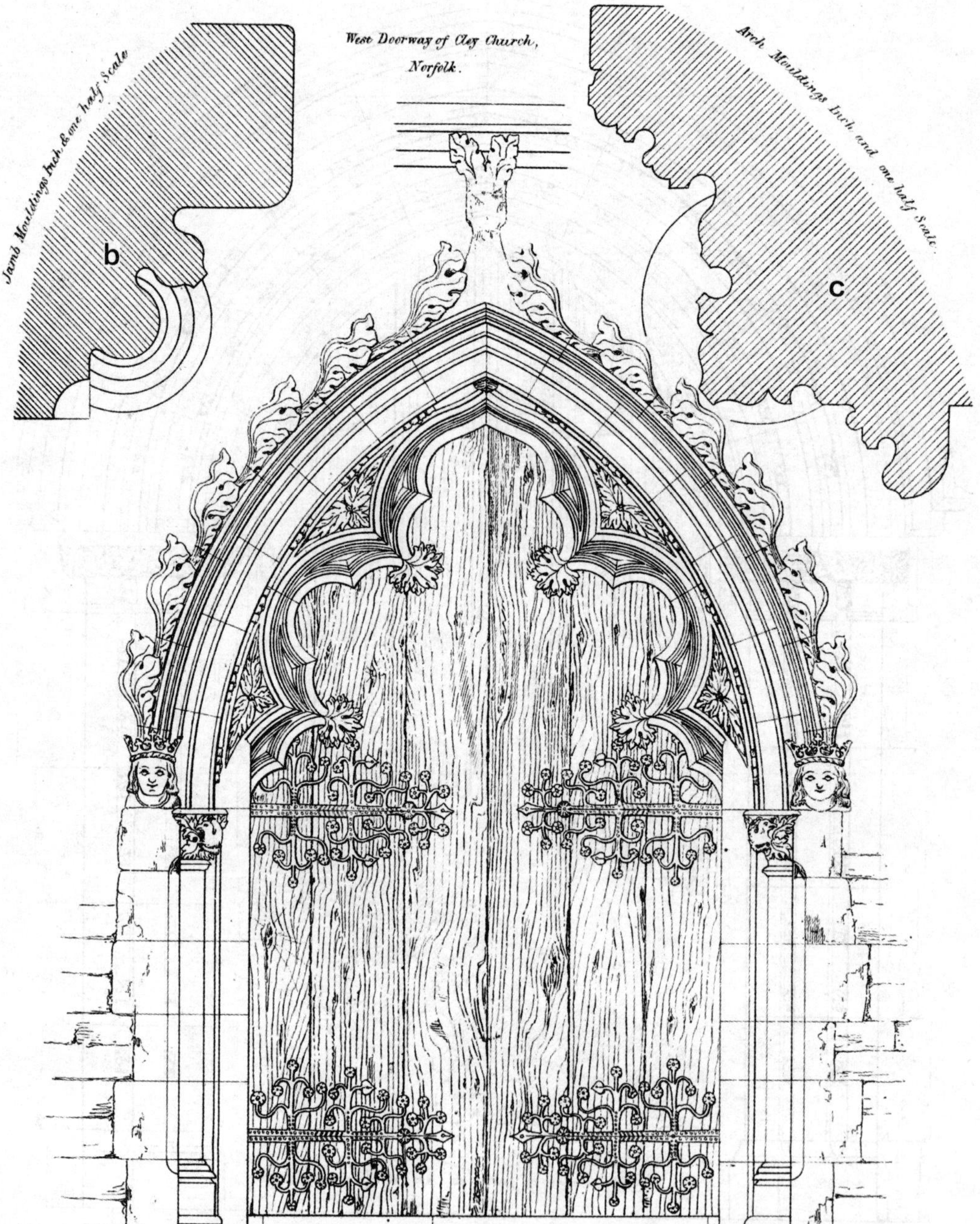

a. 诺福克郡克莱教堂西门　b. 侧壁线脚　c. 拱线脚

赫特福德郡圣阿尔班修道院南侧廊通向回廊大门

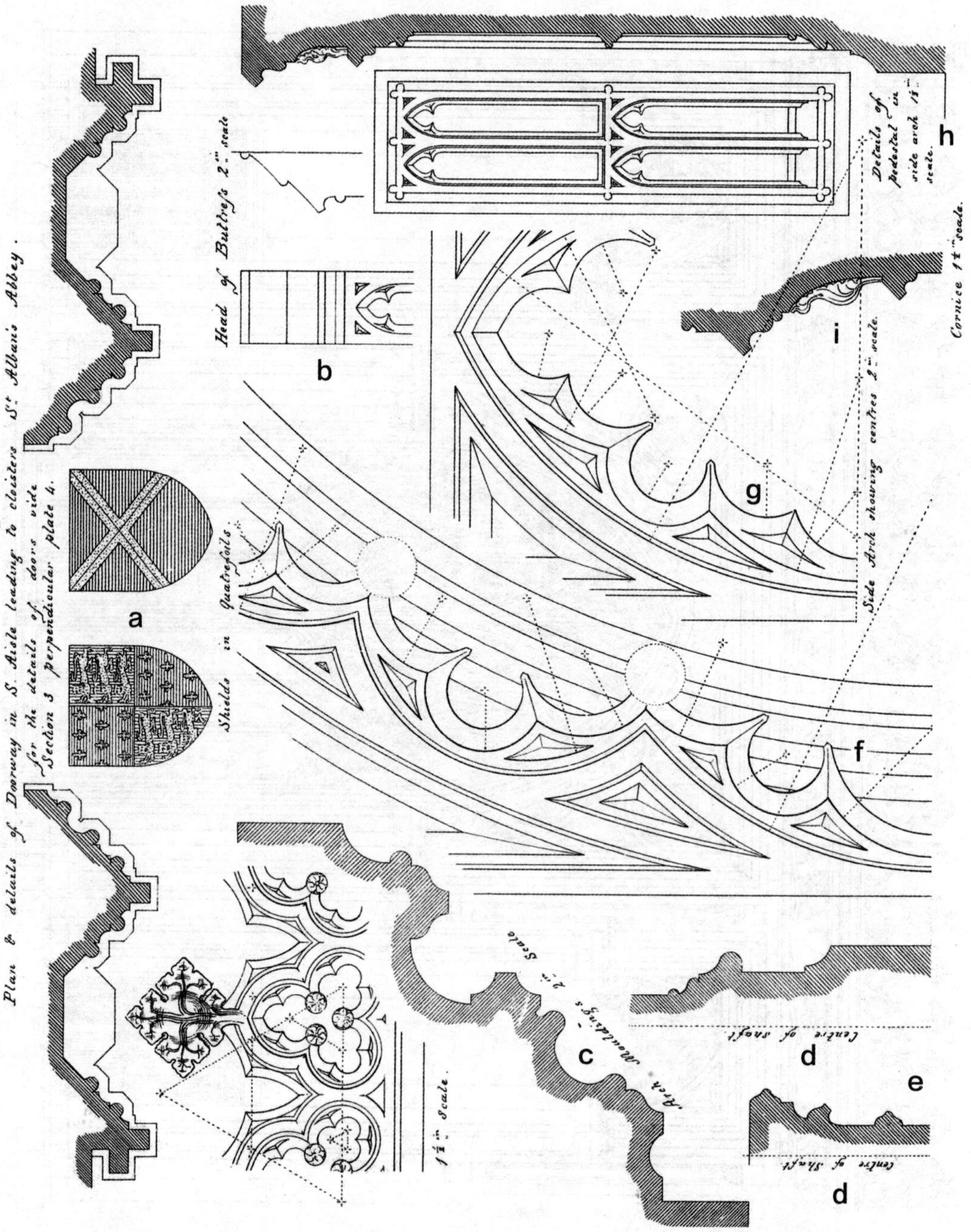

赫特福德郡圣阿尔班修道院南侧廊通向回廊大门平面及细部：a. 四叶饰内盾形纹章　b. 扶壁顶部　c. 拱线脚　d. 柱子中心线　e. 柱头及柱础剖面　f. 展示花饰中心点的中间拱局部　g. 展示花饰中心点的侧拱　h. 檐口侧拱基座细部　i. 飞檐

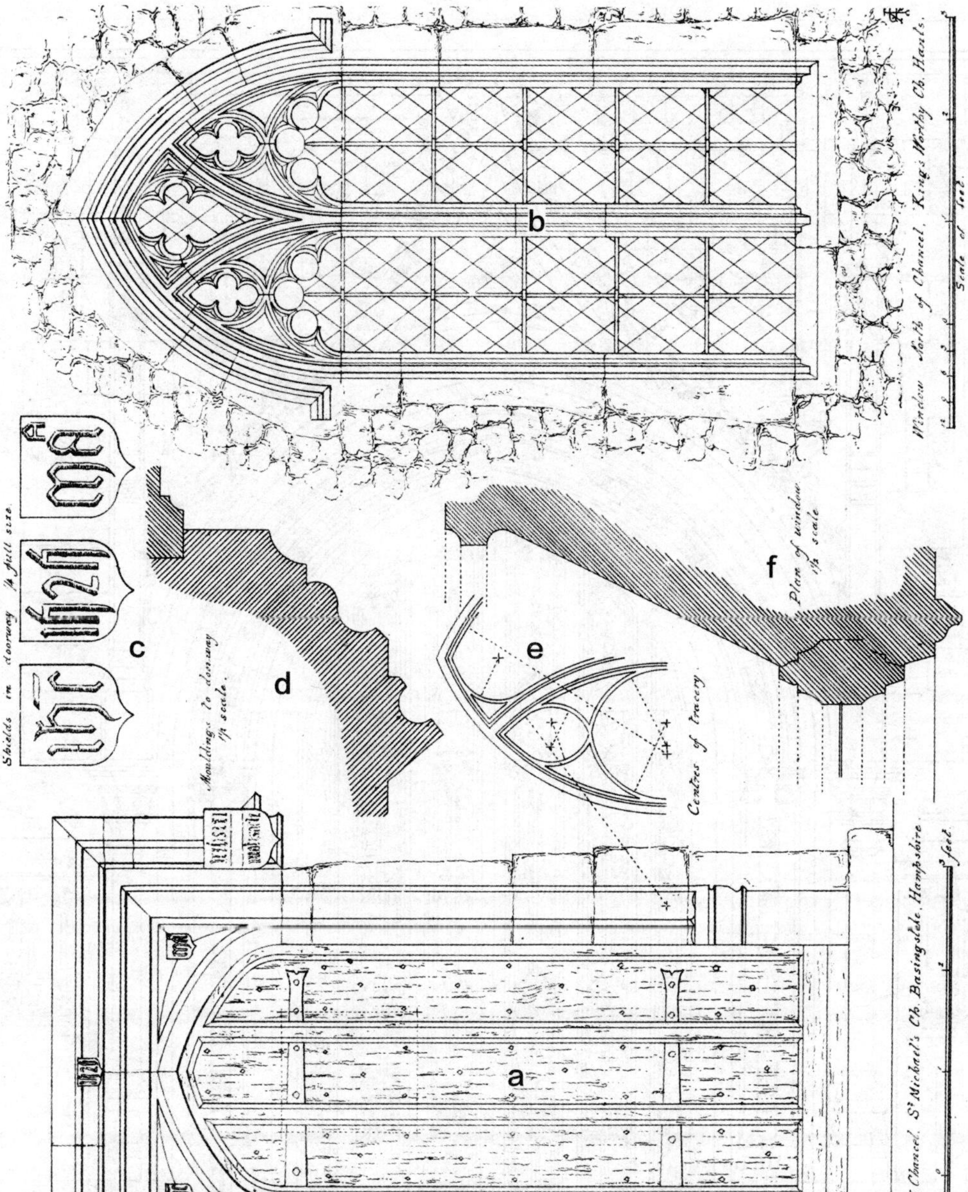

a. 汉普郡贝辛斯托克圣米迦勒教堂圣坛北侧牧师之门　b. 汉普郡金斯沃西教堂圣坛北侧窗户　c. 门上盾形纹章　d. 门上线脚　e. 花格窗饰中心点　f. 窗户平面

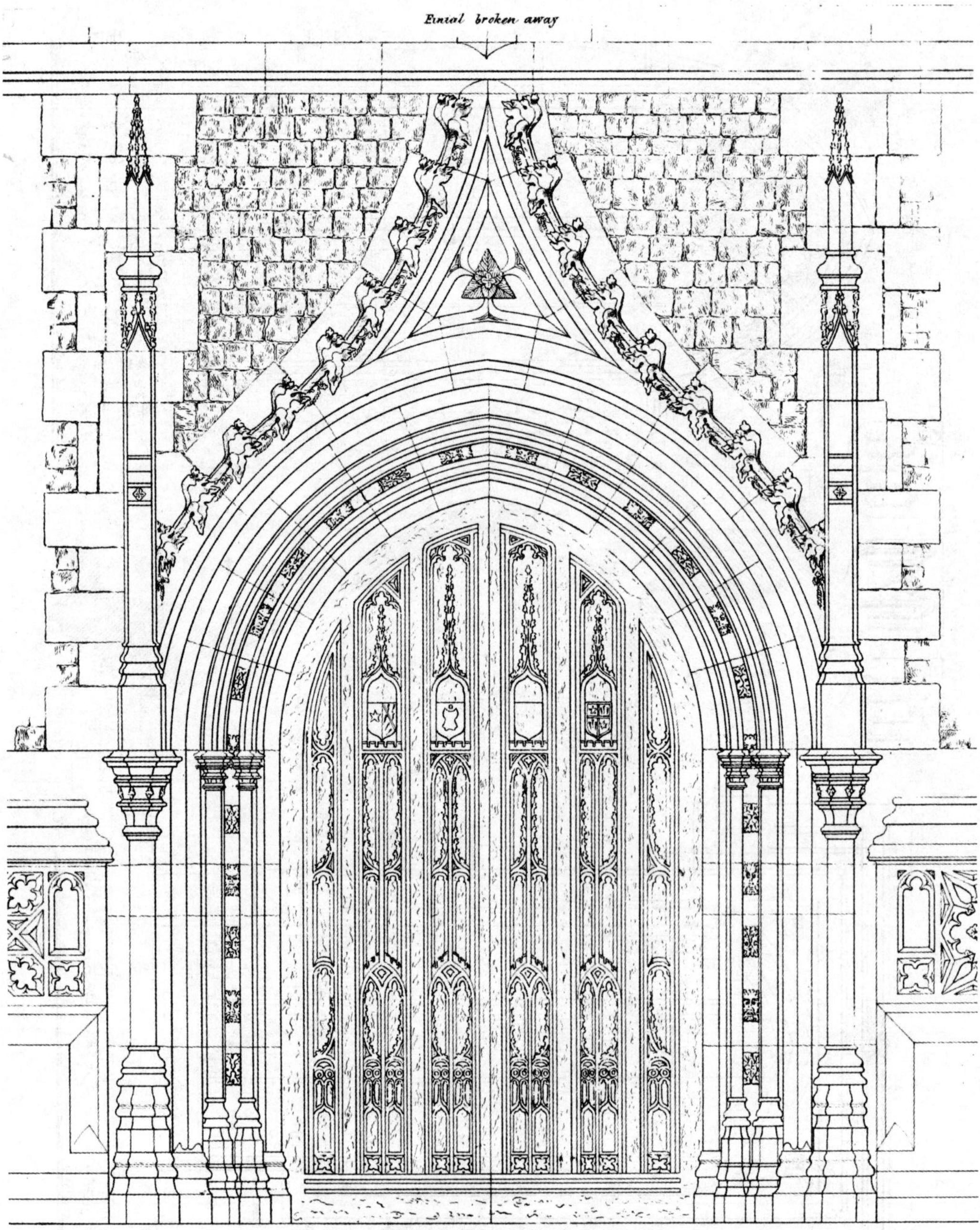

萨福克郡拉文纳姆教堂塔楼西门和尖顶饰被损坏

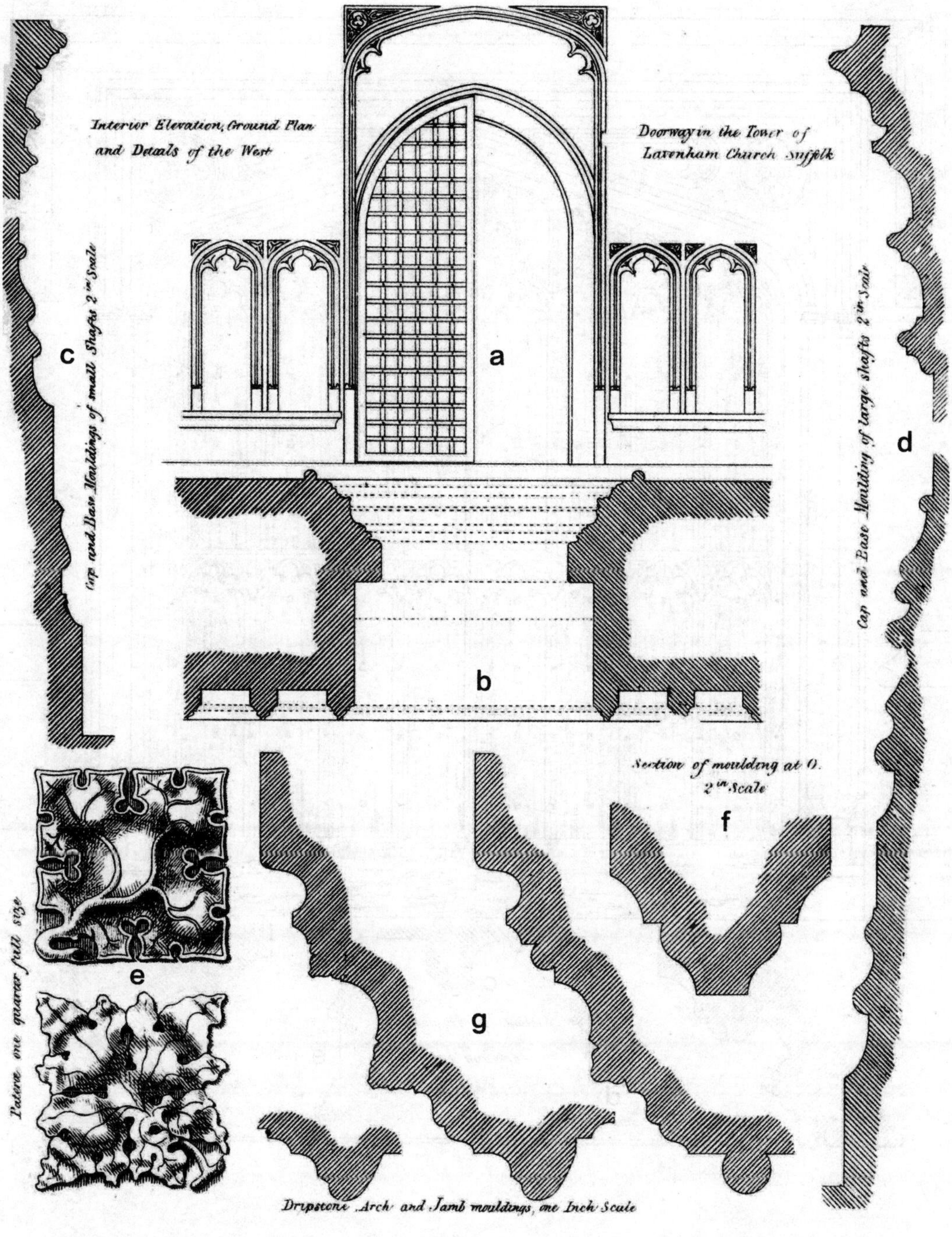

a. 萨福克郡拉文纳姆教堂塔楼西门内侧立面　b. 平面　c. 小柱子柱础及柱头线脚　d. 大柱子柱础及柱头线脚
e. 叶形饰　f.O 处的线脚剖面　g. 披水石、拱及侧壁线脚

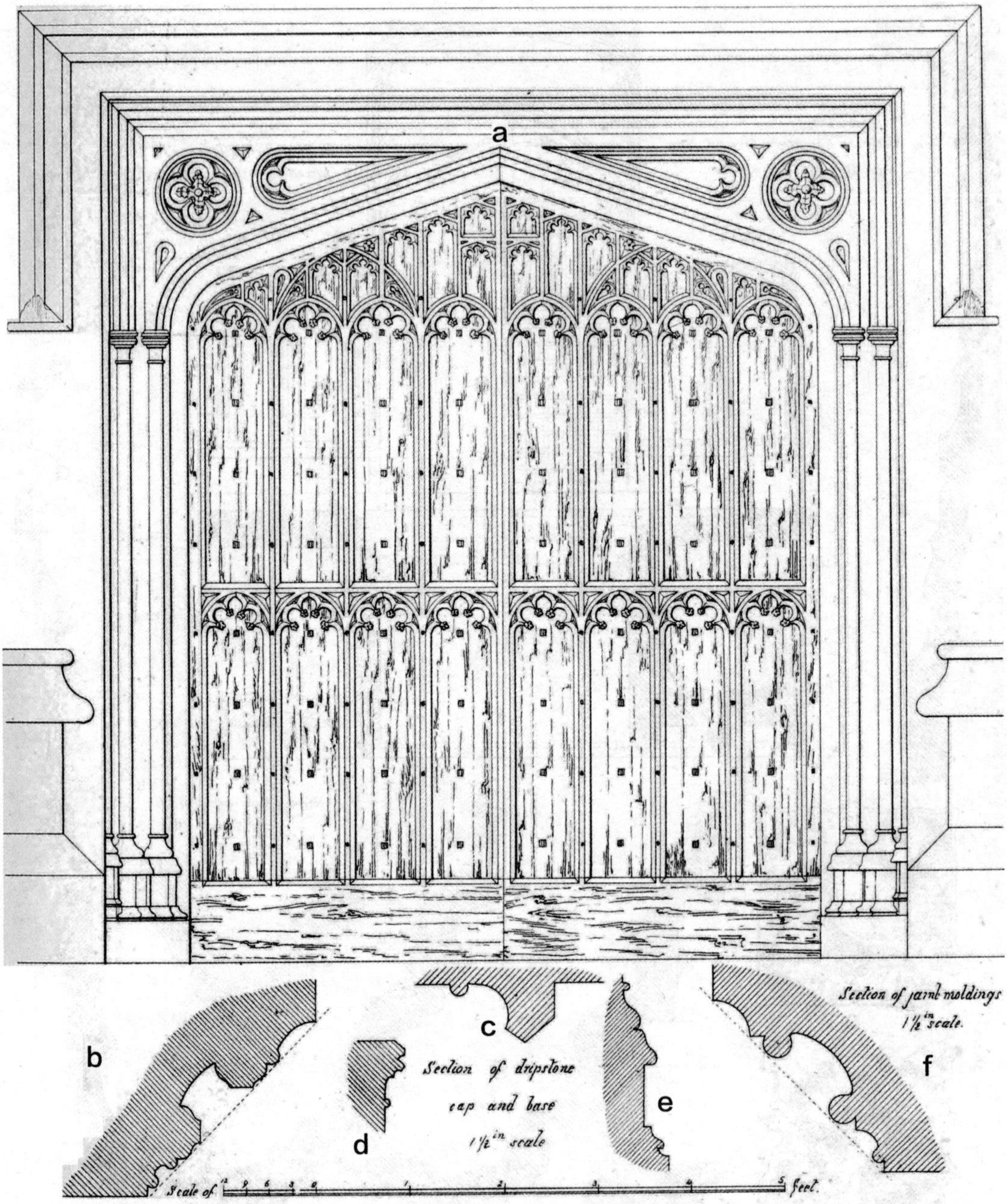

a. 巴克斯郡切舍姆教堂西门　b. 拱线脚剖面　c. 披水石　d. 柱头　e. 柱础剖面　f. 侧壁线脚剖面

a. 诺福克郡科尔蒂瑟尔教堂西门　b. 线脚剖面

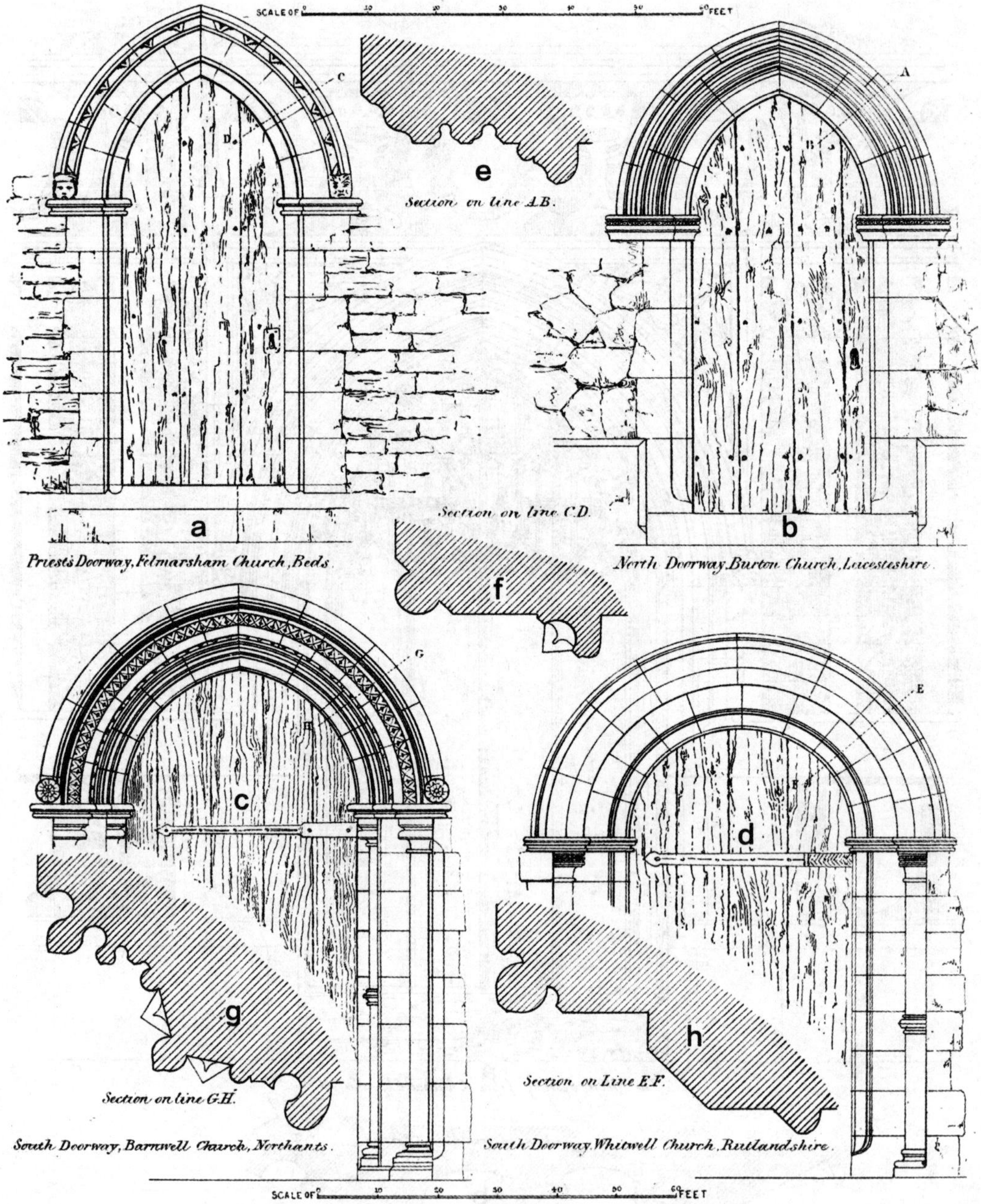

a. 贝德福德郡菲尔马什教堂牧师之门　b. 莱斯特郡伯顿教堂北门　c. 北安普敦郡巴恩韦尔教堂南门　d. 拉特兰郡惠特韦尔教堂南门　e.A–B 处的剖面　f.C–D 处的剖面　g. E–F 处的剖面　h.G–H 处的剖面

a. 赫特福德郡圣阿尔班修道院教堂门　b. 门底座线脚剖面　c. 门底座线脚　d.A–B 处的剖面

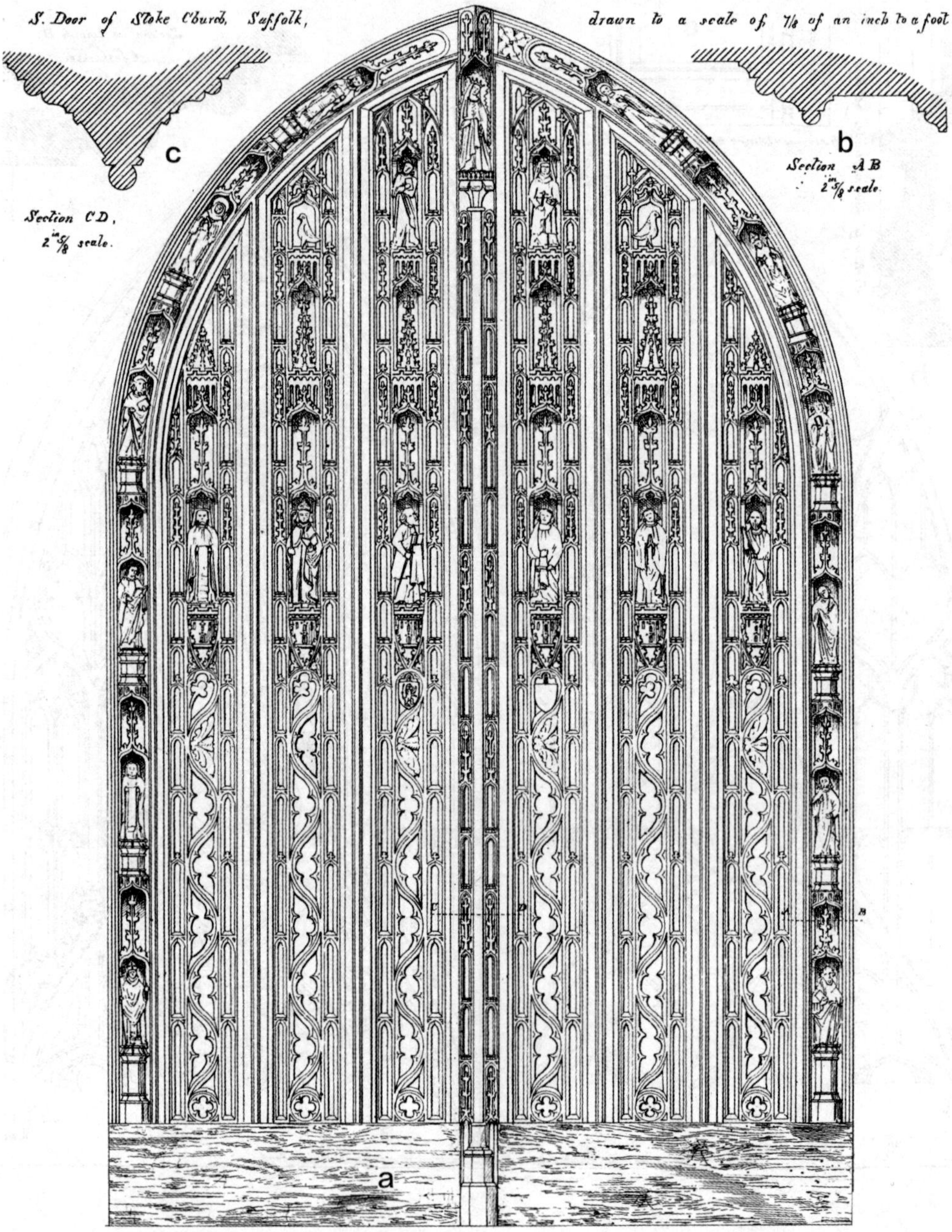

a. 萨福克郡斯托克教堂南门　b.A-B 处的剖面　c.C-D 处的剖面

第3章

窗

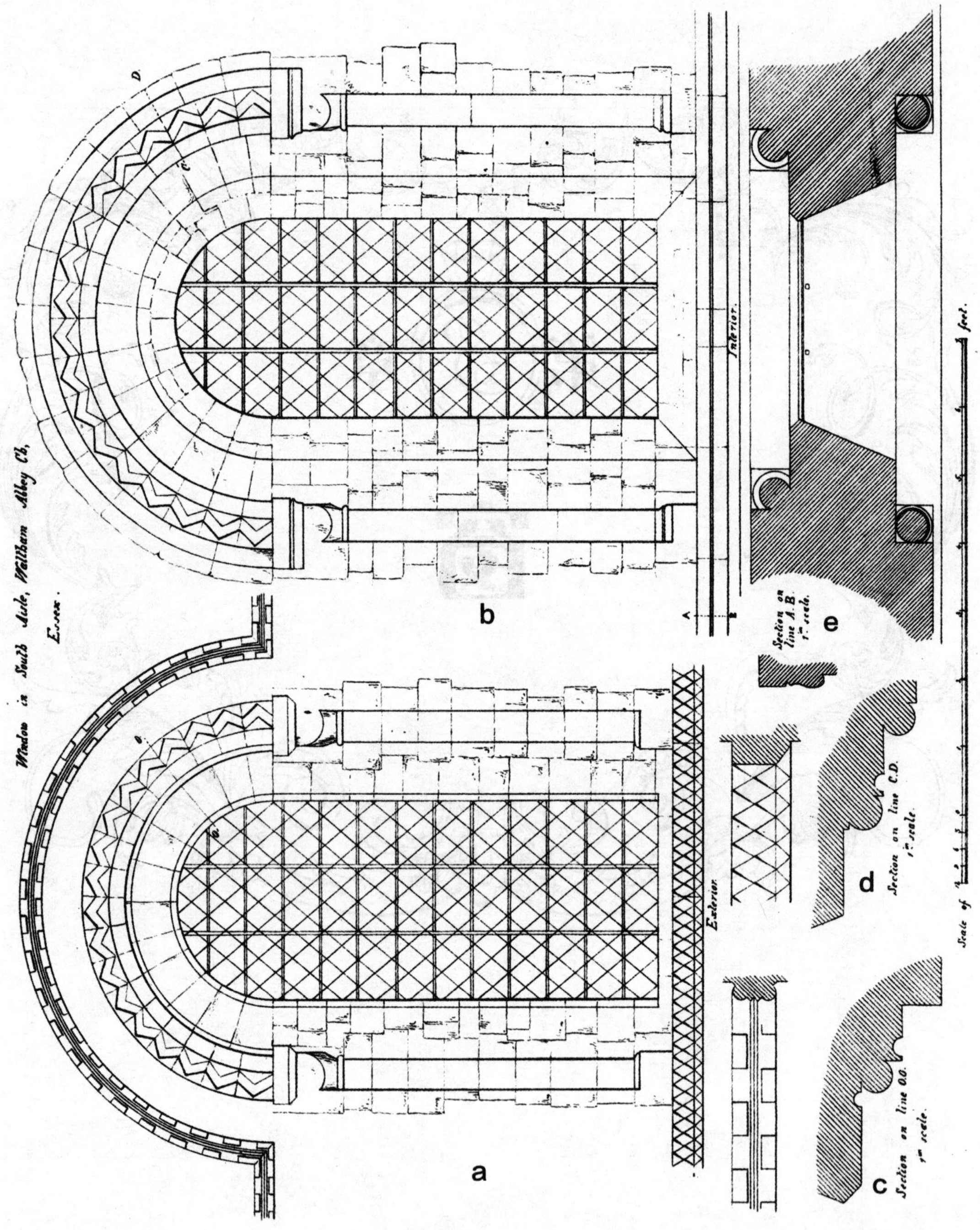

萨塞克斯郡沃尔瑟姆修道院教堂南侧廊窗户：a. 外侧　b. 内侧　c.O–O 处的剖面　d.C–D处的剖面　e. A–B处的剖面

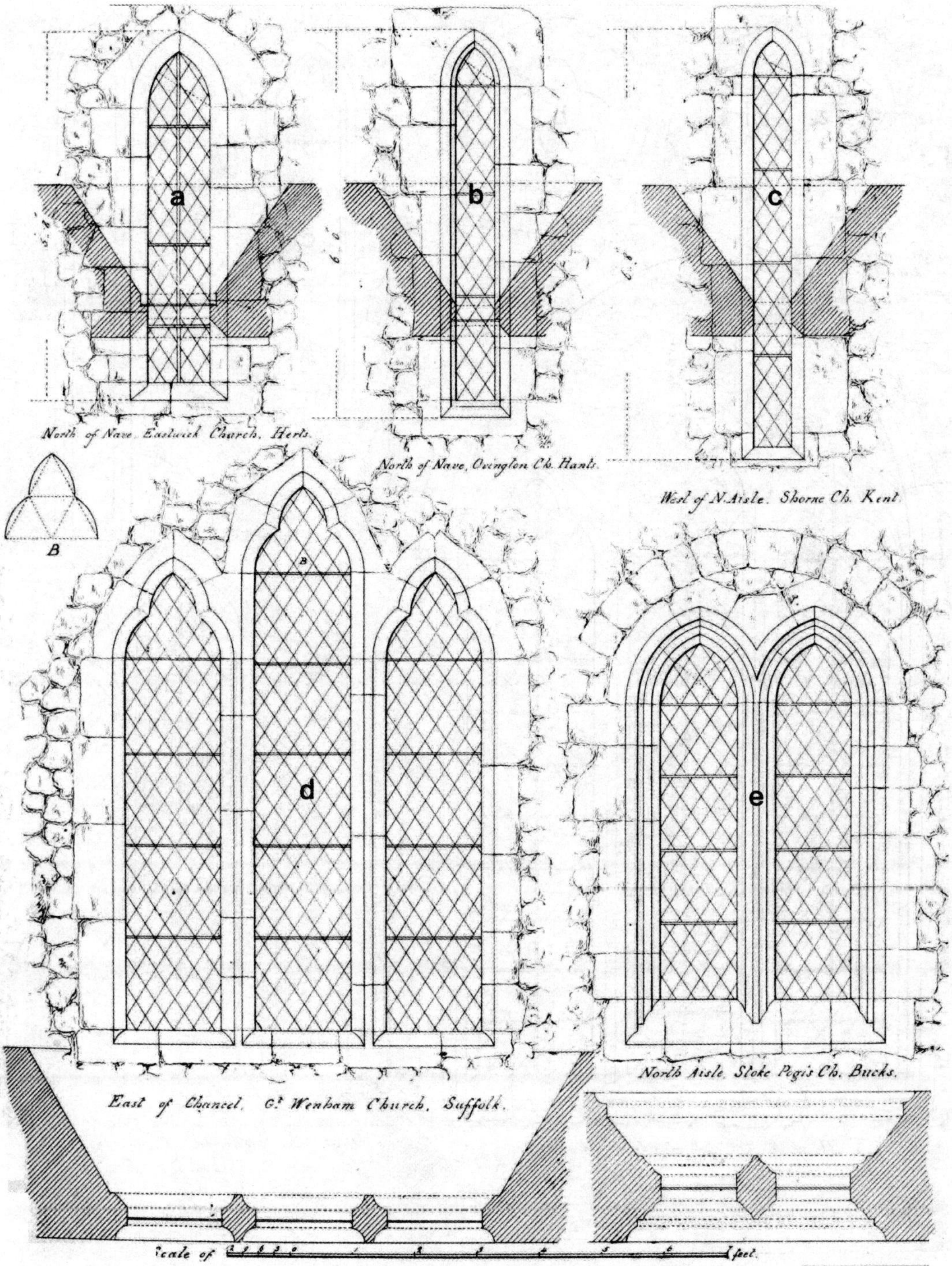

a. 赫特福德郡伊斯特威克教堂中殿北侧　b. 汉普君奥文顿教堂中殿北侧　c. 肯特郡肖勒教堂北侧廊西部　d. 萨福克郡韦纳姆大教堂圣坛东侧　e. 巴克斯郡斯托克伯吉斯教堂南侧廊

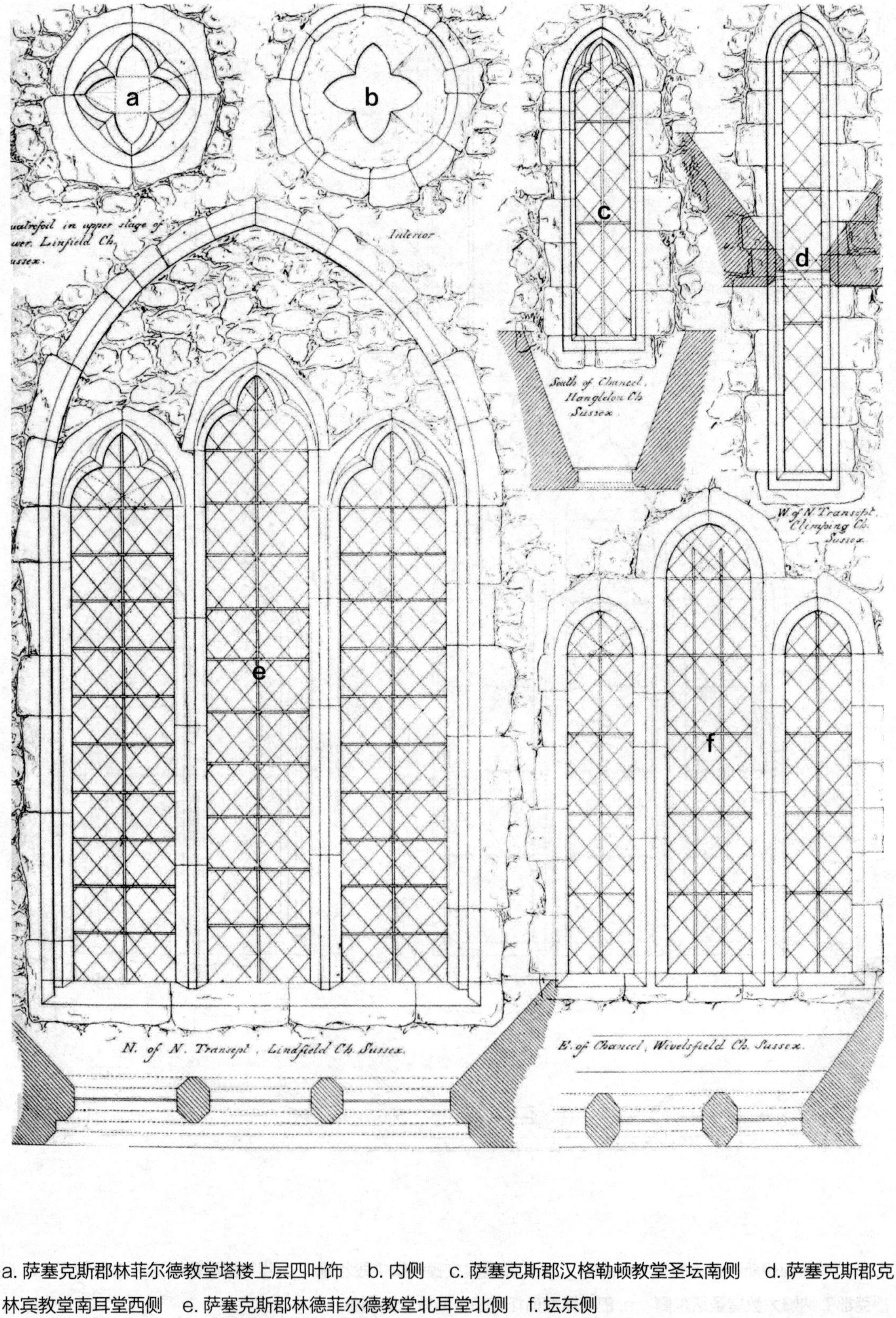

a. 萨塞克斯郡林菲尔德教堂塔楼上层四叶饰　b. 内侧　c. 萨塞克斯郡汉格勒顿教堂圣坛南侧　d. 萨塞克斯郡克林宾教堂南耳堂西侧　e. 萨塞克斯郡林德菲尔德教堂北耳堂北侧　f. 坛东侧

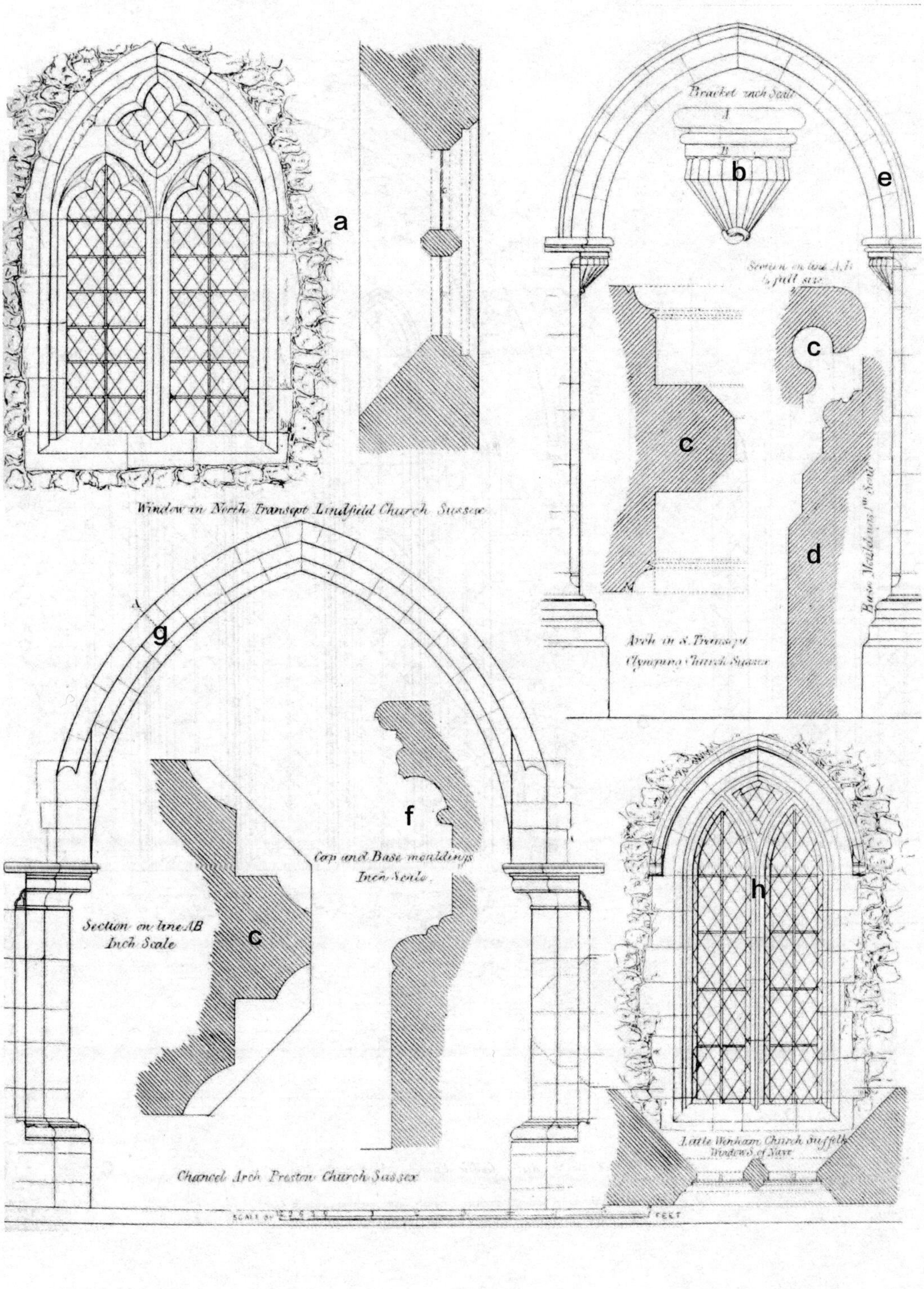

a. 萨塞克斯郡林德菲尔德教堂北耳堂窗户　b. 托架　c.A-B 处的剖面　d. 柱础　e. 萨塞克斯郡克林宾教堂南耳堂内拱　f. 柱头及柱础线脚　g. 萨塞克斯郡普勒斯顿教堂圣坛内拱　h. 萨福克郡韦纳姆小教堂中殿窗户

a. 北安普敦郡昂德尔教堂南侧廊西侧窗户　b. 窗台剖面　c. 披水石剖面　d. 竖框剖面

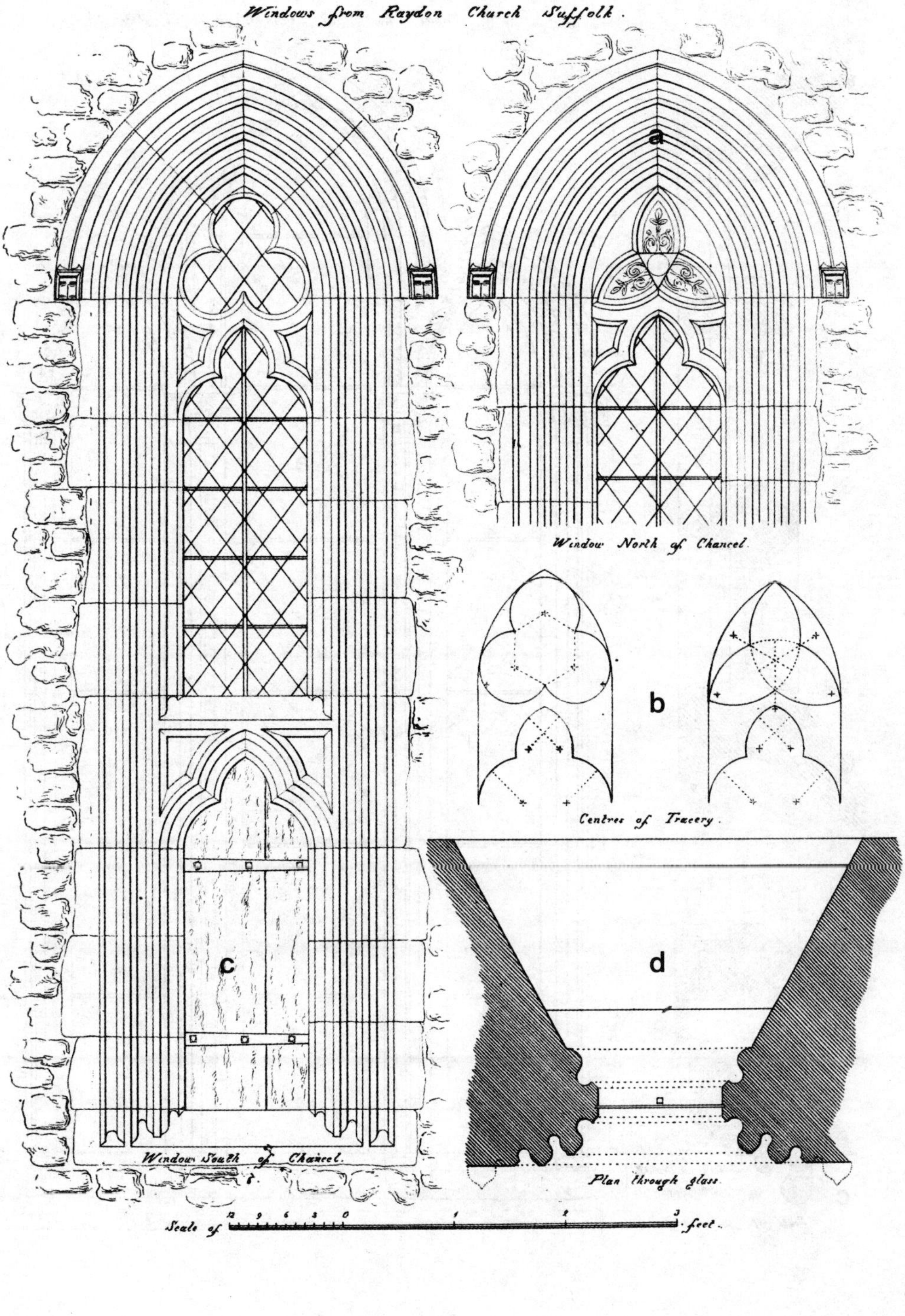

萨福克雷东教堂窗户：a. 圣坛北侧窗户　b. 花饰窗格中心　c. 圣坛南侧窗户　d. 玻璃处剖面

a. 威尔特郡威利教堂圣坛东侧三联窗　b. 窗台剖面　c.A–B 处的剖面

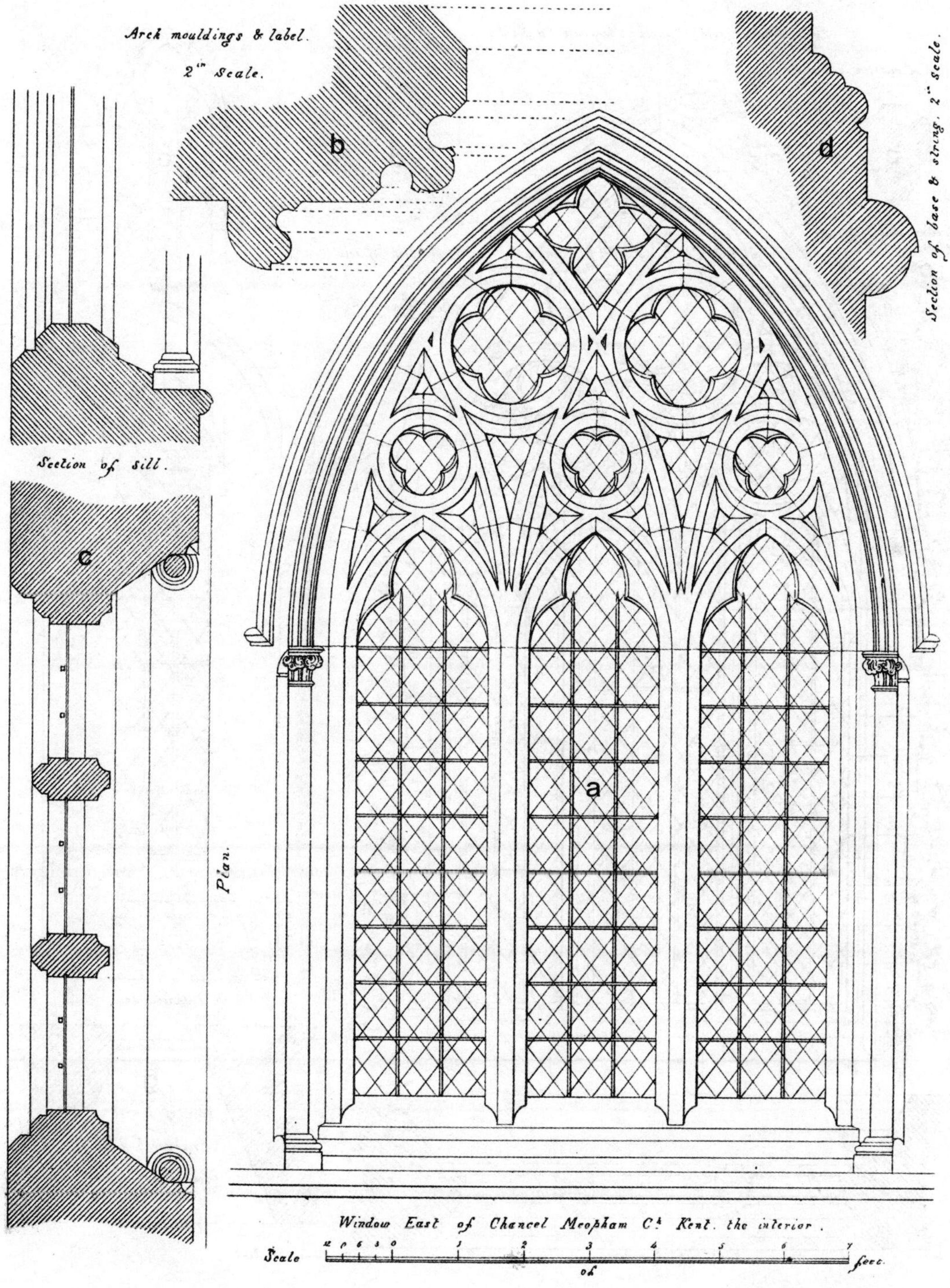

a. 肯特郡梅珀姆教堂圣坛东侧窗户内侧　b. 披水石及拱线脚　c. 窗台剖面　d. 底座及束带层剖面

a. 威尔特郡威利教堂圣坛东侧窗户　b. 竖框剖面　c. A-B 处的剖面　d. 窗台剖面　e. 外侧　f. 内侧　g. 格洛斯特郡布罗克沃斯教堂尖顶窗平面

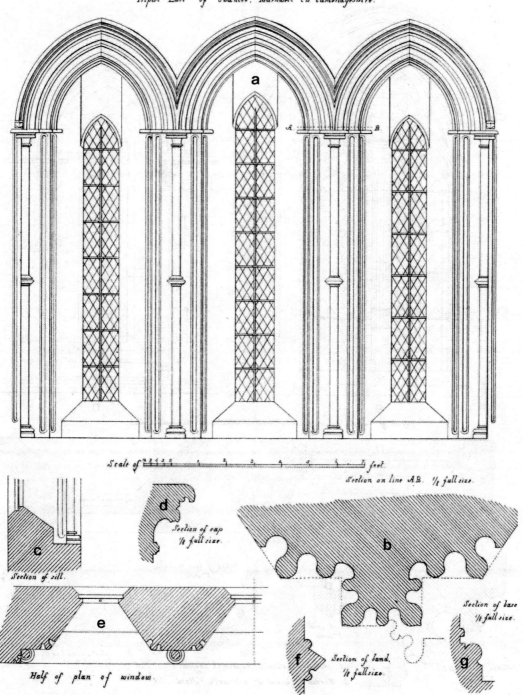

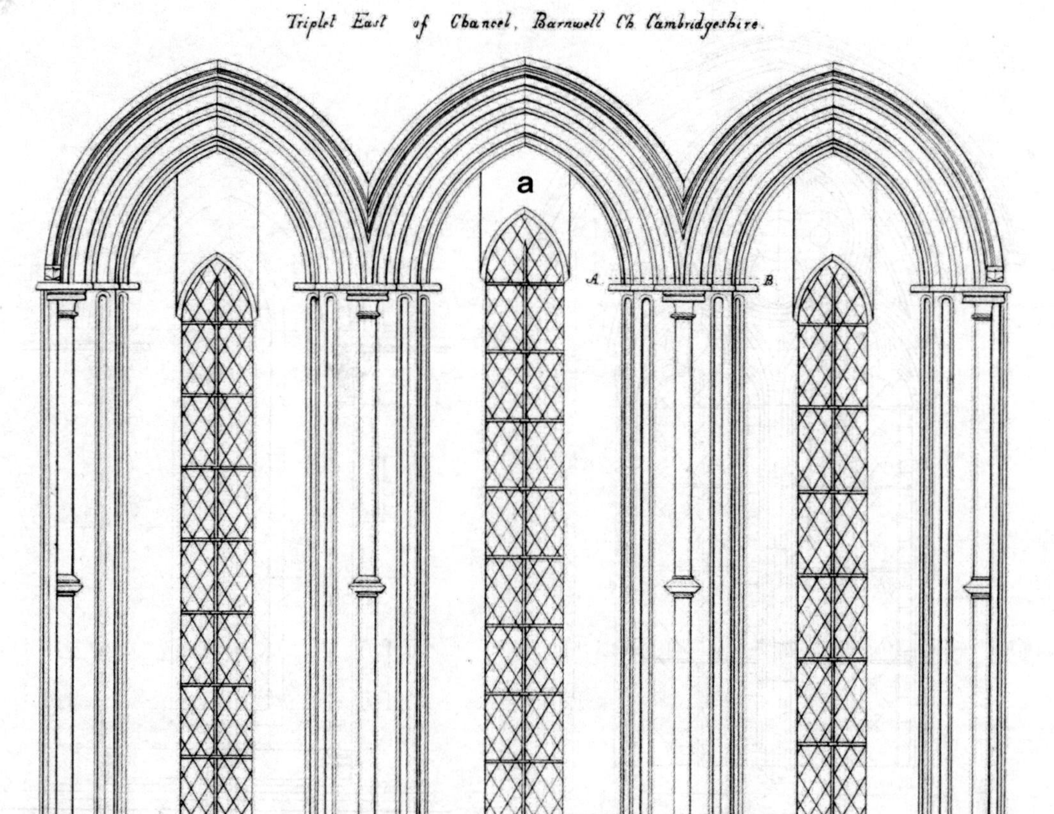

a. 剑桥郡巴恩韦尔教堂圣坛东侧三联窗　b.A–B 处的剖面　c. 窗台剖面　d. 柱头剖面　e. 窗户半平面　f. 束带剖面　g. 柱础剖面

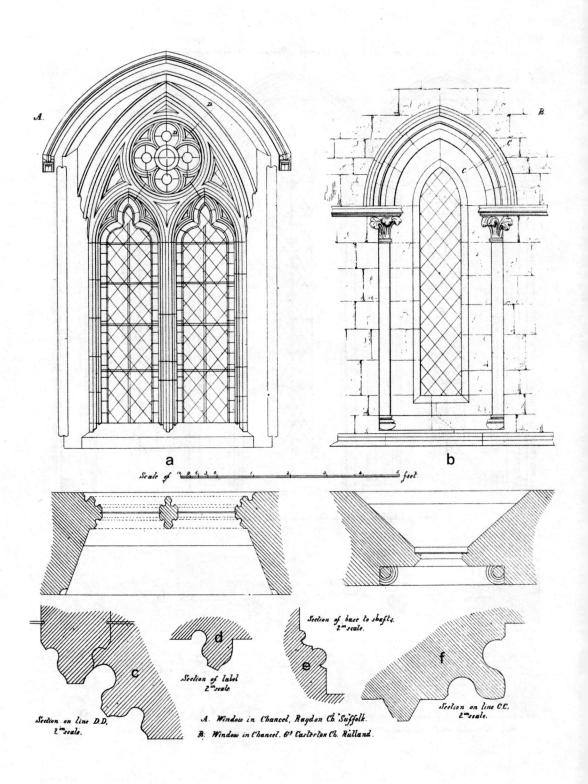

a. 萨福克雷东教堂圣坛窗户　b. 拉特兰郡卡斯特顿教堂圣坛窗户　c.D-D 处的剖面　d. 披水石剖　e. 柱础剖面　f.C-C 的剖面

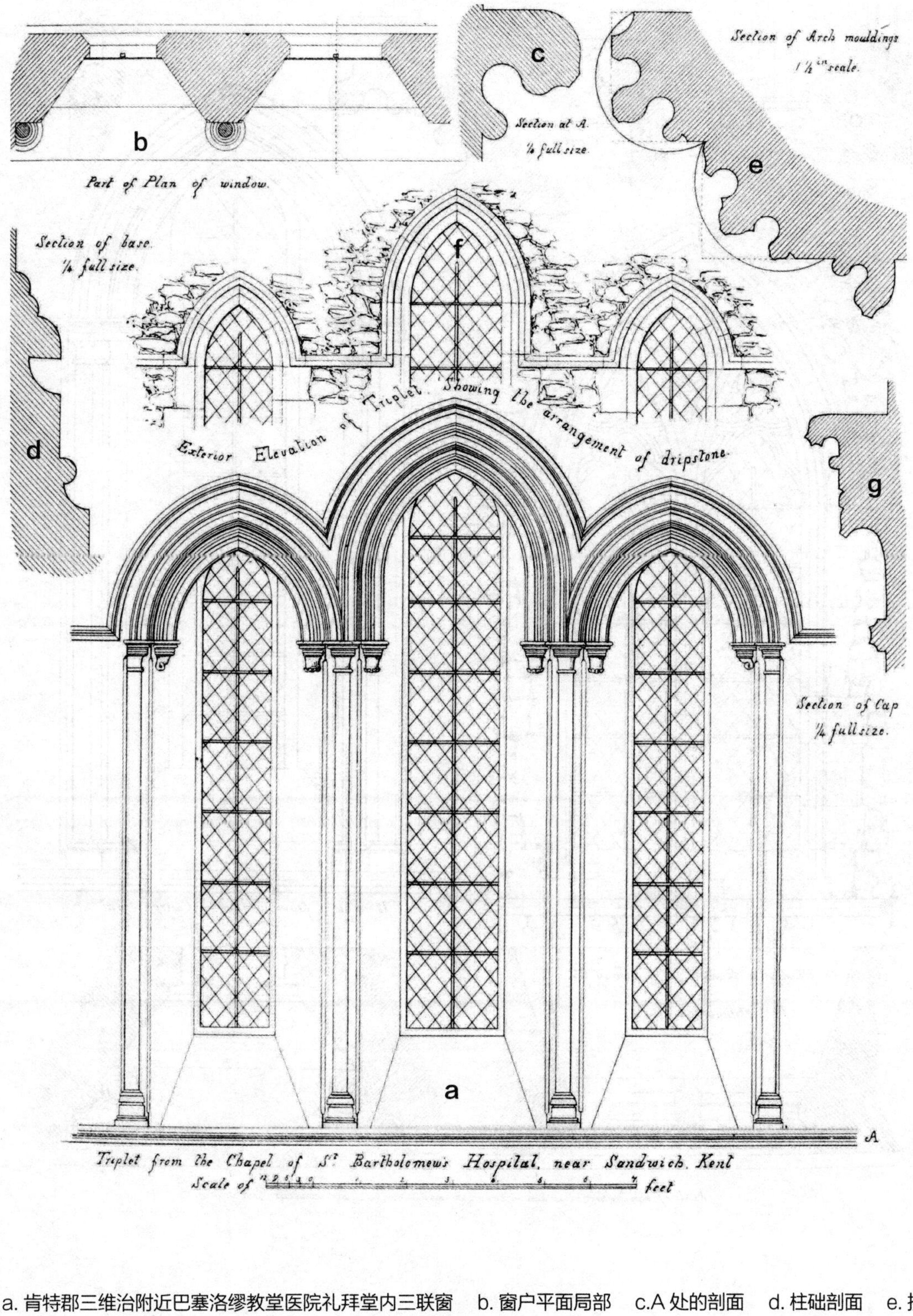

a. 肯特郡三维治附近巴塞洛缪教堂医院礼拜堂内三联窗　b. 窗户平面局部　c.A 处的剖面　d. 柱础剖面　e. 拱线脚剖面　f. 展示披水石布局的三联窗外侧　g. 柱头剖面

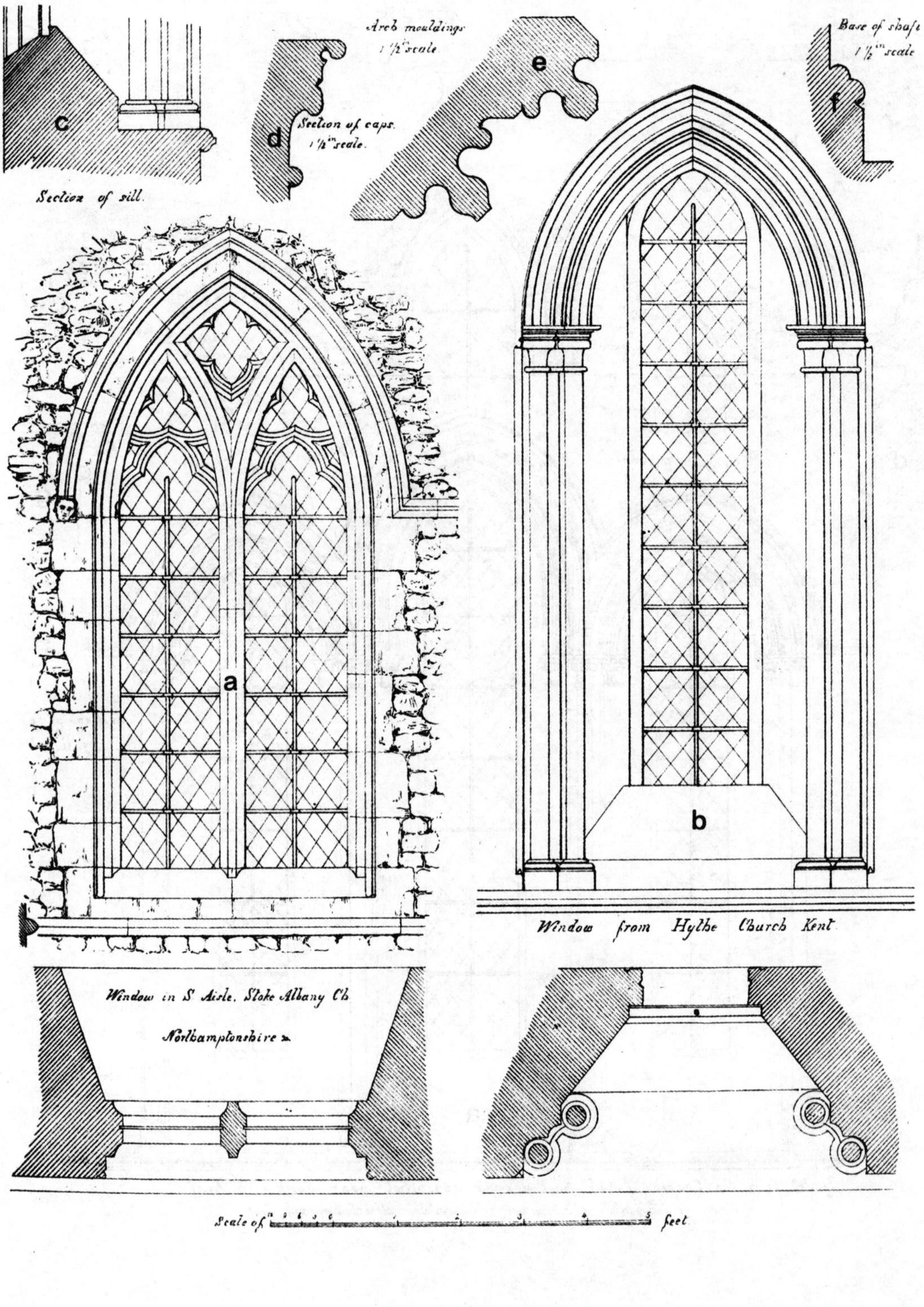

a. 北安普敦郡斯托克奥尔巴尼教堂南侧廊窗户　b. 肯特郡海斯教堂窗户　c. 窗台剖面　d. 柱头剖面　e. 拱线脚　f. 柱础

第3章 窗

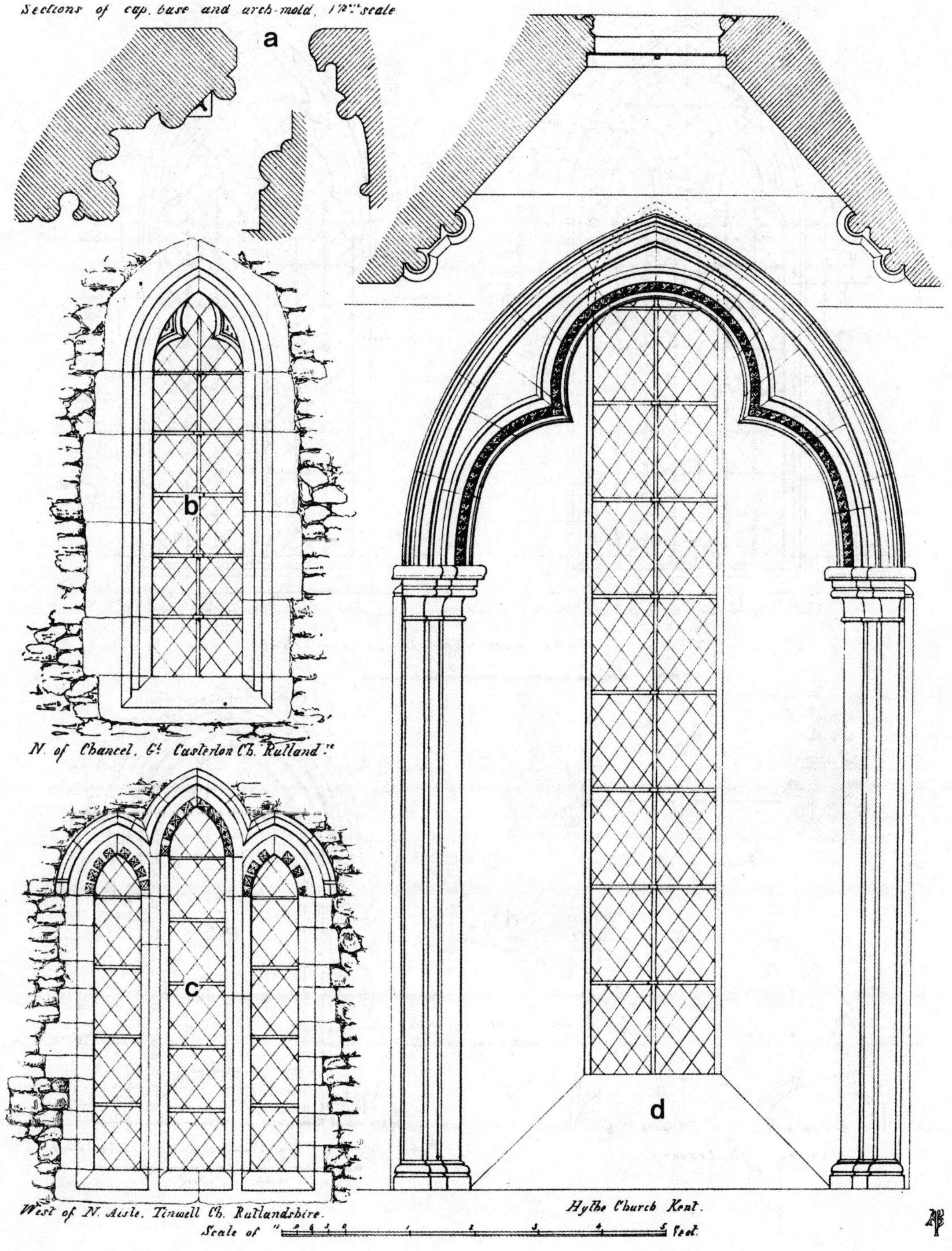

a. 柱头、柱础和拱线脚剖面 b. 拉特兰郡卡斯特顿教堂圣坛北侧 c. 拉特兰郡廷韦尔教堂北侧廊 d. 西部肯特郡海斯教堂

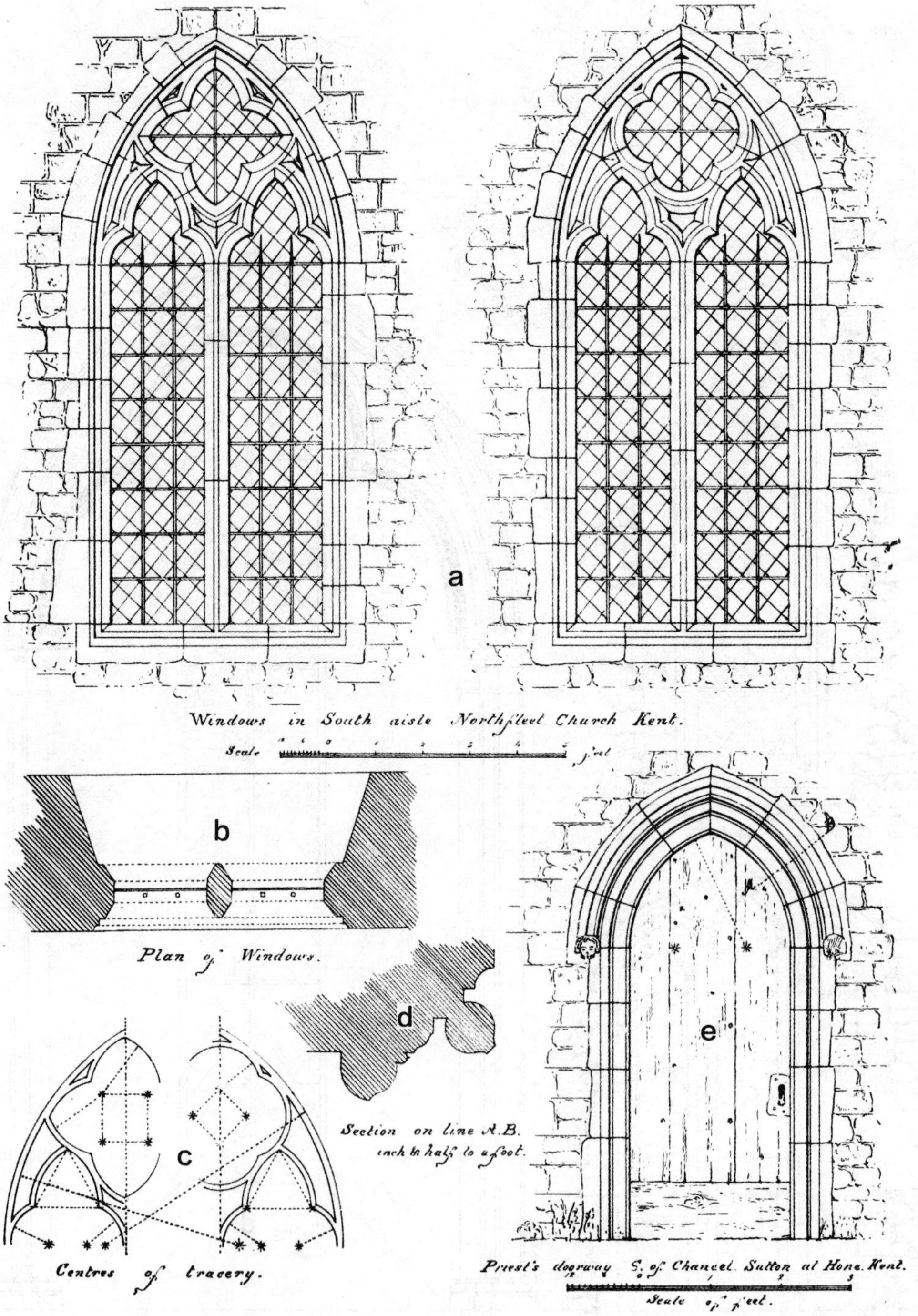

a. 肯特郡诺思弗利特教堂南侧廊窗户　b. 窗户平面　c. 花格窗饰中心点　d.A–B 处的剖面　e. 肯特郡霍恩萨顿教堂圣坛南侧牧师之门

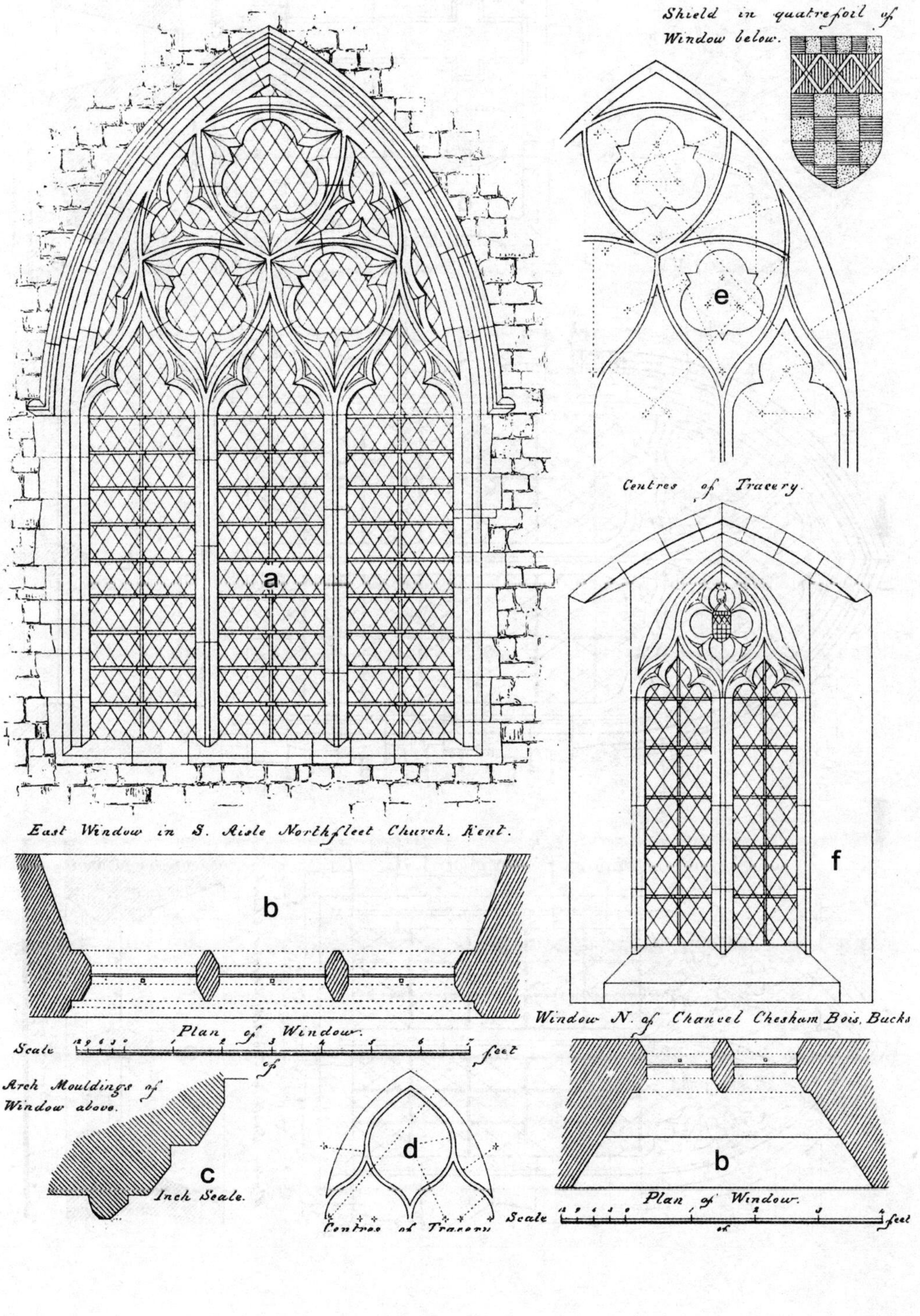

a. 肯特郡诺思弗利特教堂南侧廊东部窗户　b. 窗户平面　c. 窗户拱线脚　d. 花格窗饰中心点　e. 窗户四叶饰内盾形纹章　f. 巴克斯郡切舍姆博伊斯教堂圣坛北侧窗户

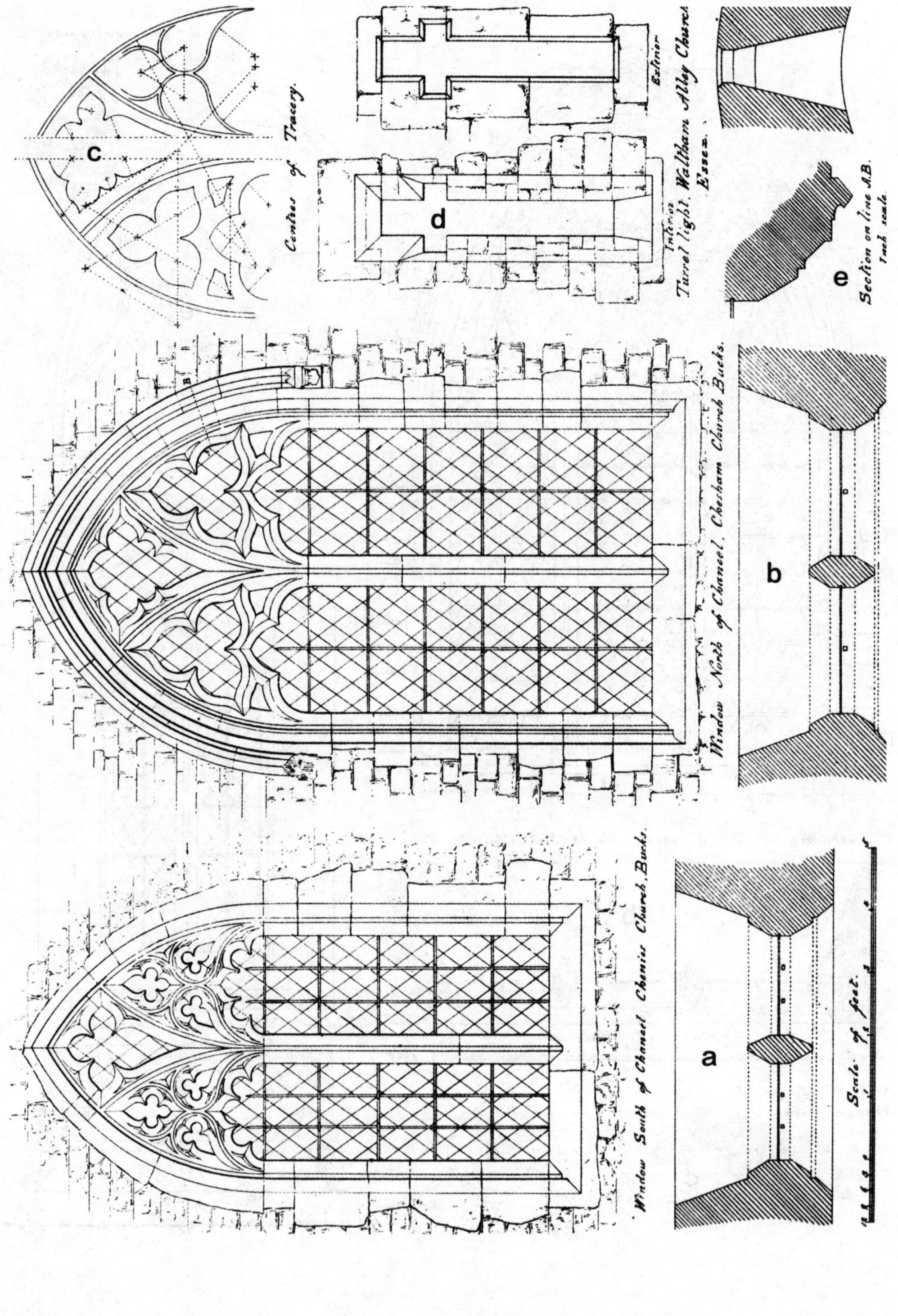

a. 巴克斯郡切尼斯教堂圣坛南侧窗户　b. 巴克斯切舍姆教堂圣坛北侧窗户　c. 花格窗饰中心点　d. 埃塞克斯郡沃尔瑟姆修道院教堂塔楼窗户　e.A–B 处的剖面

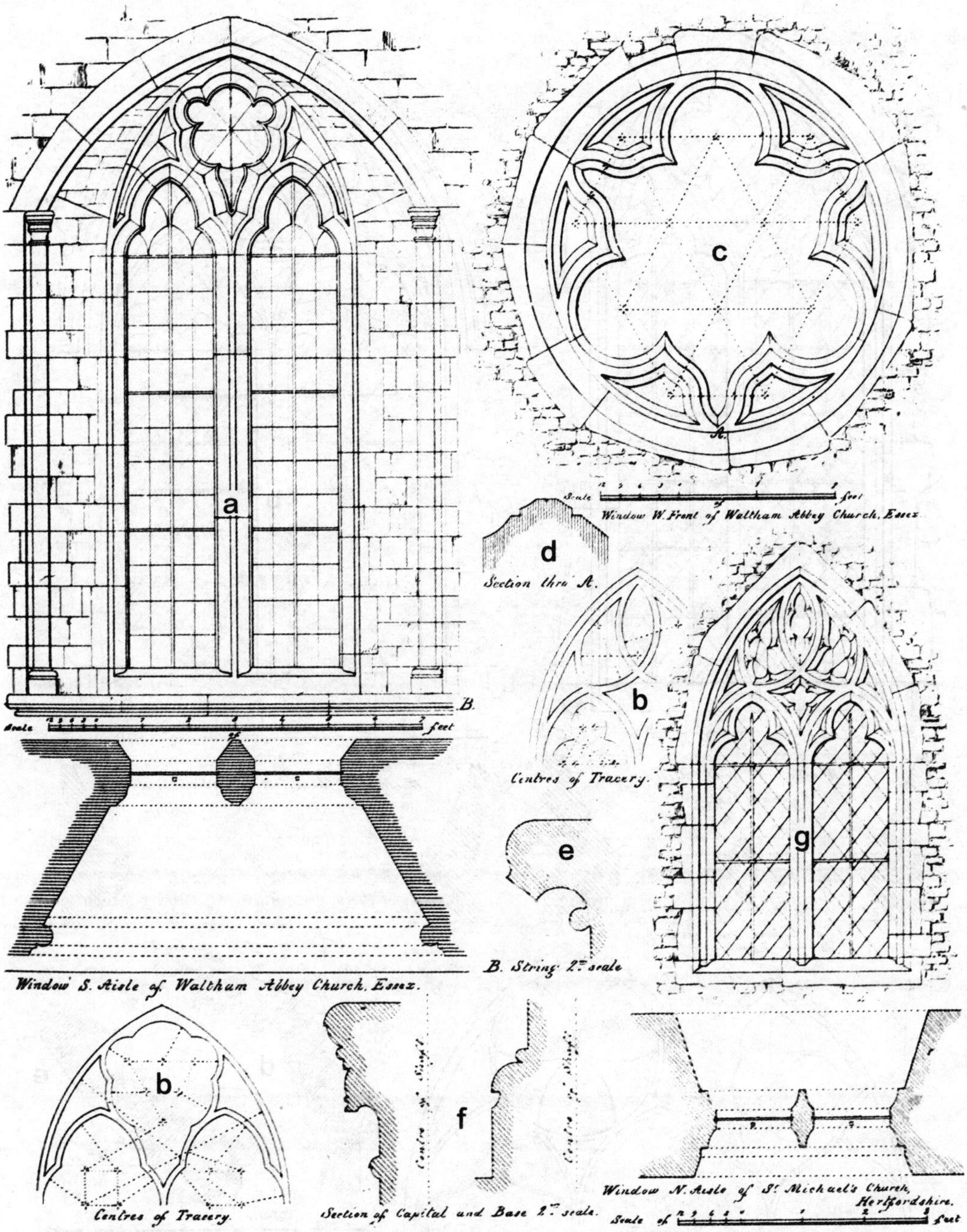

a. 埃塞克斯郡沃尔瑟姆修道院教堂南侧廊窗户　b. 花格窗饰中心点　c. 埃塞克斯郡沃尔瑟姆修道院教堂西侧正面窗户　d.A 处的剖面　e.B 处的束带　f. 柱头及柱础剖面　g. 赫特福德郡圣米迦勒教堂北侧廊窗户

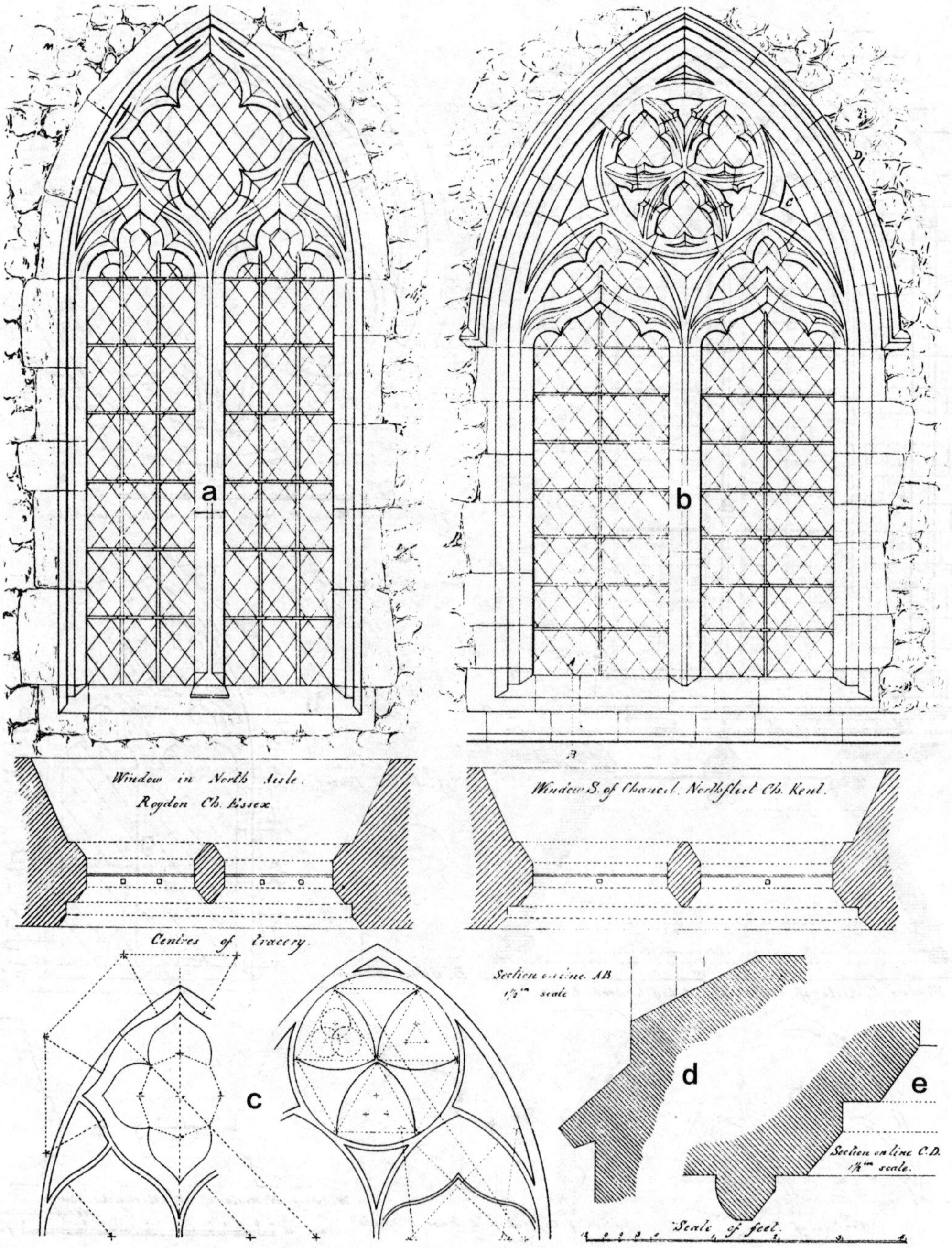

a. 埃塞克斯郡雷东教堂北侧廊窗户　b. 肯特郡诺思弗利特教堂圣坛南侧窗户　c. 花格窗饰中心点　d.A–B 处的剖面　e. C–D 处的剖面

a. 塞萨克斯郡林德菲尔德教堂圣坛东侧窗户　b. A–B 处的剖面　c. 内侧花格窗饰中心点

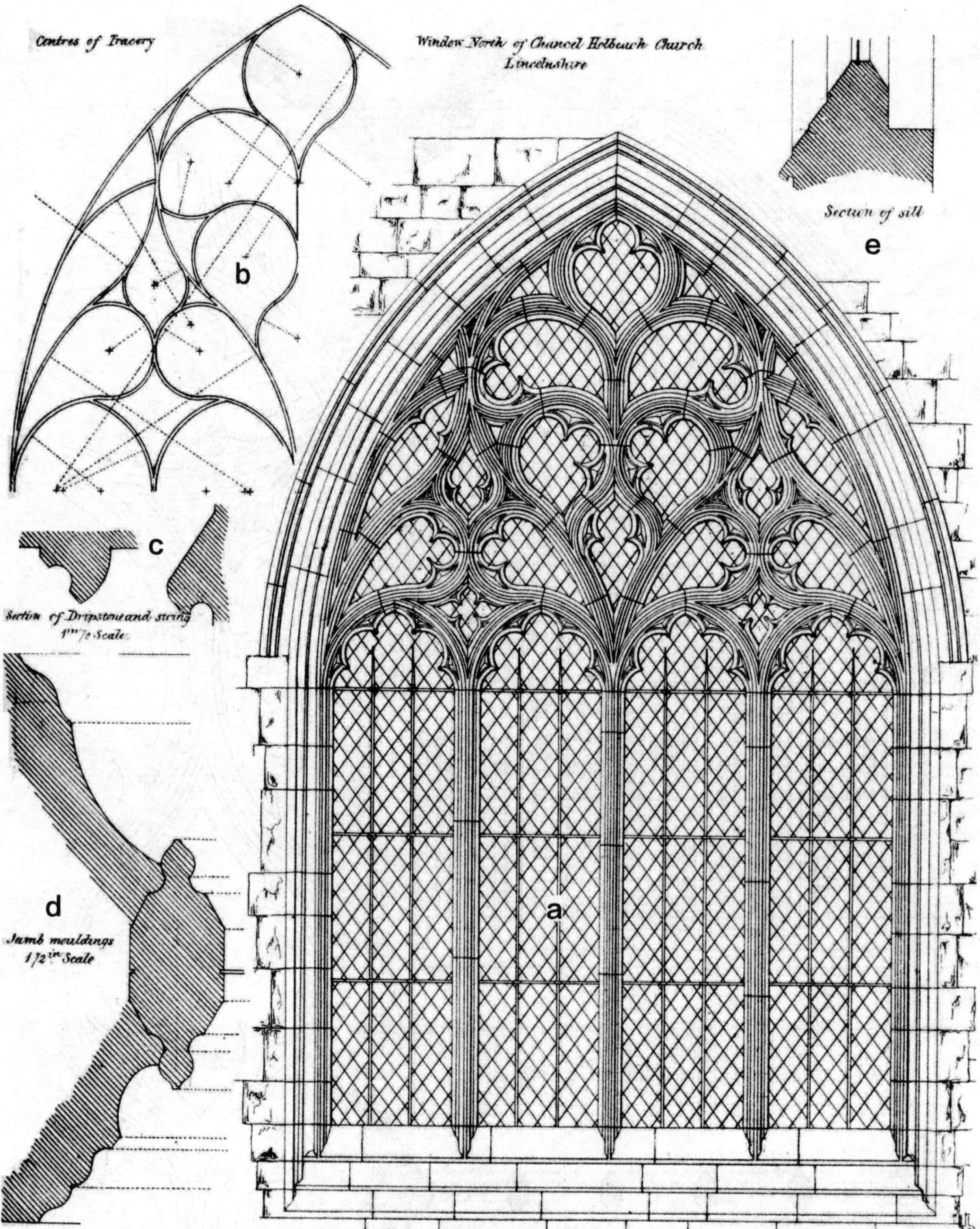

a. 林肯郡霍尔比奇教堂圣坛北侧窗户　b. 花格窗饰中心　c. 披水石剖面　d. 侧壁线脚　e. 窗台剖面

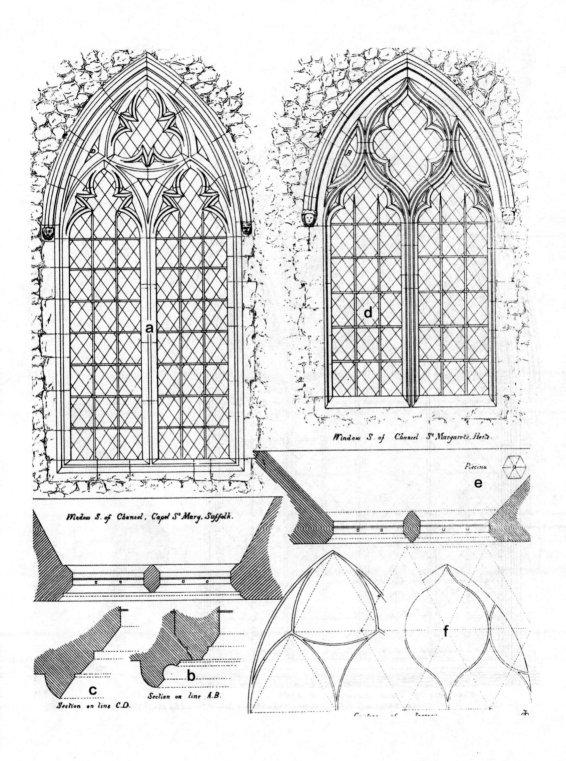

a. 萨福克郡卡佩尔圣玛丽教堂圣坛南侧窗户　b.A–B 处的剖面　c. C–D 处的剖面　d. 赫特福德郡圣玛格丽特教堂圣坛南侧窗户　e. 排水石盆　f. 花格窗饰中心点

a. 萨福克郡斯特拉特福德圣玛丽教堂圣坛东侧窗户　b. 束带层剖面　c. 花格窗饰中心点　d. 侧壁线脚

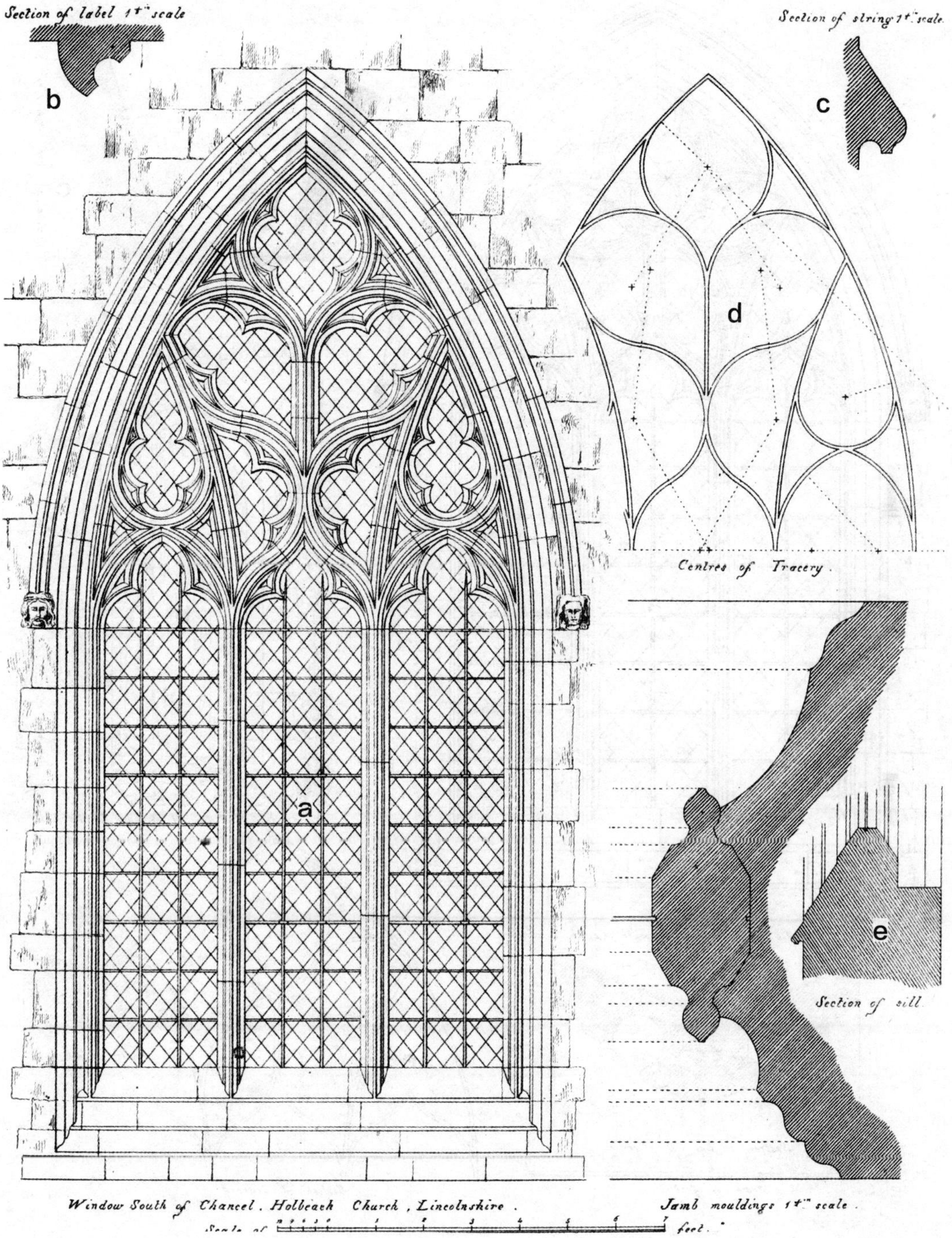

a. 林肯郡霍尔比奇教堂圣坛南侧窗户　b. 披水石剖面　c. 束带层剖面　d. 花格窗饰中心点　e. 窗台剖面侧壁线脚

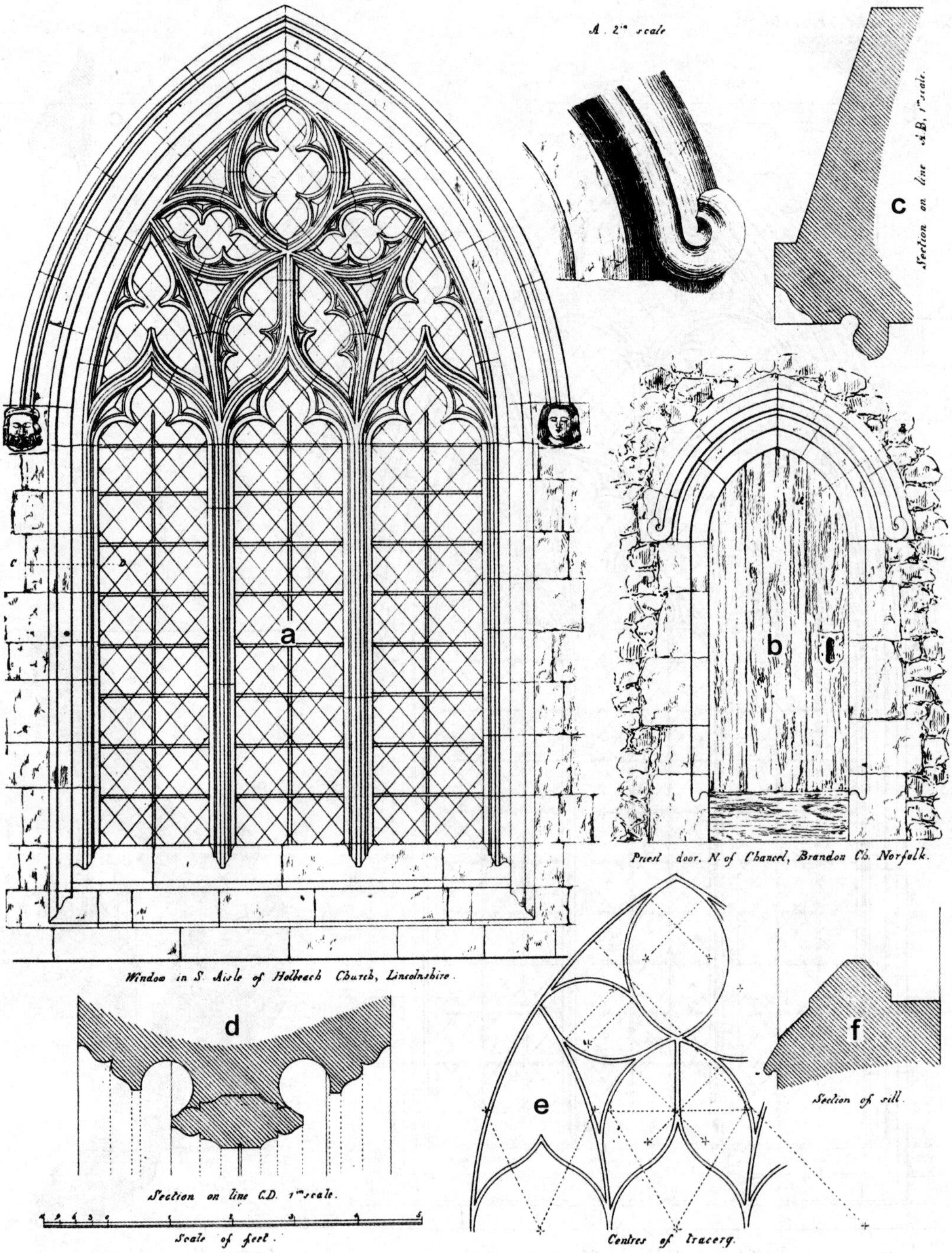

a. 林肯郡霍尔比奇教堂南侧廊窗户　b. 诺福克郡布兰登教堂圣坛北侧牧师之门　c.A–B 处的剖面　d.C–D 处的剖面　e. 花格窗饰中心点　f. 窗台剖面

a. 埃塞克斯郡雷东教堂北侧廊窗户　b. 赫特福德郡圣玛格丽特教堂圣坛东侧窗户　c. 花格窗饰中心点

a. 肯特郡鲍顿阿鲁教堂西侧窗户　b. 花格窗饰中心点　c. 窗户平面

a. 肯特郡赫恩教堂窗户　b. 窗户平面　c. 侧壁线脚剖面

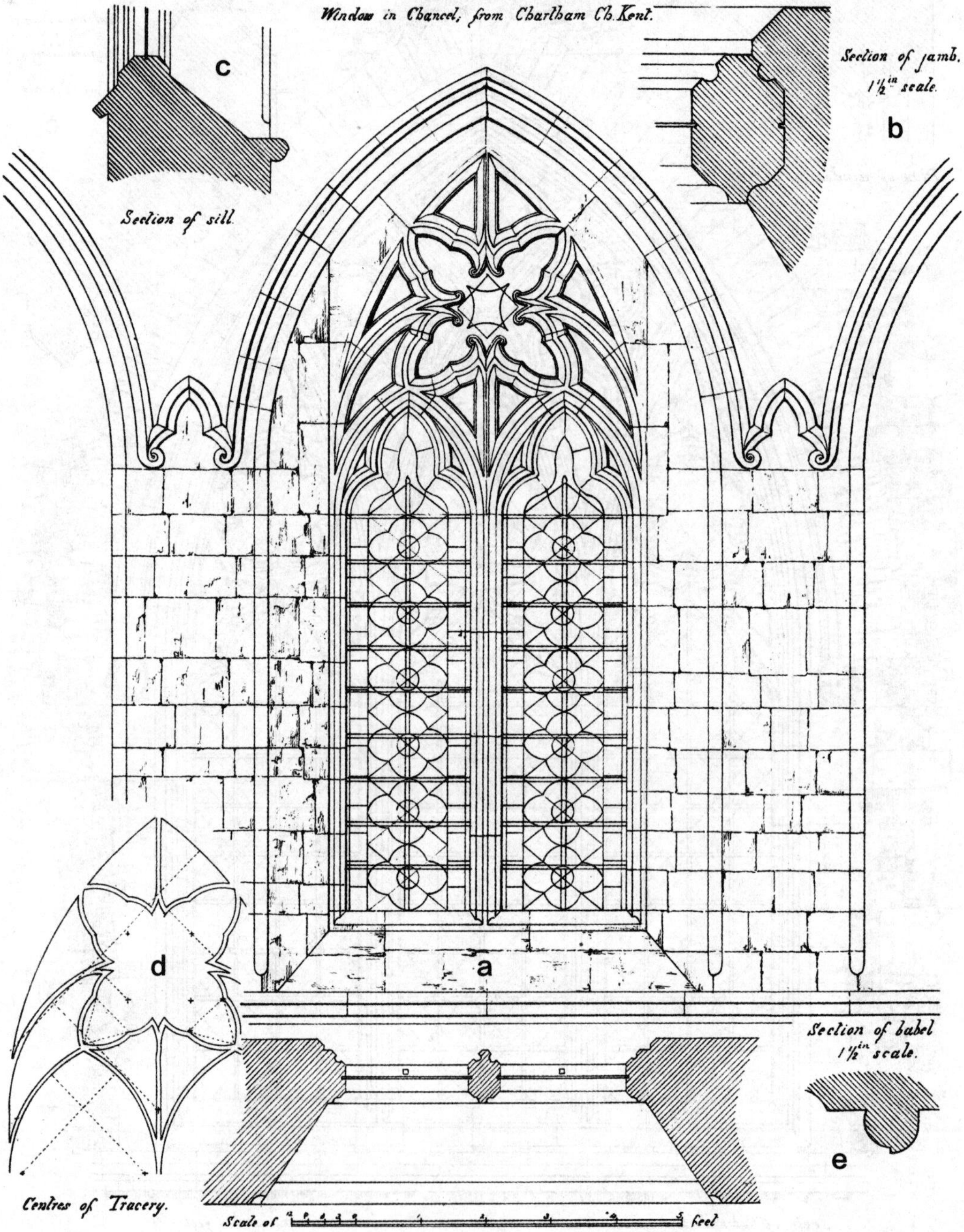

a. 肯特郡查塔姆教堂圣坛窗户　b. 侧壁剖面　c. 窗台剖面　d. 花格窗饰中心点　e. 披水石剖面

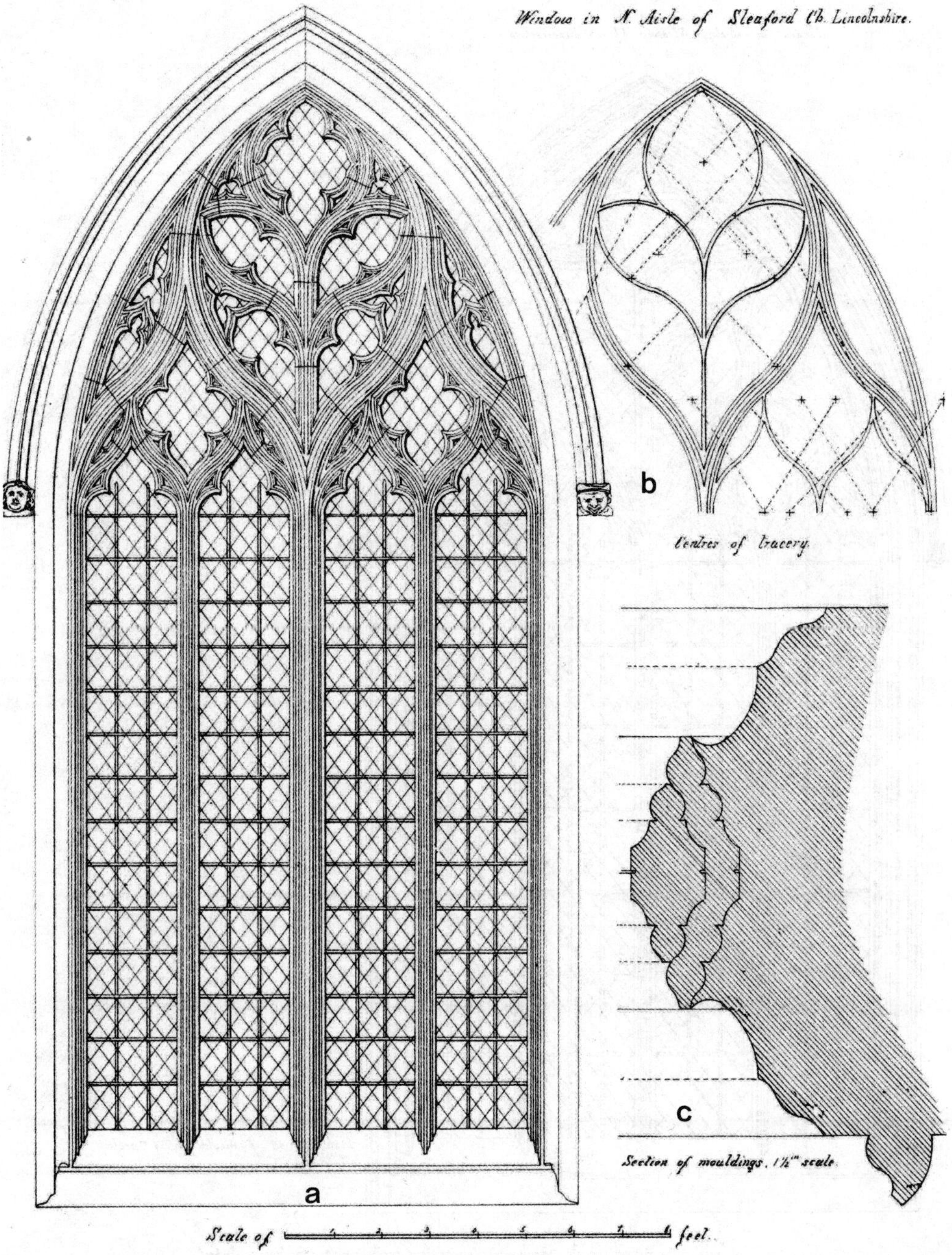

a. 林肯郡斯利福德教堂北侧廊窗户 b. 花格窗饰中心点 c. 线脚剖面

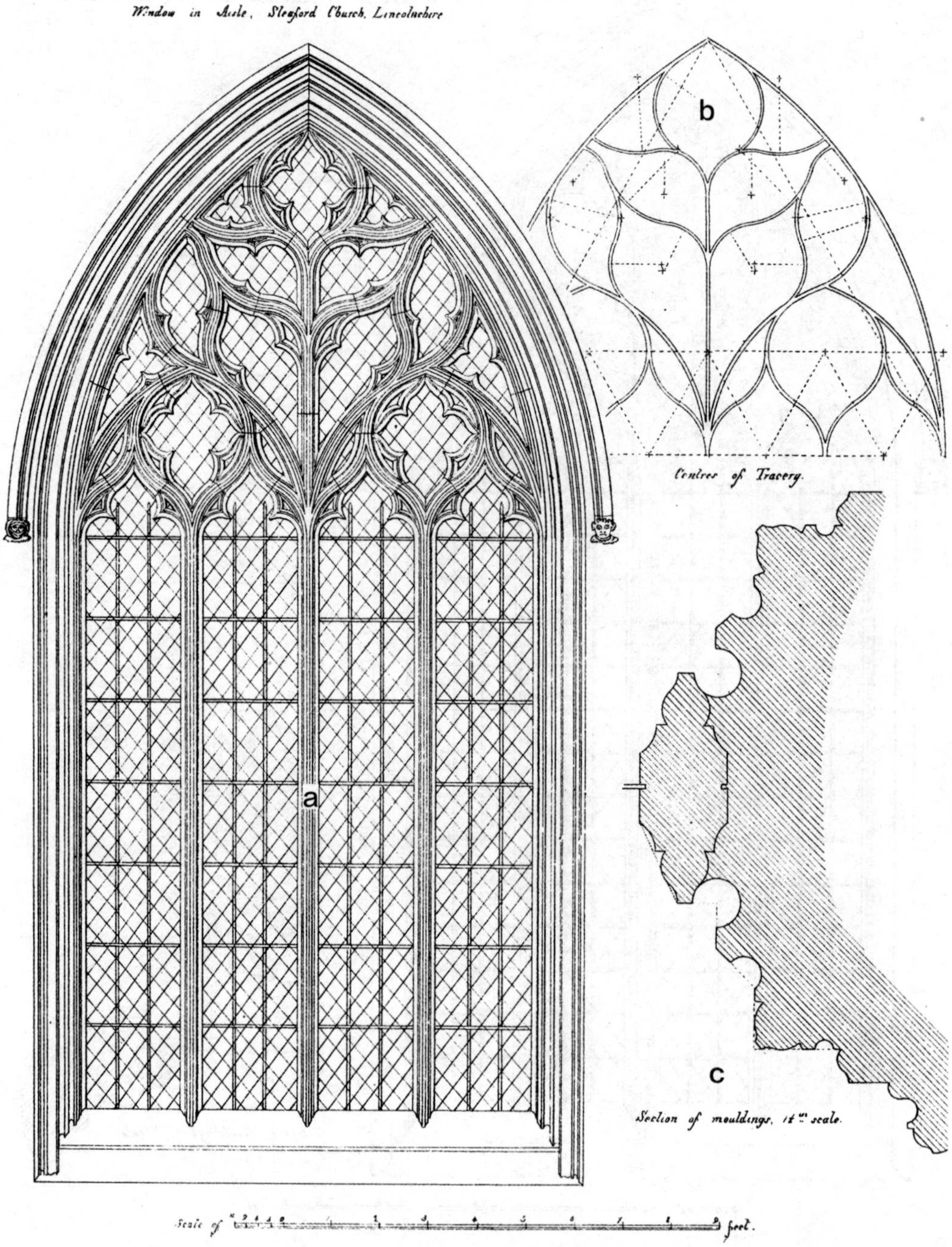

a. 林肯郡斯利福德教堂侧廊窗户　b. 花格窗饰中心点　c. 线脚剖面

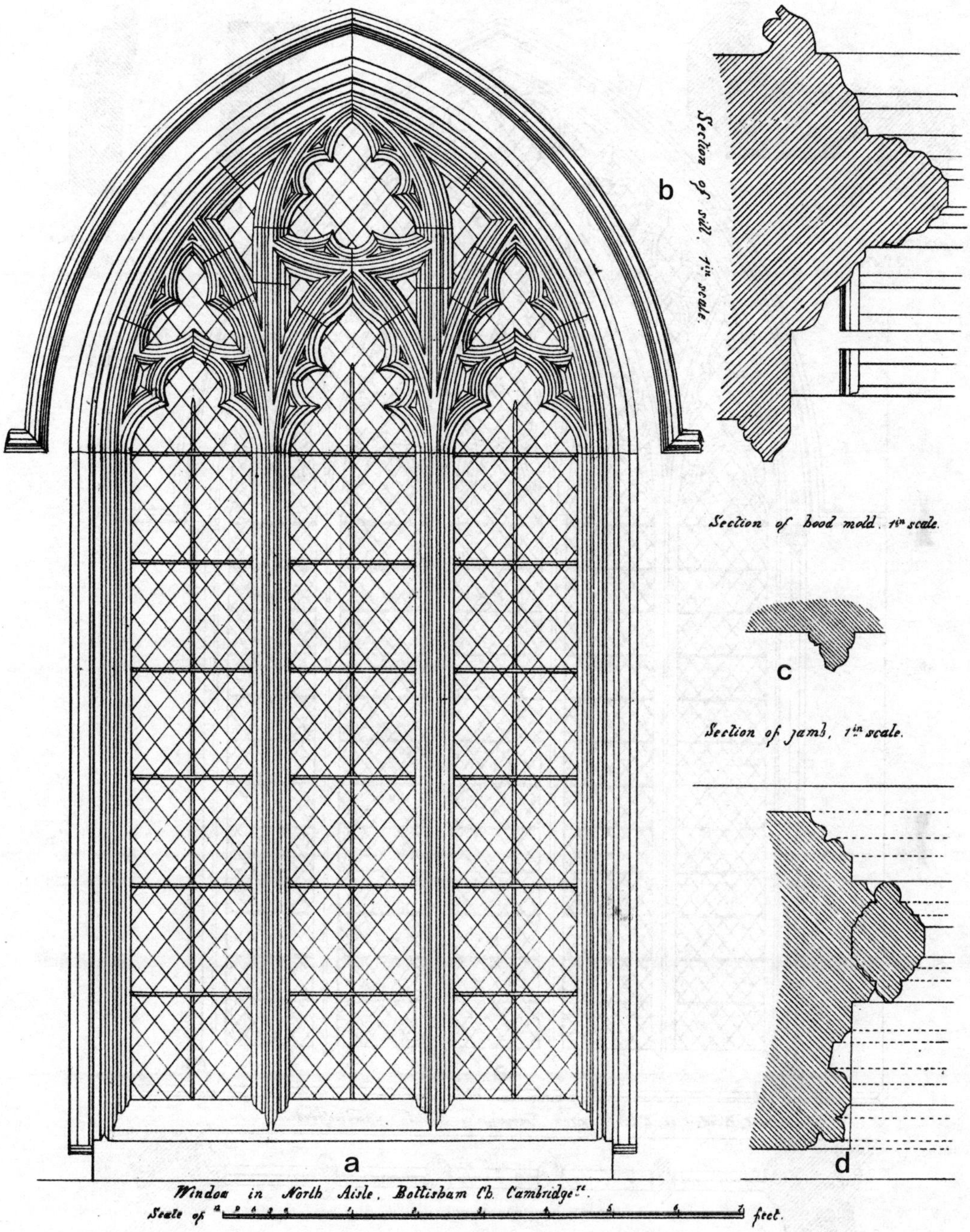

a. 剑桥郡博蒂舍姆教堂北侧廊窗户　b. 窗台剖面　c. 出檐线脚剖面　d. 侧壁剖面

a. 剑桥郡特兰平顿教堂圣坛东侧窗户　b.A-B 处的剖面　c. 柱头及柱础剖面

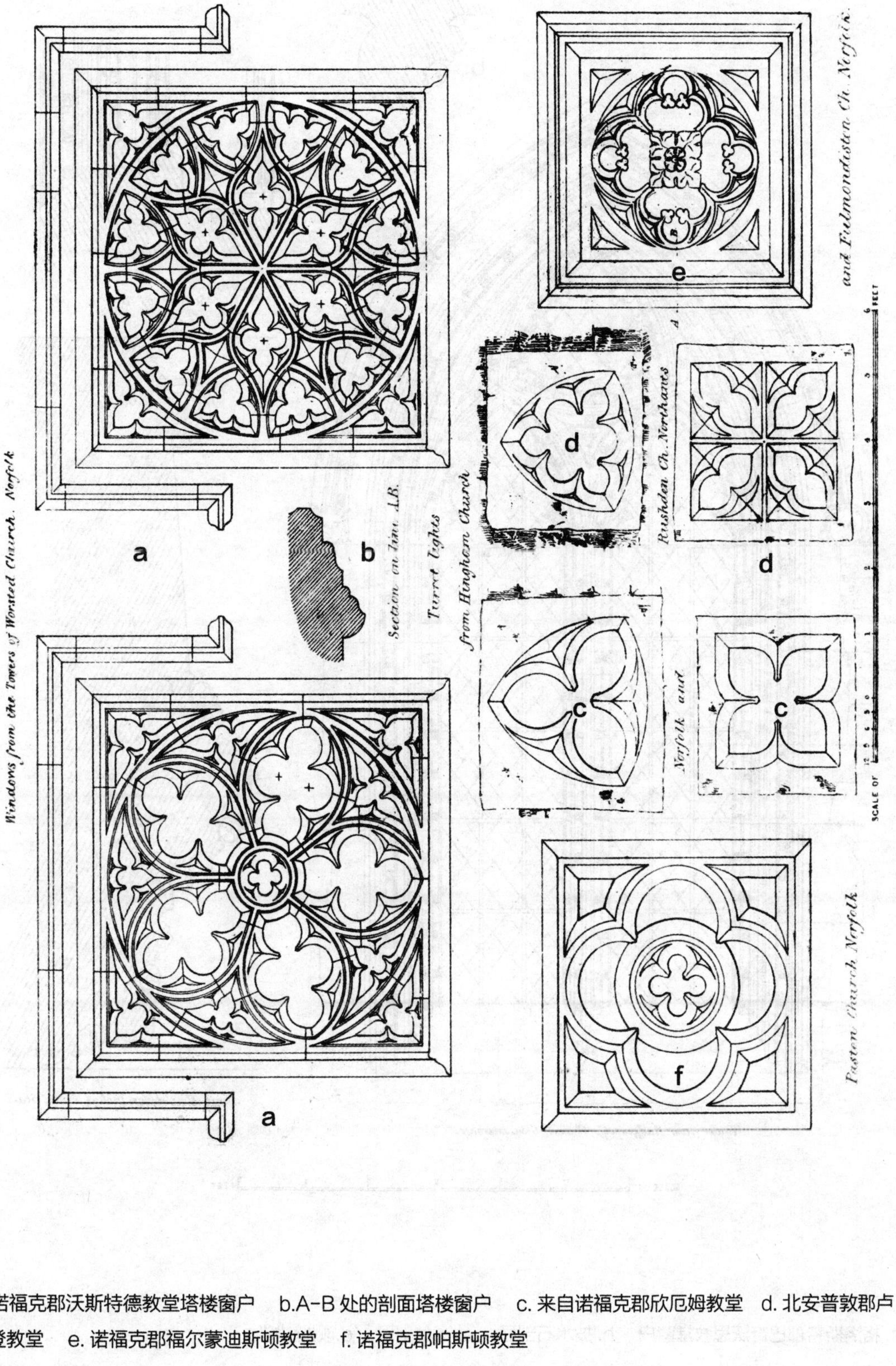

a. 诺福克郡沃斯特德教堂塔楼窗户　b.A–B 处的剖面塔楼窗户　c. 来自诺福克郡欣厄姆教堂　d. 北安普敦郡卢斯登教堂　e. 诺福克郡福尔蒙迪斯顿教堂　f. 诺福克郡帕斯顿教堂

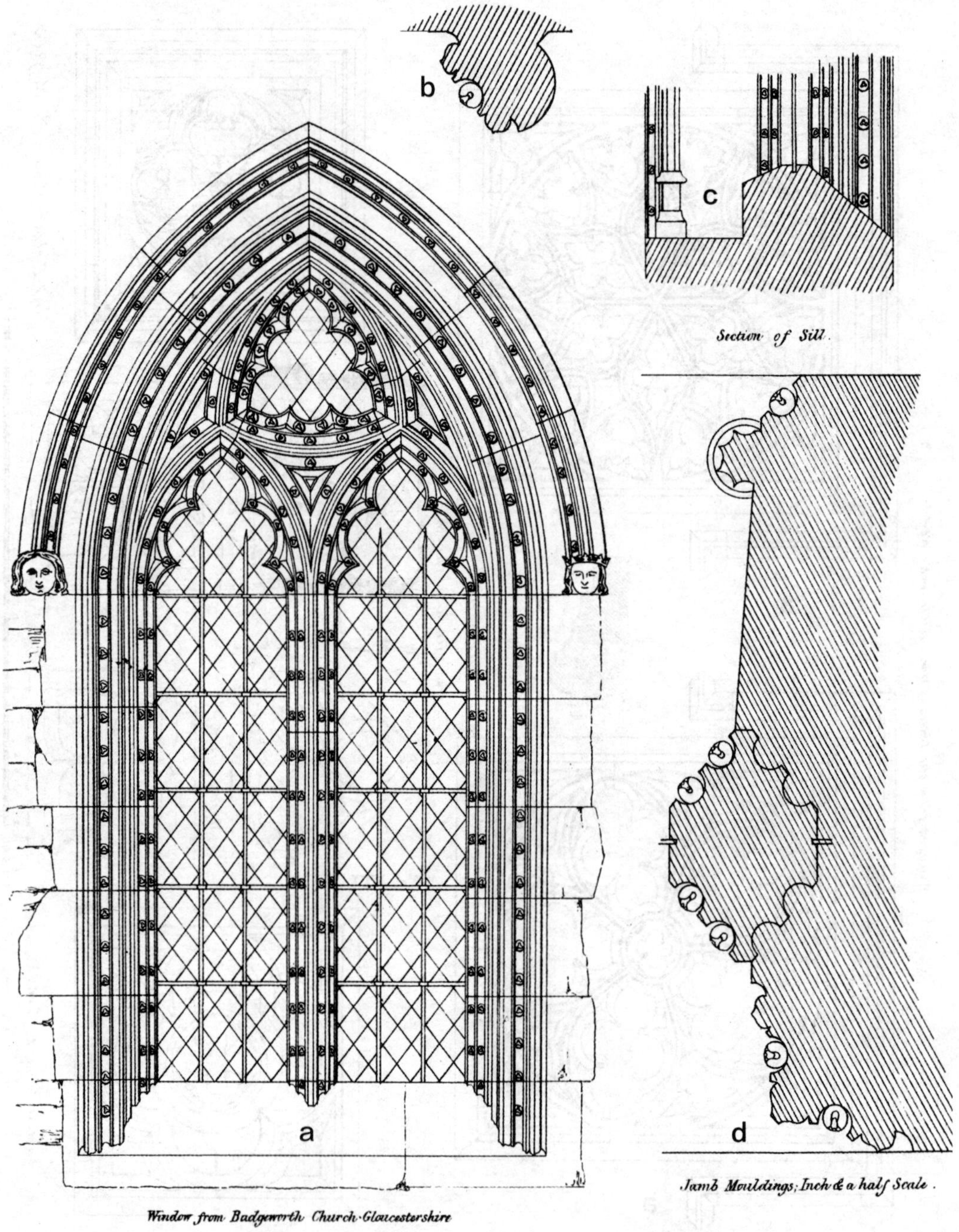

a. 格洛斯特郡巴奇沃思教堂窗户　b. 披水石剖面　c. 窗台剖面　d. 侧壁线脚

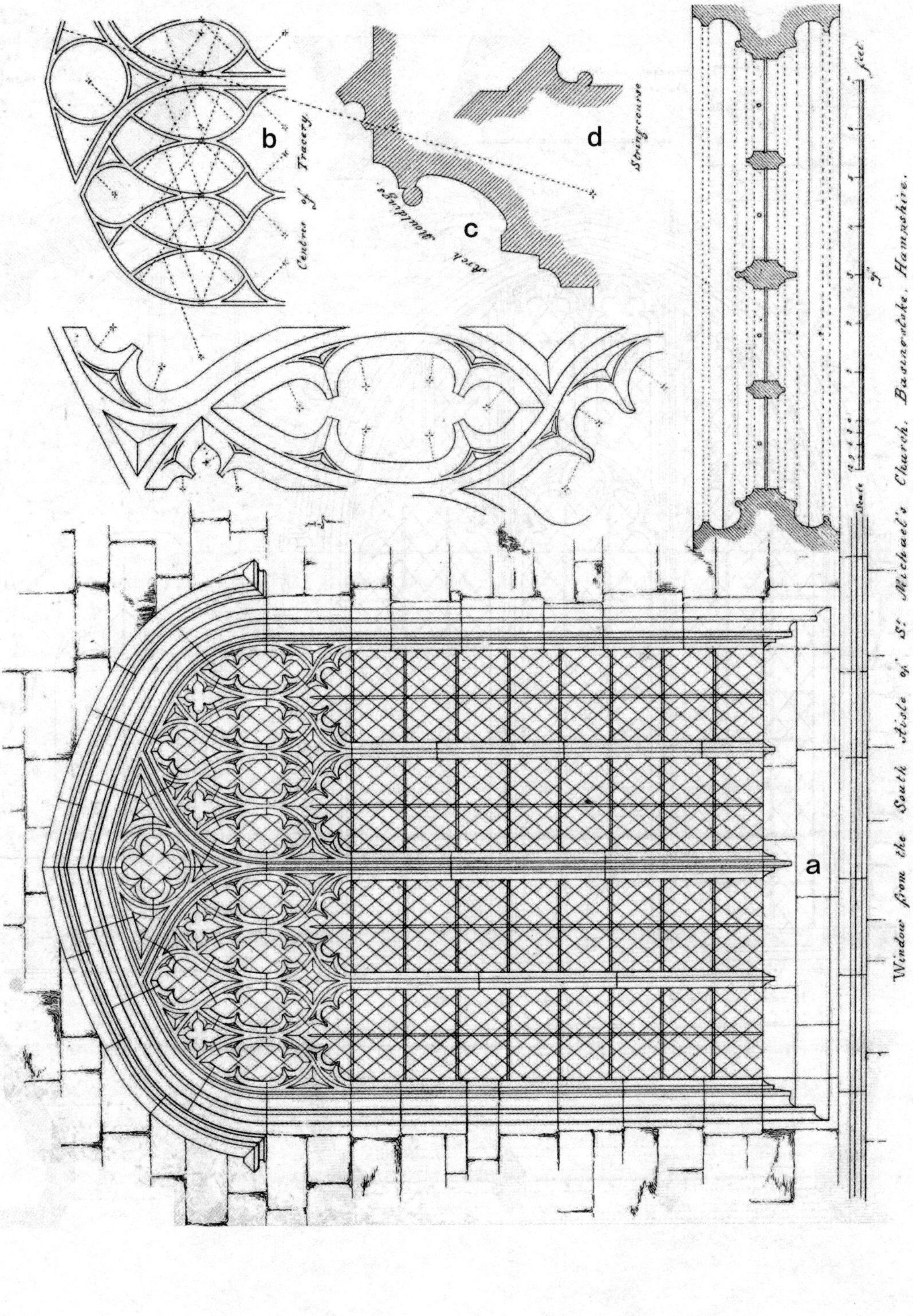

a. 汉普郡贝辛斯托克圣米迦勒教堂南侧廊窗户　b. 花格窗饰中心点　c. 拱线脚　d. 束带层

a. 肯特郡诺思弗利教堂北侧廊窗户 b. 汉普郡达默教堂西门廊窗户 c.A-B 处的剖面 d. 花格窗饰中心点 e. 肯特郡乔克教堂西门廊窗户

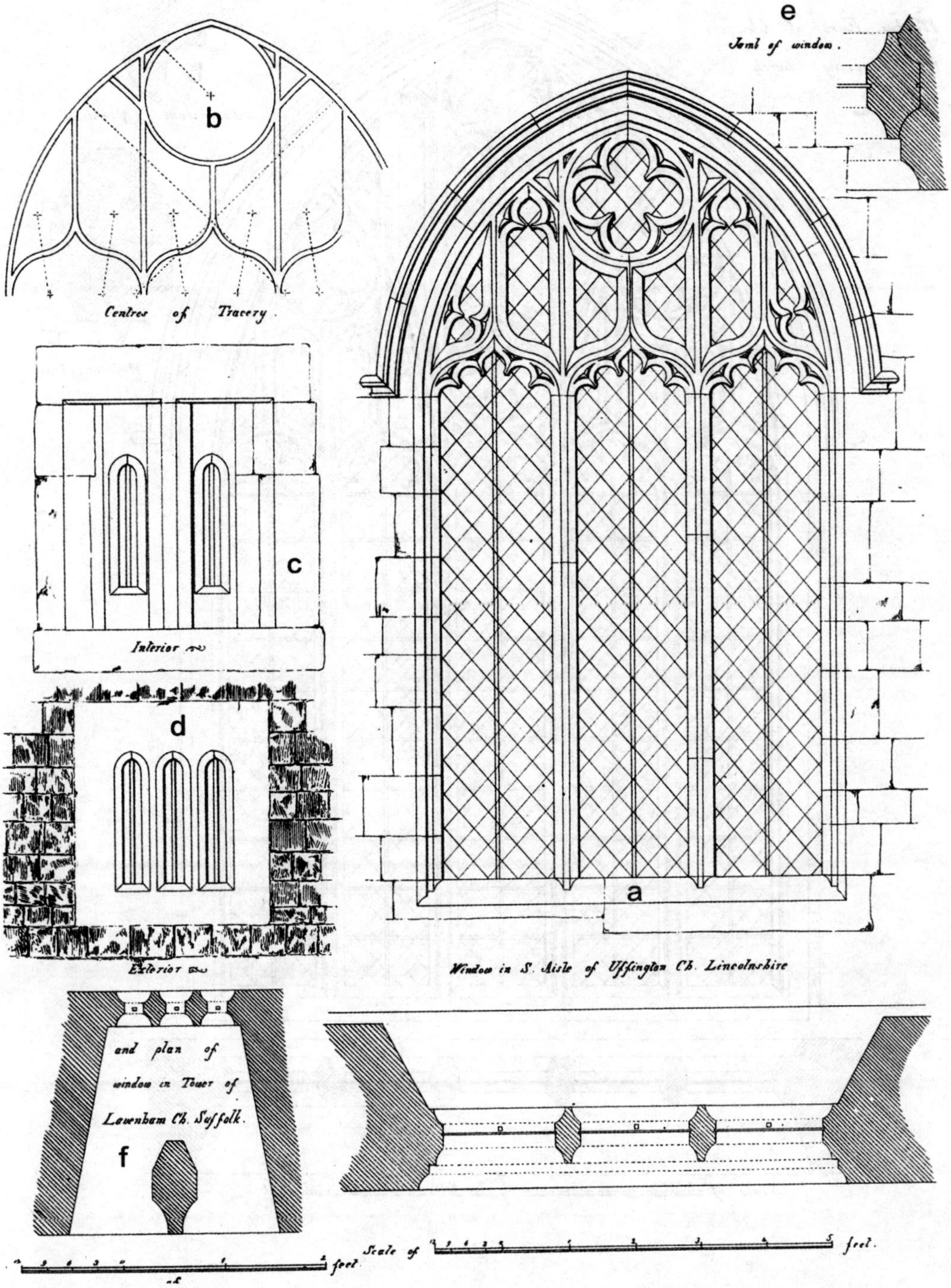

a. 林肯郡阿芬顿教堂南侧廊窗户　b. 花格窗饰中心点　c. 内侧　d. 外侧　e. 窗户侧壁　f. 萨福克郡拉文纳姆教堂塔楼窗户平面

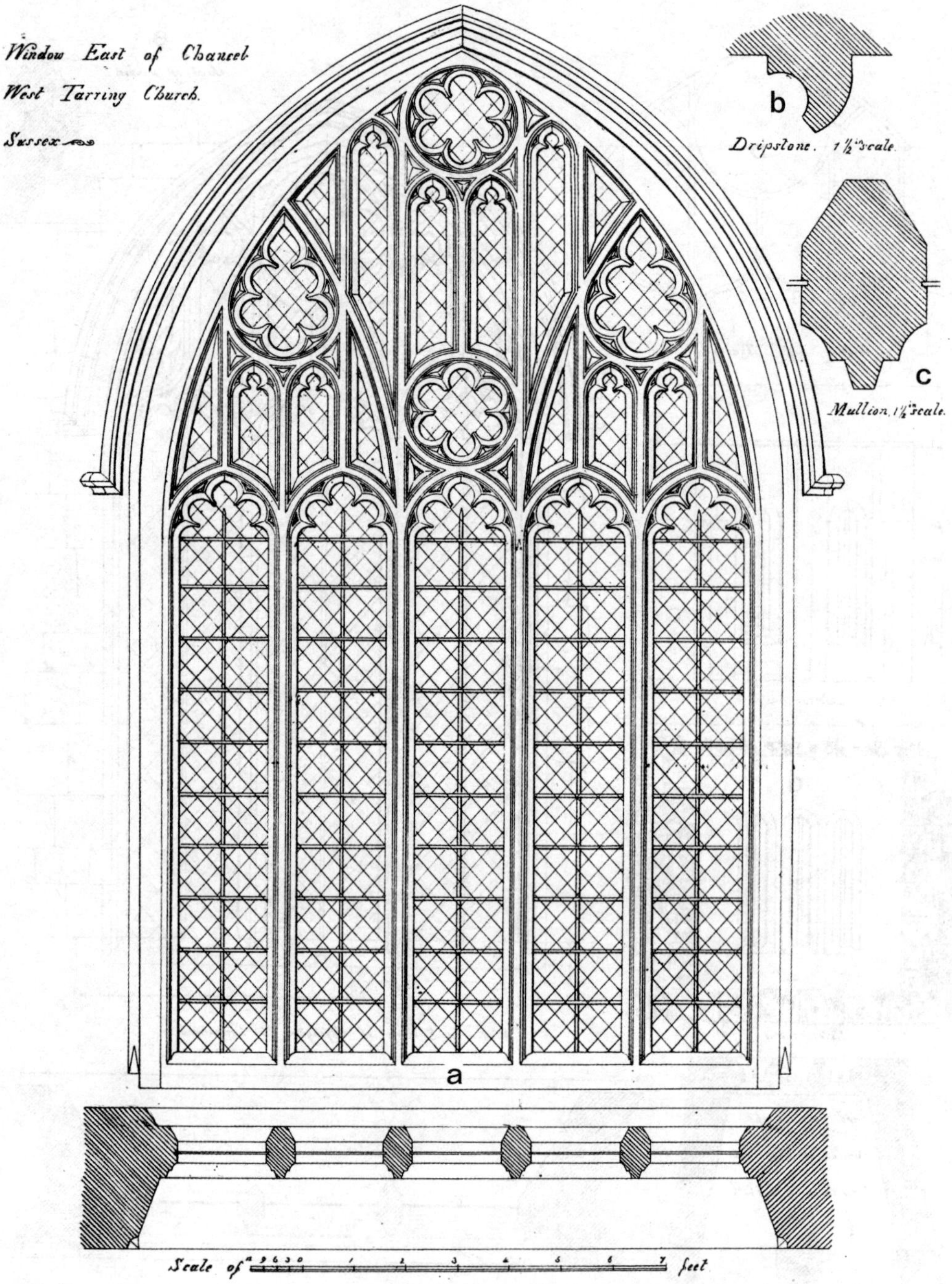

a. 塞萨克斯郡西塔灵教堂圣坛东侧窗户　b. 披水石　c. 竖框

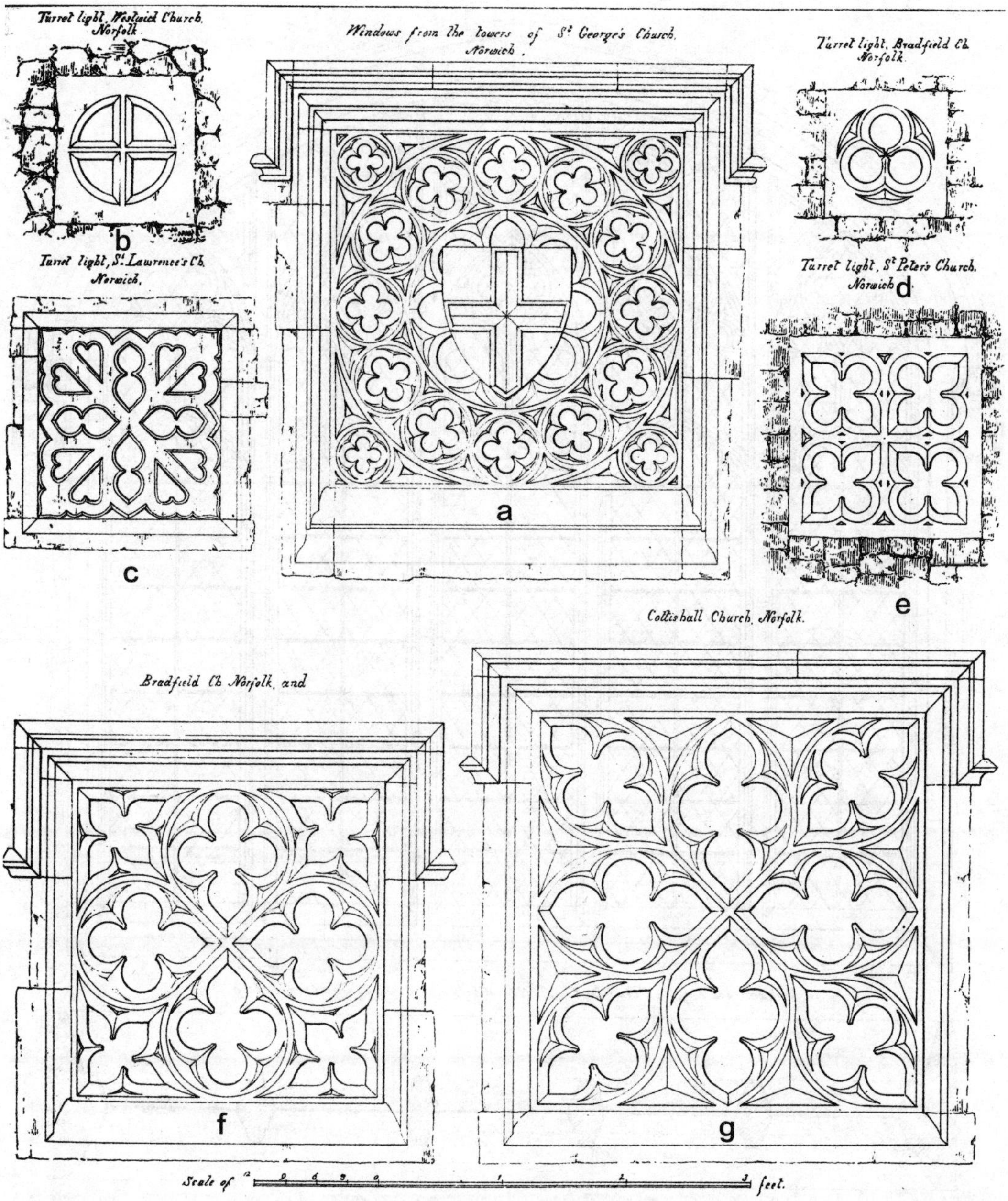

a. 诺维奇圣乔治教堂塔楼窗户　b. 诺福克郡维斯特维克教堂塔楼窗户　c. 诺维奇圣劳伦斯教堂塔楼窗户　d. 诺福克郡布拉德菲尔德教堂塔楼窗户　e. 诺维奇圣彼得教堂塔楼窗户　f. 诺福克郡布拉德菲尔德教堂　g. 诺福克郡科尔蒂瑟教堂

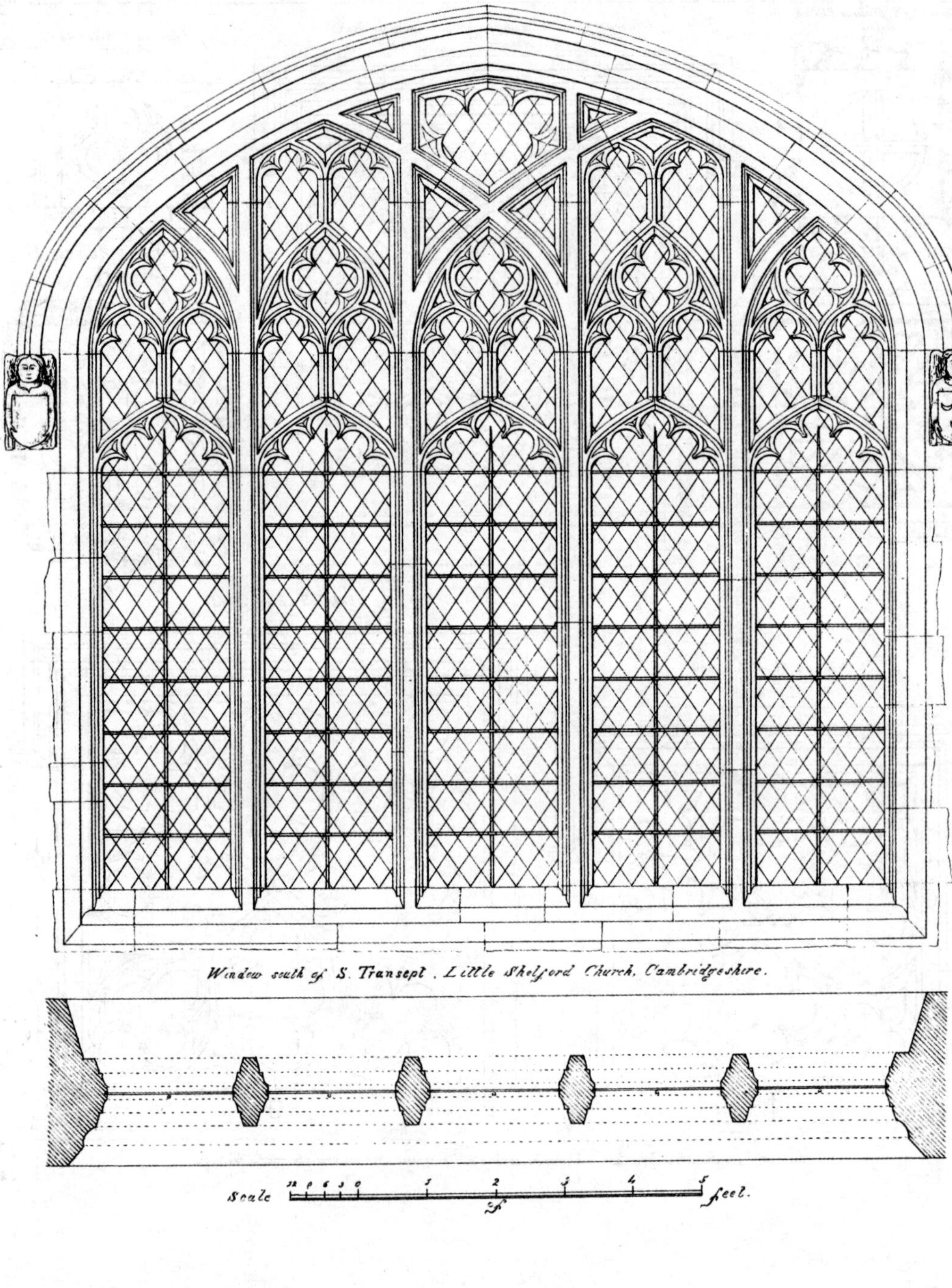

剑桥郡小谢尔福德教堂南耳堂南侧窗户

第4章

扶壁及女儿墙

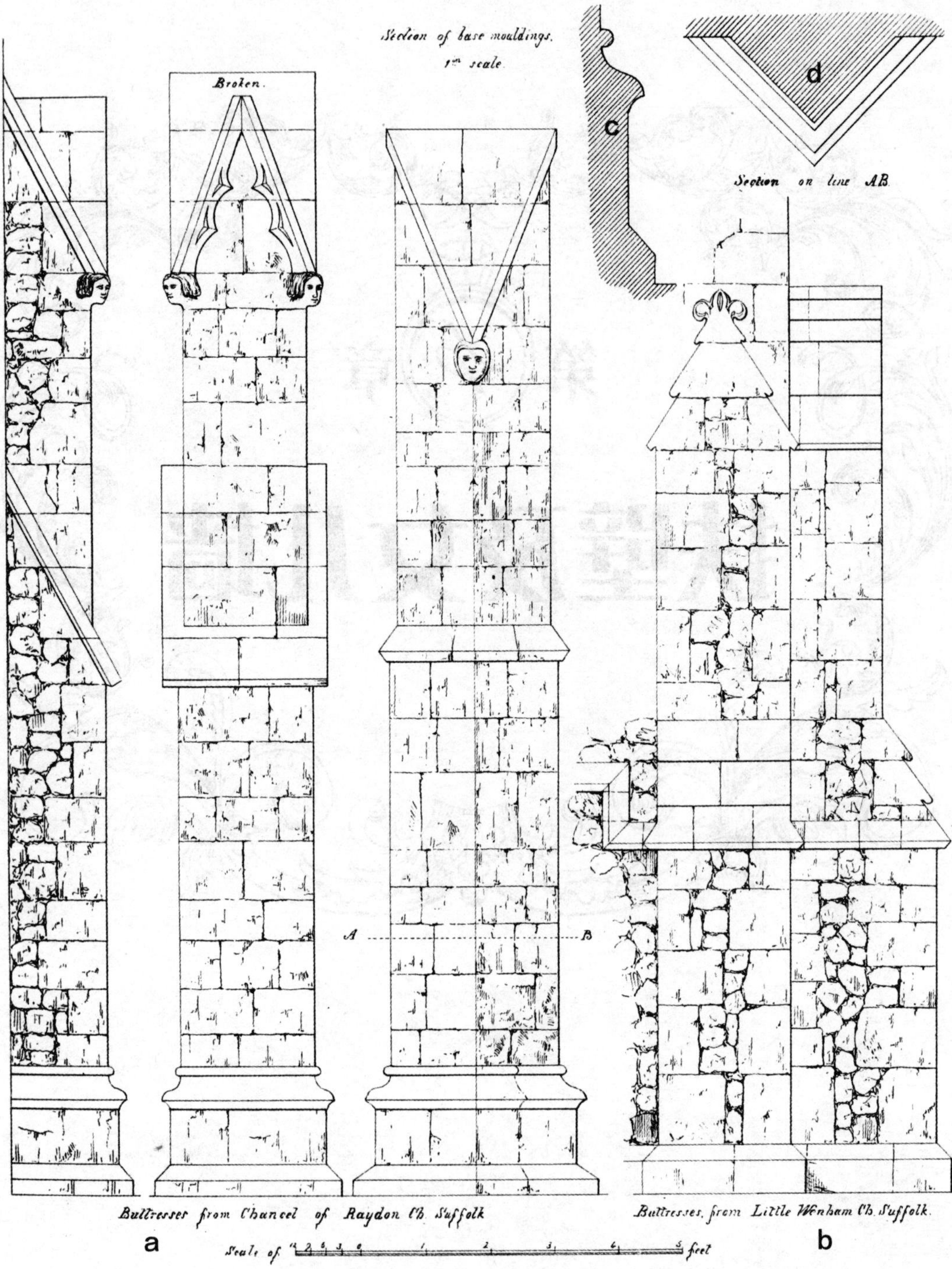

a. 萨福克雷东教堂圣坛扶壁 b. 萨福克韦纳姆小教堂扶壁 c. 底座线脚剖面 d. A–B 处的剖面

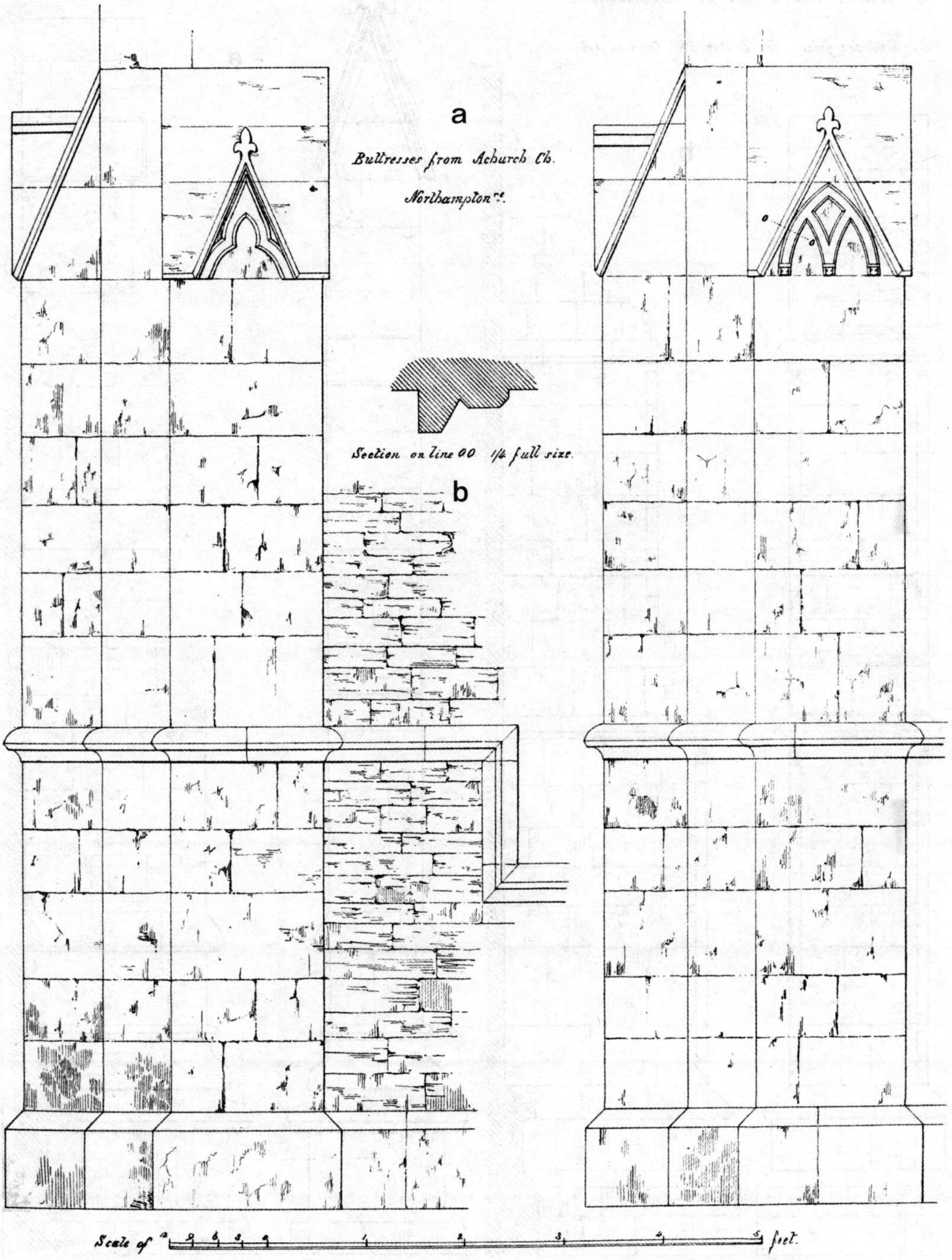

a. 北安普敦郡阿克切奇教堂扶壁　b.O-O 处的剖面

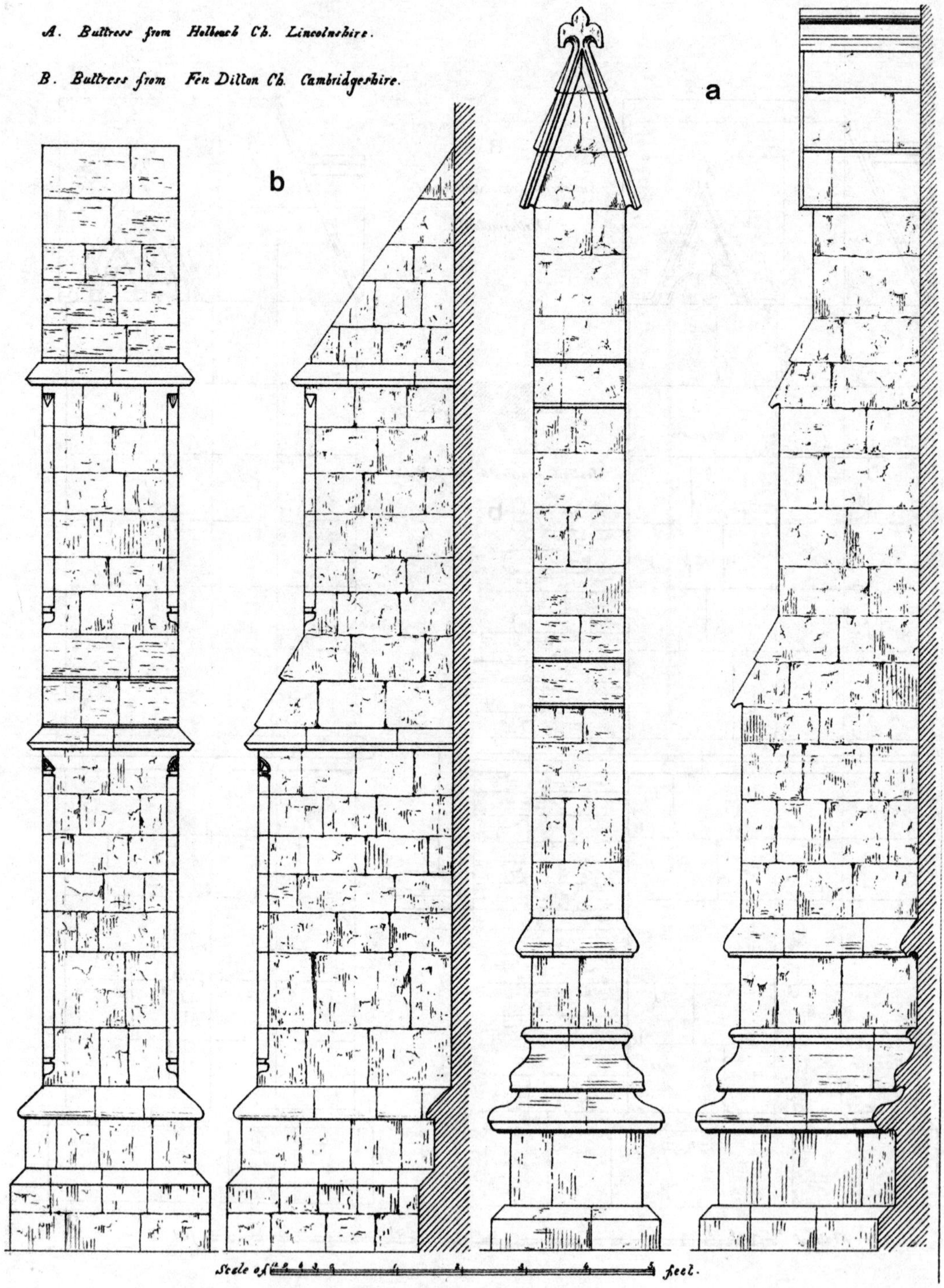

a. 林肯郡霍尔比奇教堂扶壁　b. 剑桥郡芬迪顿教堂扶壁

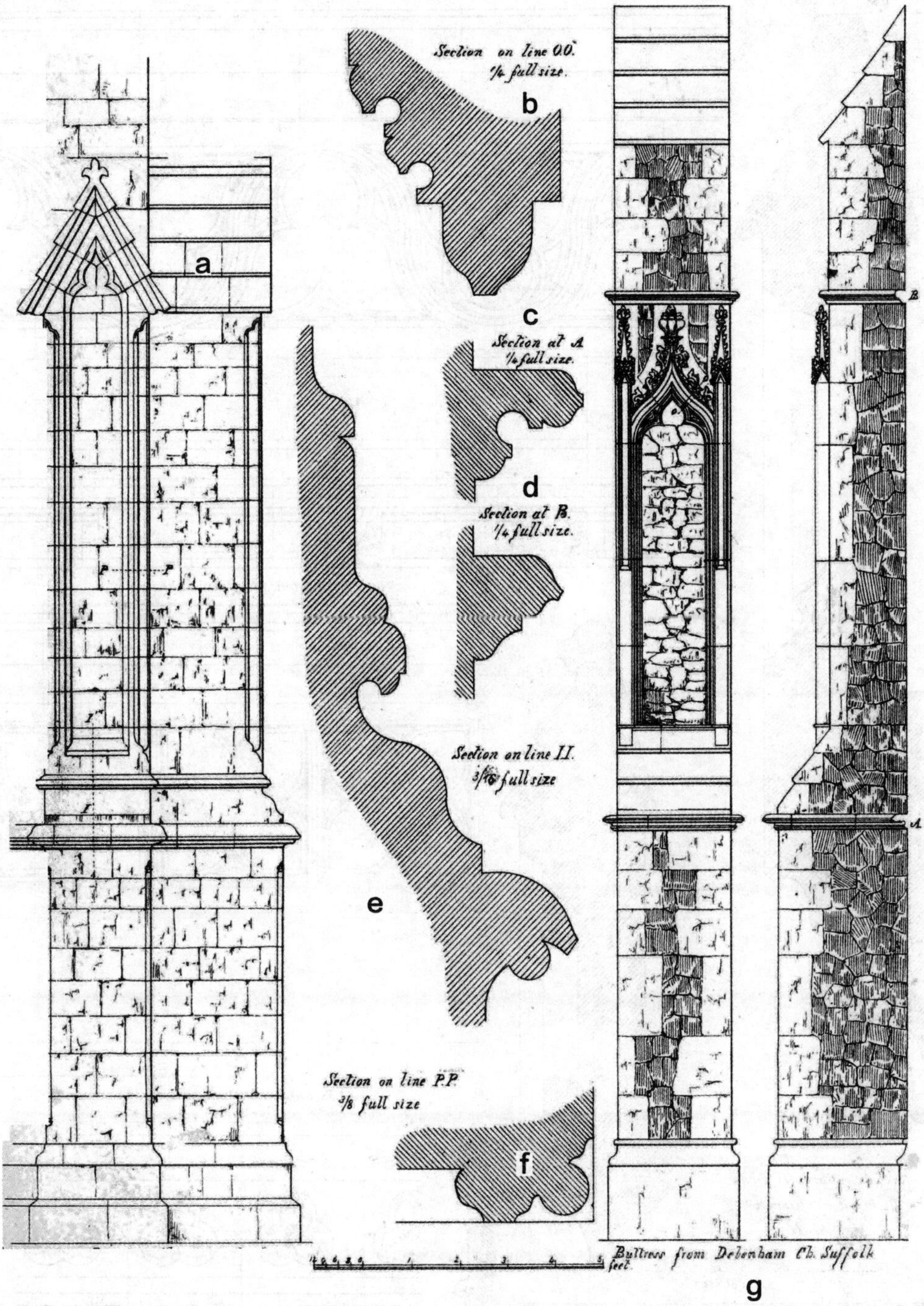

a. 剑桥郡博蒂舍姆教堂扶壁　b. O–O 处的剖面　c.A 处的剖面　d.B 处的剖面　e.I–I 处的剖面　f .P–P 处的剖面　g. 萨福克郡德贝纳姆教堂扶壁

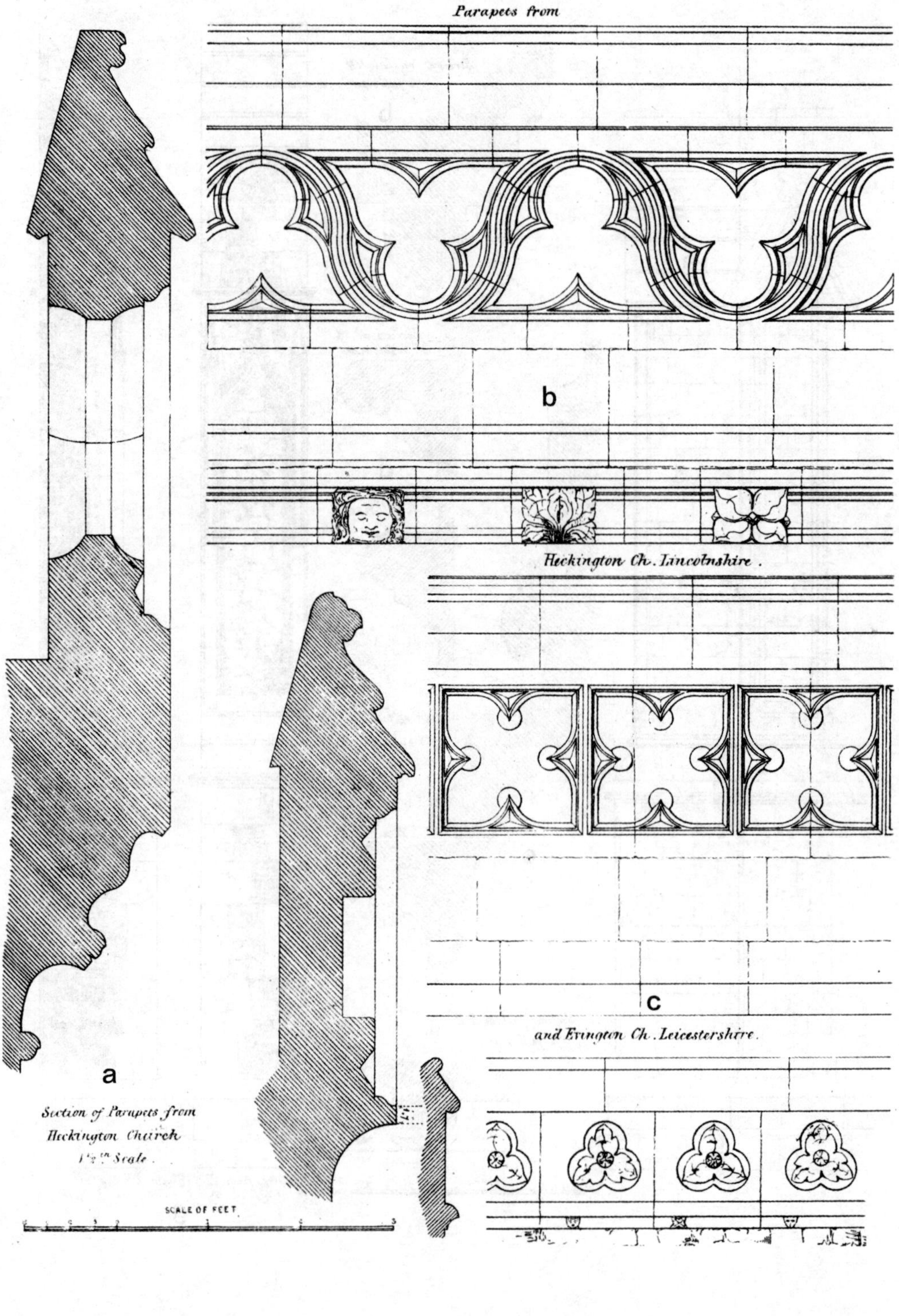

女儿墙来自于：a. 林肯郡赫金顿教堂　b. 莱斯特郡埃文顿教堂　c. 赫金顿教堂女儿墙剖面

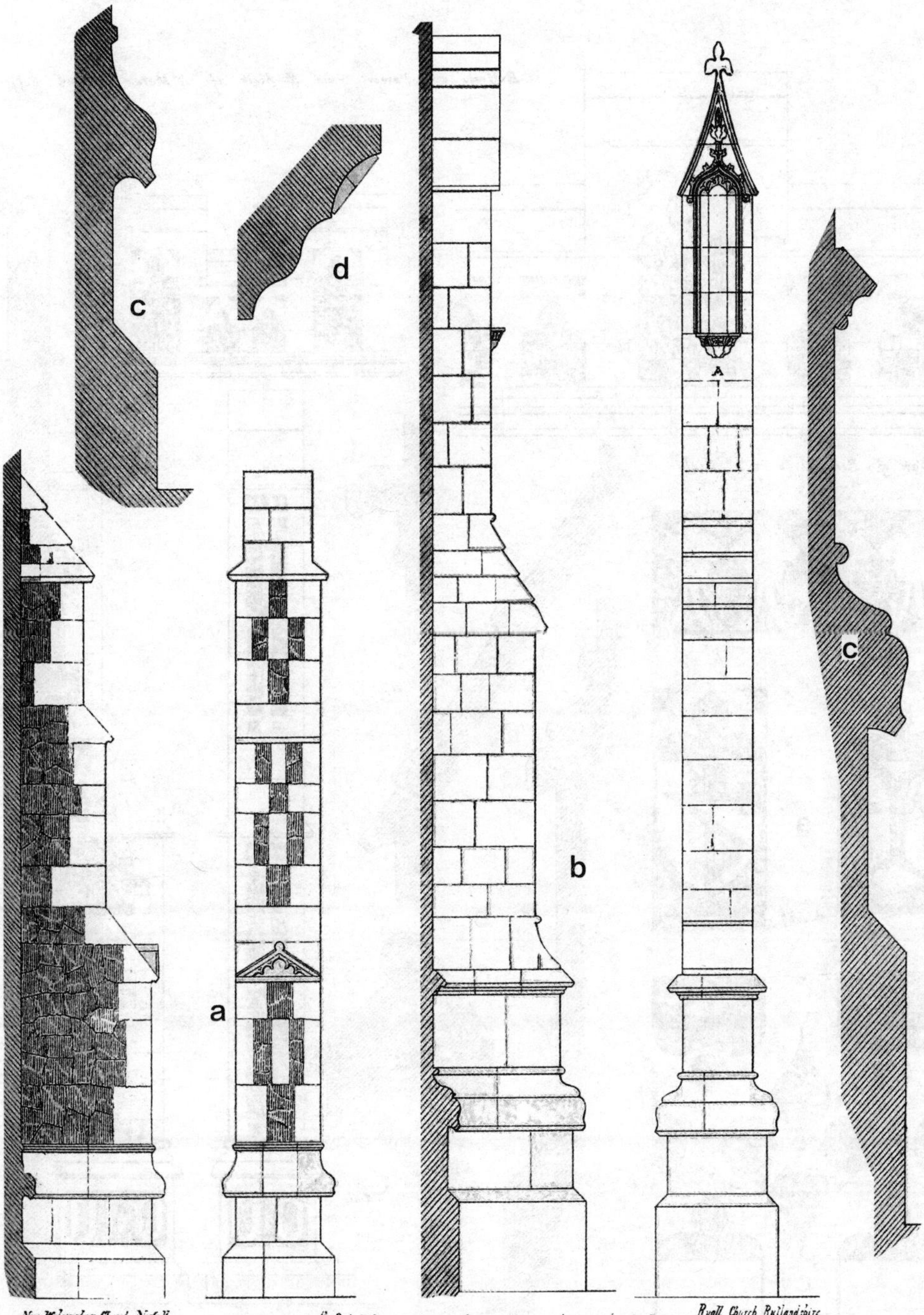

a. 诺福克郡新沃尔辛厄姆教堂　b. 拉特兰郡赖亚尔教堂　c. 底座剖面线脚　d.A 处的剖面

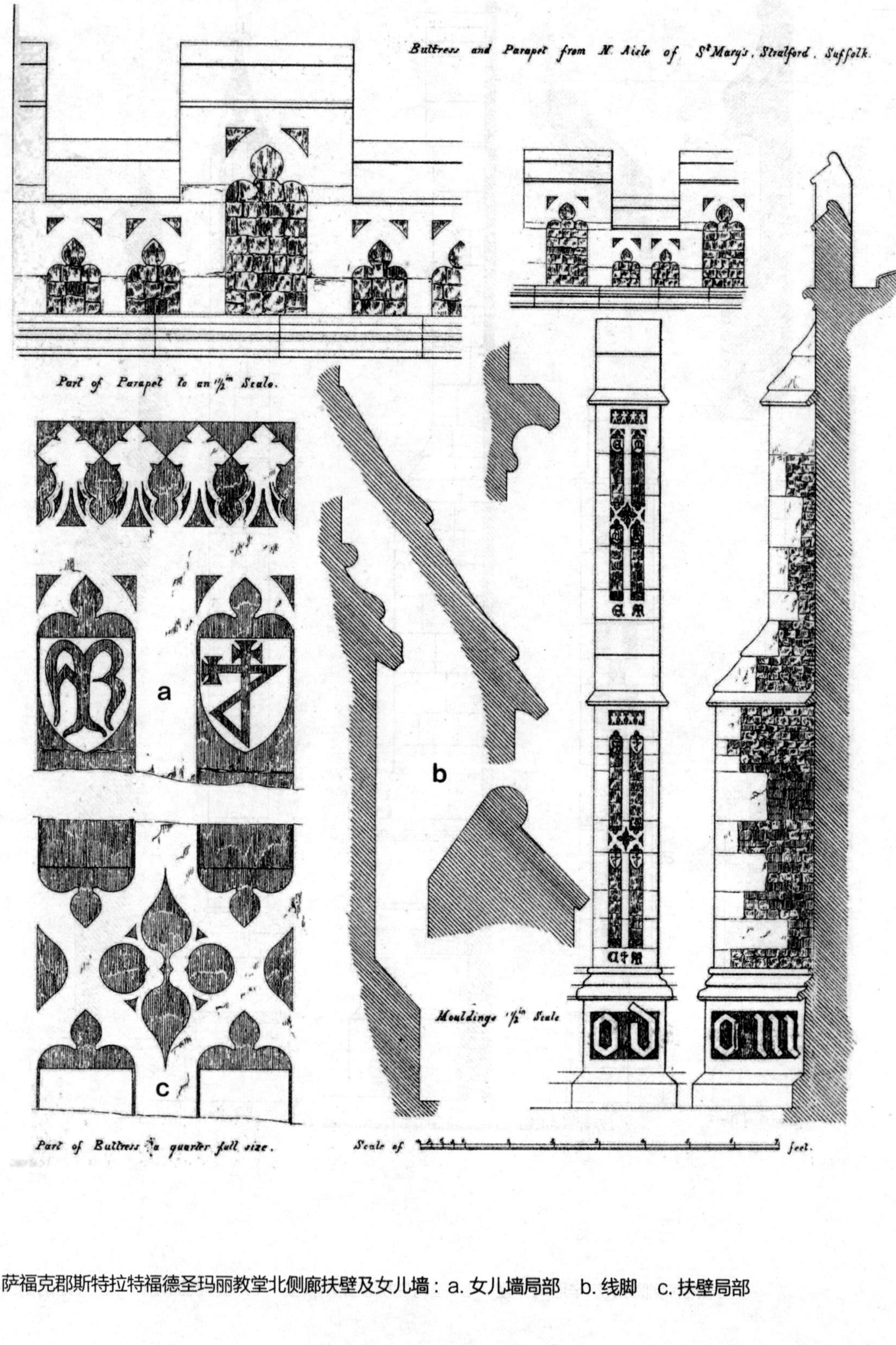

萨福克郡斯特拉特福德圣玛丽教堂北侧廊扶壁及女儿墙：a. 女儿墙局部　b. 线脚　c. 扶壁局部

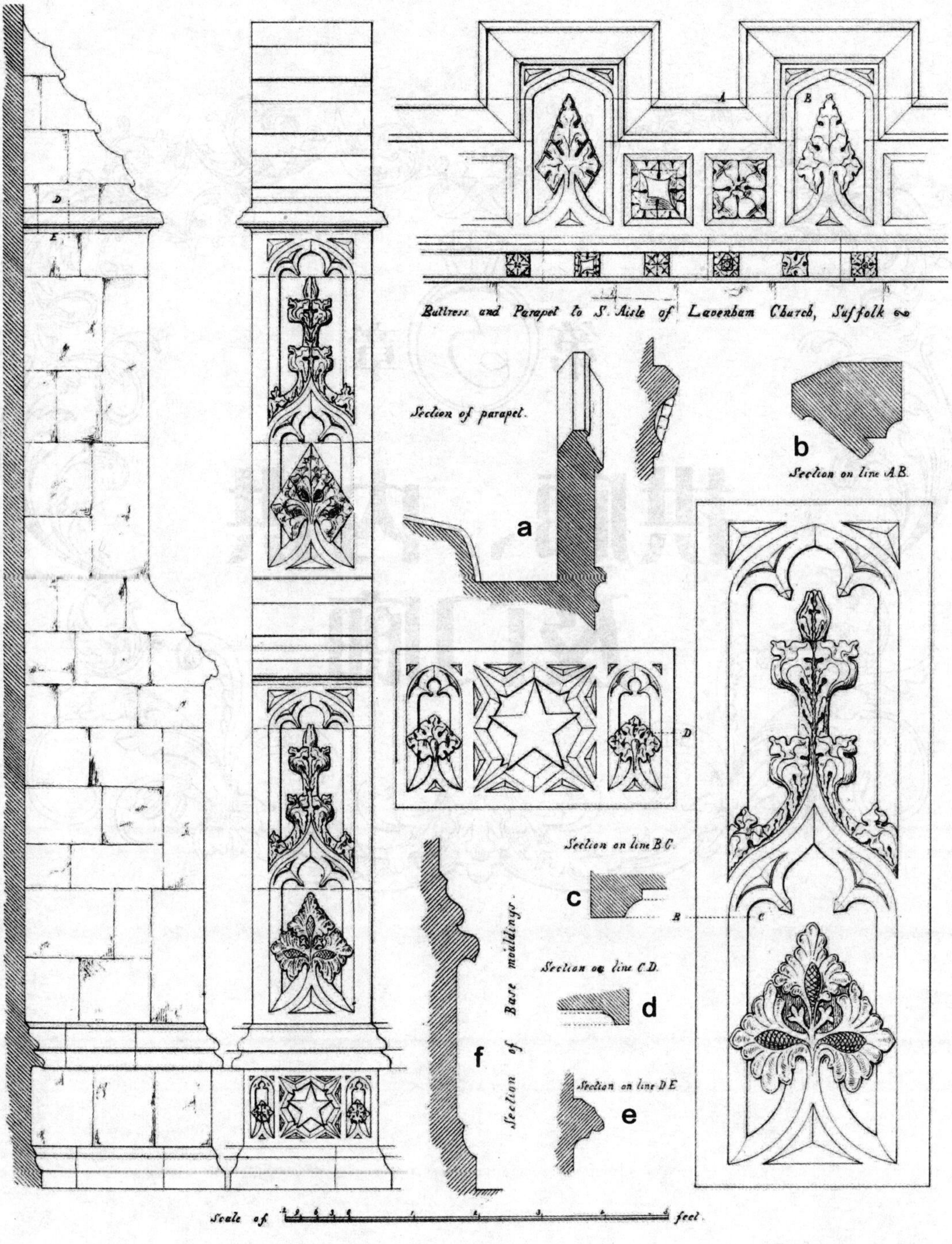

萨福克郡拉文纳姆教堂南侧廊扶壁及女儿墙：a. 女儿墙剖面　b. A–B 处的剖面　c.B–C 处的剖面　d.C–D 处的剖面　e. D–E 面　f. 底座线脚剖面

第5章

拱廊、内拱及门廊

萨塞克斯郡新肖勒姆教堂拱廊：a. 北侧廊拱　b. 南侧廊拱　c. 山形纹样装饰　d.A–B 处的线脚剖面　e. 柱础平面　f.C 处的剖面　g. 柱础剖面　h. 拱局部　i. 束带层局部

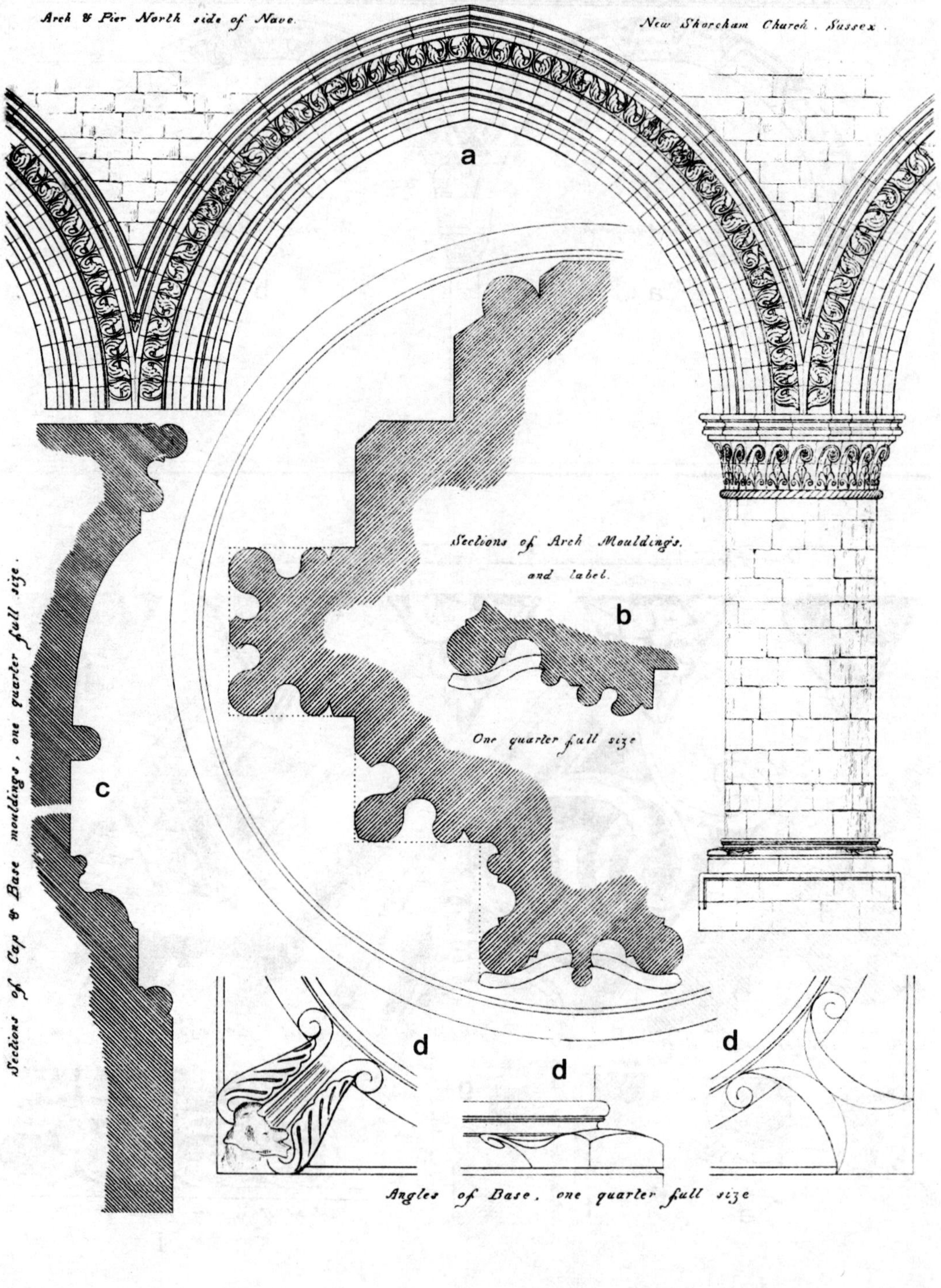

萨塞克斯郡新肖勒姆教堂：a. 中殿北侧拱及柱子　b. 拱线脚及披水石剖面　c. 柱头及柱础剖面　d. 柱础转角处

a. 圣阿尔班修道院教堂拱廊　b. 拱廊平面　c. 柱头剖面　d. A–B 处的剖面

威尔斯教堂：a. 环绕牧师会礼堂拱廊一间　b. 牧师会礼堂转角柱子

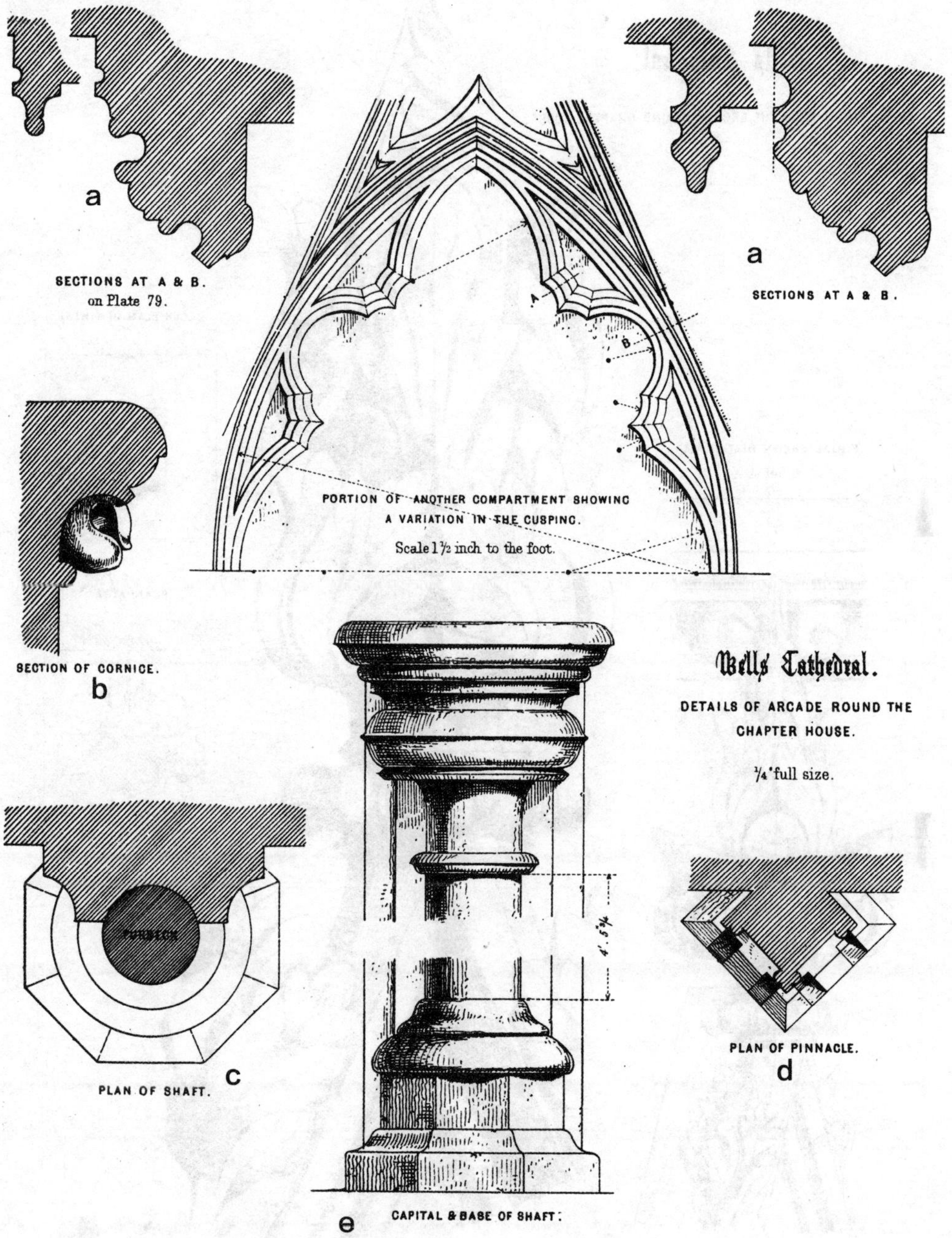

威尔斯教堂环绕牧师会礼堂拱廊细部：a.A 和 B 处的剖面　b. 檐口剖面　c. 柱身平面　d. 小尖塔平面　e. 柱子柱头和柱础

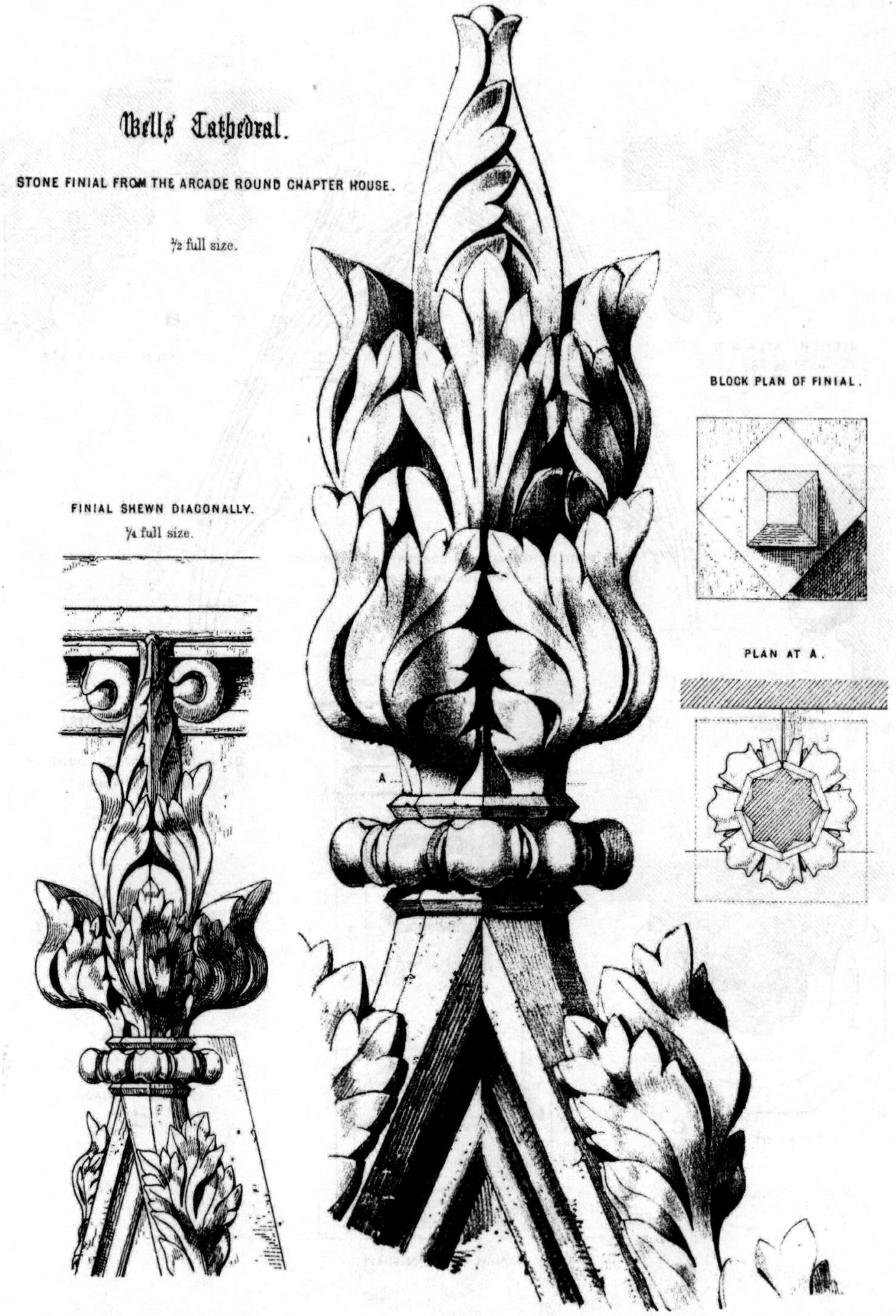

威尔斯教堂：环绕牧师会礼堂拱廊细部　A 和 B 处的剖面、檐口剖面、柱身平面 、小尖塔平面柱子柱头和柱础

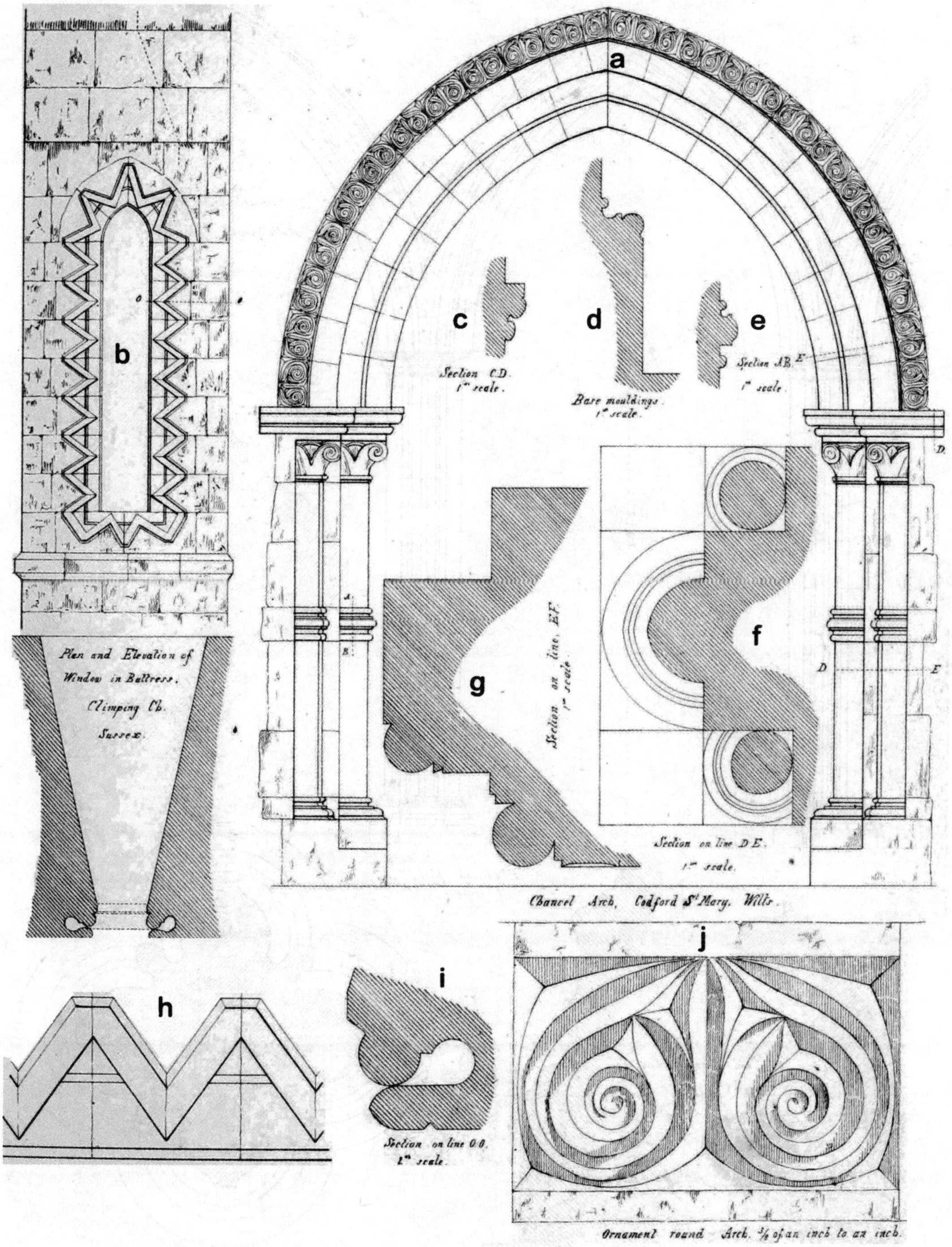

a. 威尔特郡考德福德圣玛丽教堂圣殿内拱　b. 萨塞克斯郡克林宾教堂扶壁内窗户平面及剖面　c.C–D 处的剖面　d. 柱础线脚　e.A–B 处的剖面　f.D–E 处的剖面　g.E–F 处的剖面　h. 环绕窗户的山形纹样　i.O–O 处的剖面　j. 环绕拱的装饰

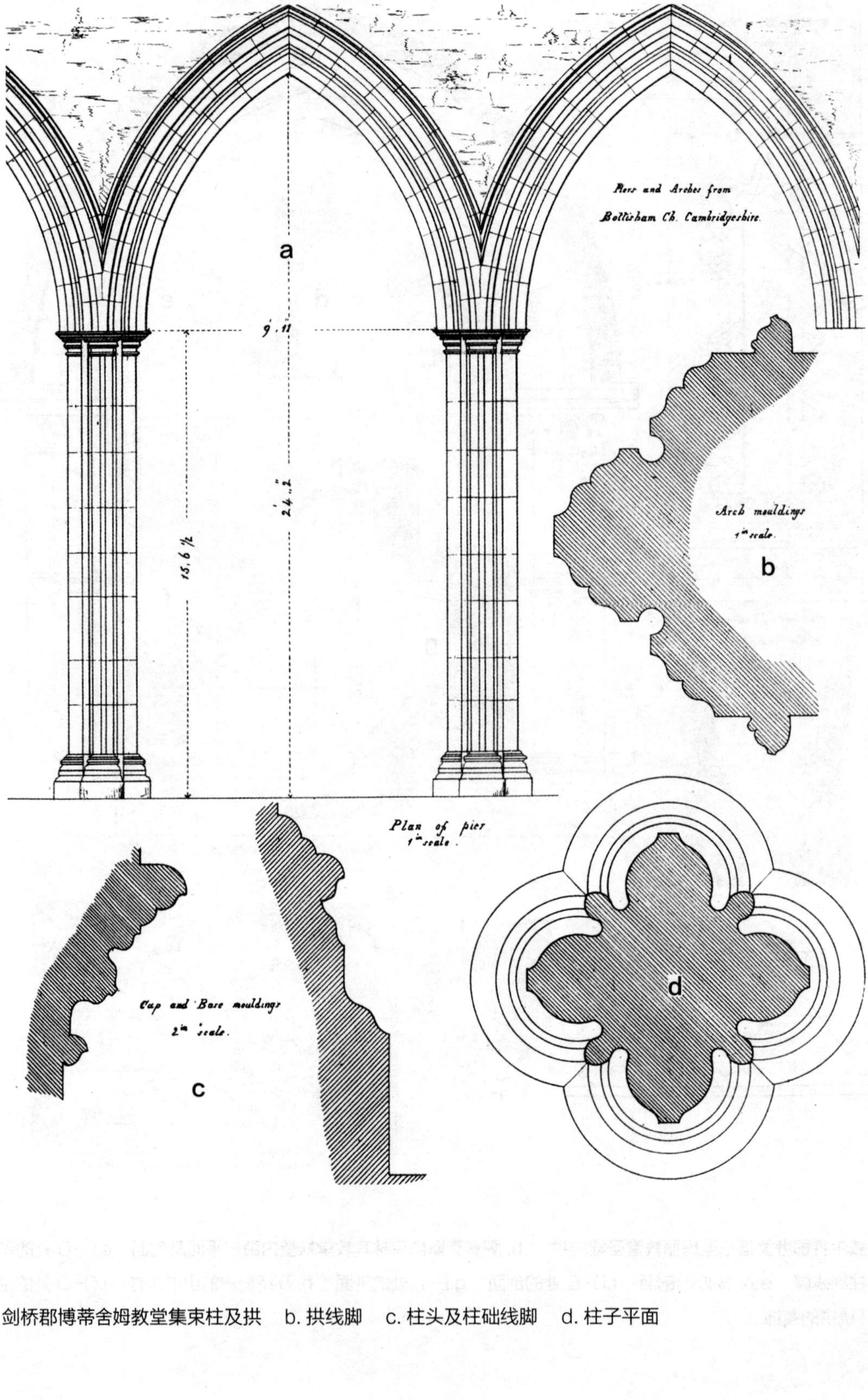

a. 剑桥郡博蒂舍姆教堂集束柱及拱　b. 拱线脚　c. 柱头及柱础线脚　d. 柱子平面

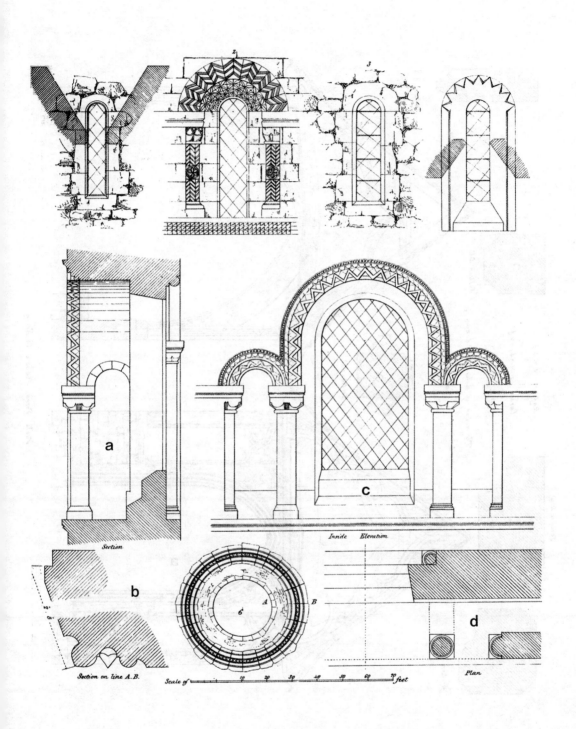

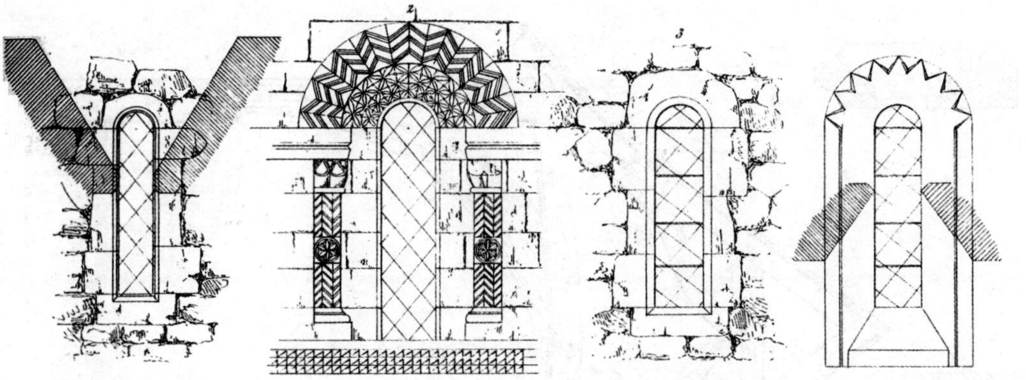

a. 剖面　b. A-B 处的剖面　c. 内侧立面　d. 平面

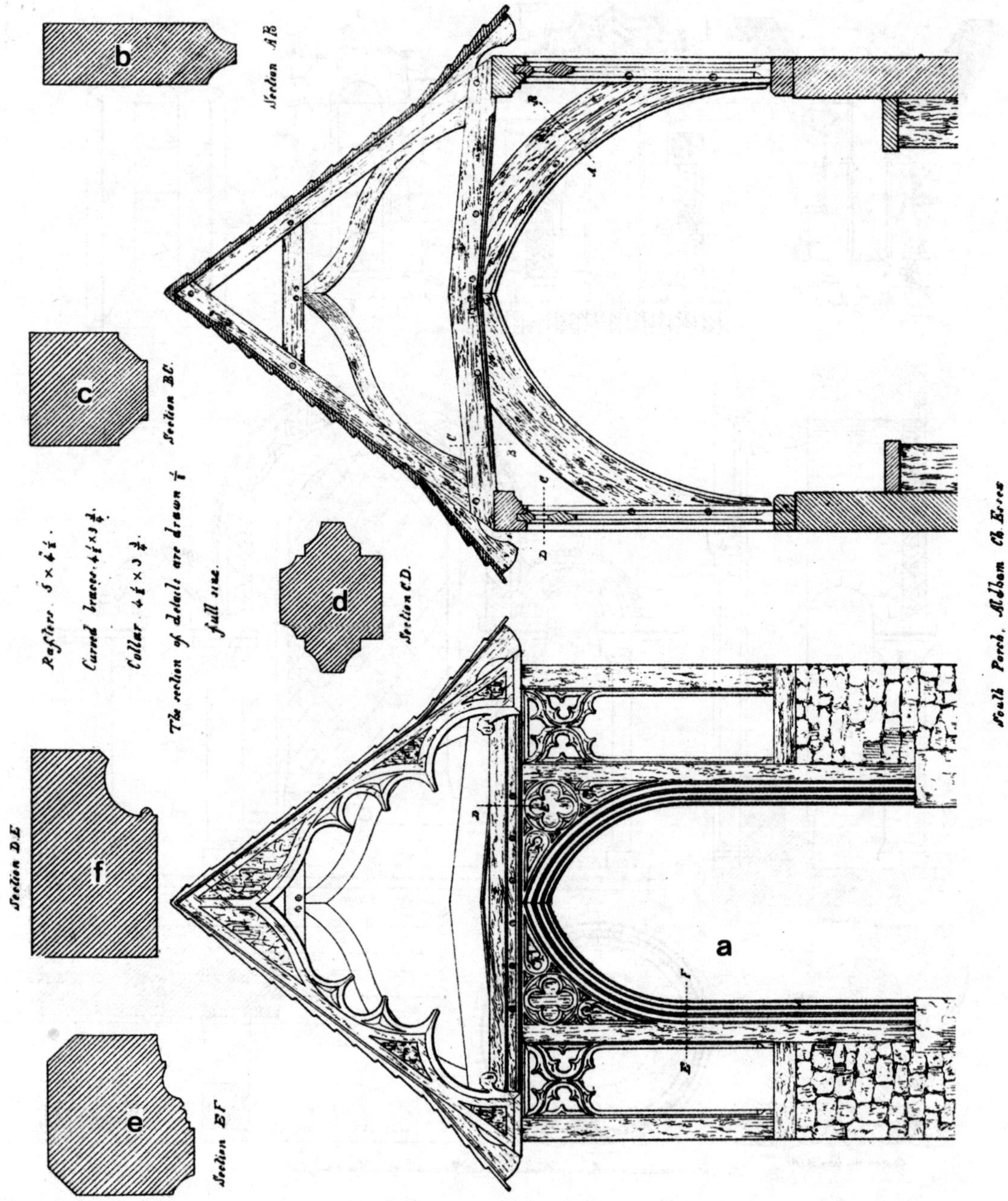

a. 埃塞克斯郡奥尔德姆教堂南门廊　b.A-B 处的剖面　c.B-C 处的剖面　d.C-D 处的剖面　e.E-F 处的剖面　f.D-E 处的剖面

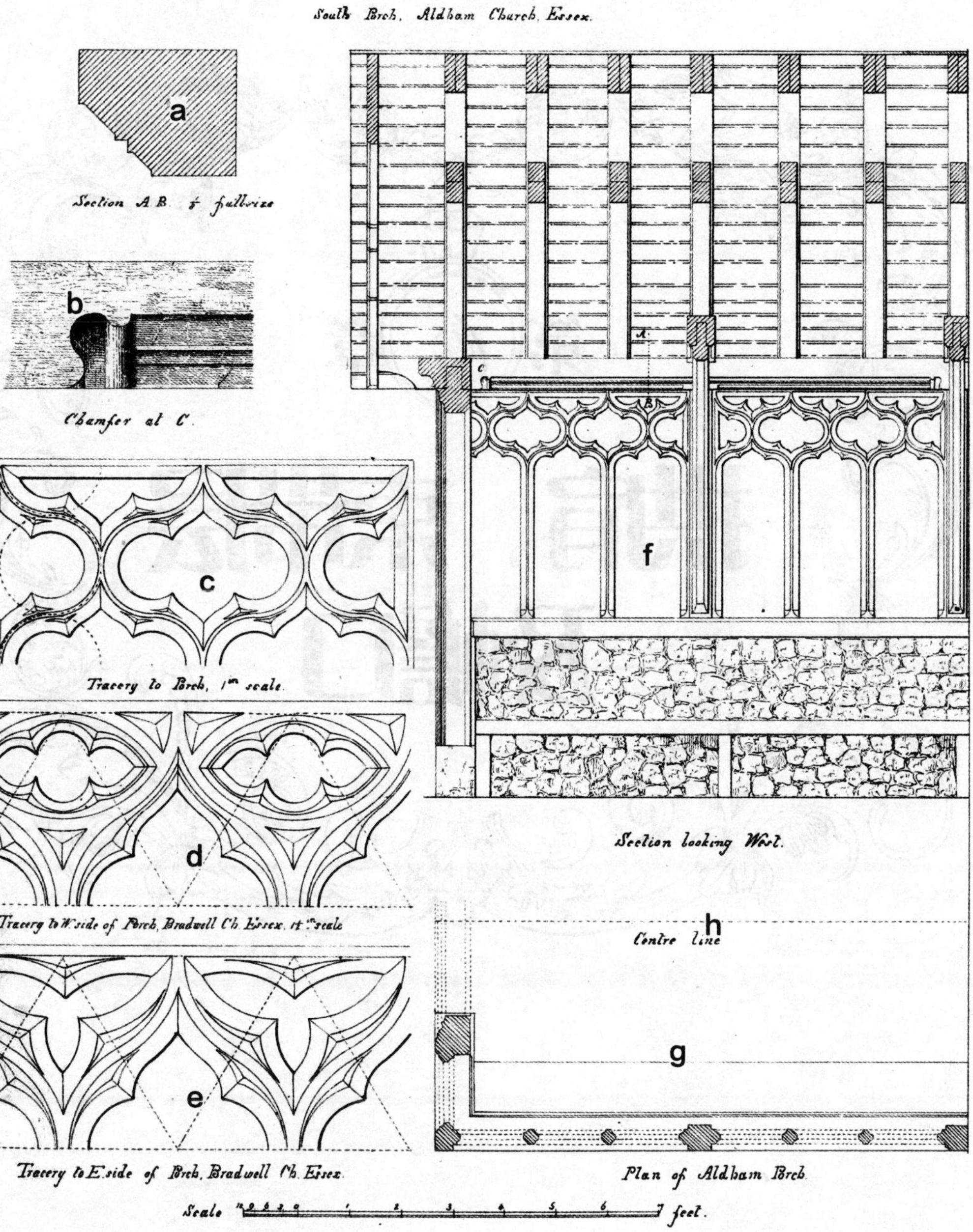

埃塞克斯郡奥尔德姆教堂南门廊：a.A–B 处的剖面　b.C 处的切角　c. 门廊内花格饰　d. 埃塞克斯郡布　拉德韦尔教堂门廊西侧花格饰　e. 埃塞克斯郡布　拉德韦尔教堂门廊东侧花格饰　f. 向西看剖面　g. 奥尔德姆教堂南门廊平面　h. 中轴线

第6章

拱肩、束带层及檐口

第 6 章 拱肩、束带层及檐口

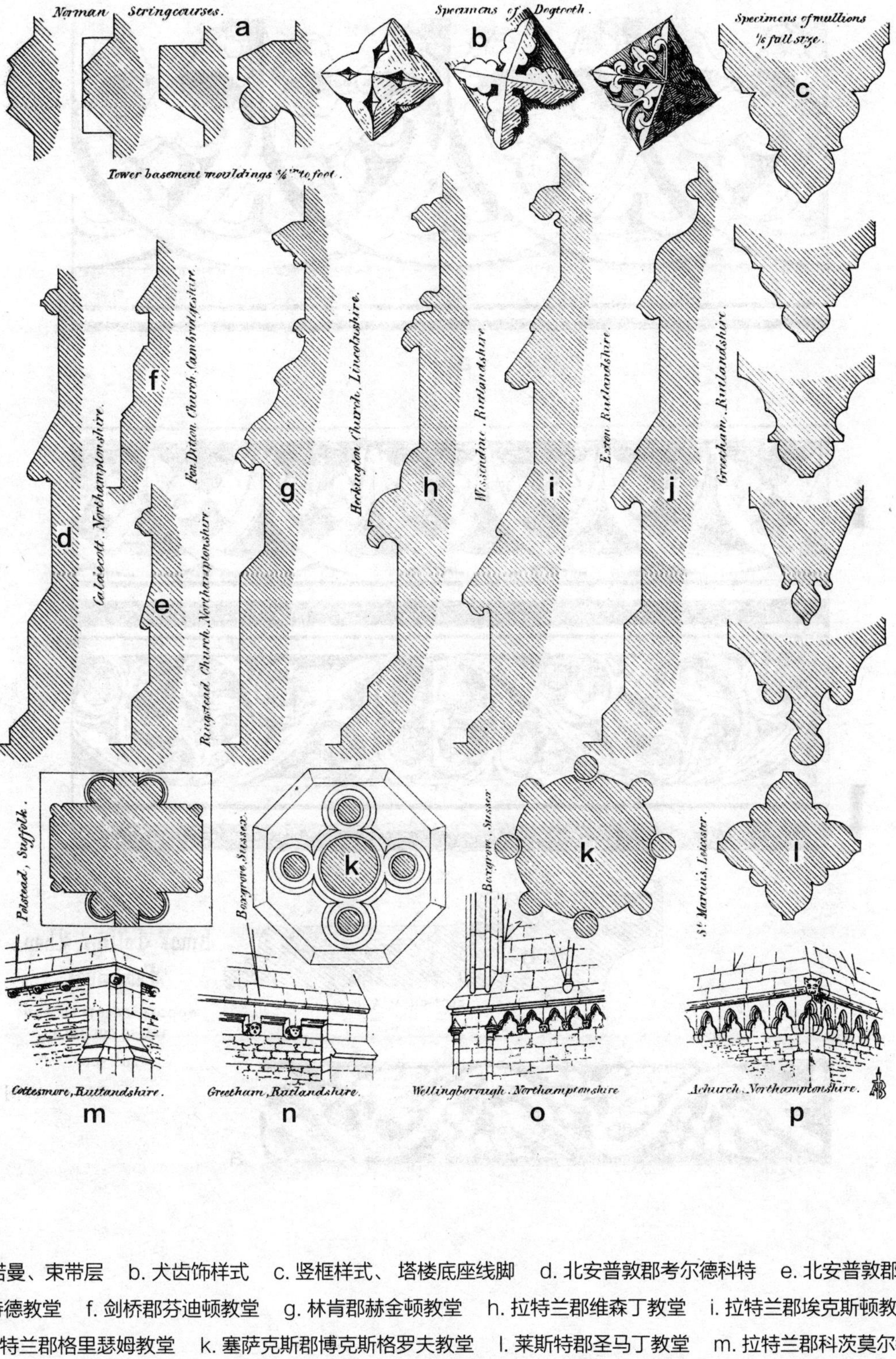

a. 诺曼、束带层　b. 犬齿饰样式　c. 竖框样式、塔楼底座线脚　d. 北安普敦郡考尔德科特　e. 北安普敦郡灵斯特德教堂　f. 剑桥郡芬迪顿教堂　g. 林肯郡赫金顿教堂　h. 拉特兰郡维森丁教堂　i. 拉特兰郡埃克斯顿教堂　j. 拉特兰郡格里瑟姆教堂　k. 塞萨克斯郡博克斯格罗夫教堂　l. 莱斯特郡圣马丁教堂　m. 拉特兰郡科茨莫尔教堂　n. 拉特兰郡格里瑟姆教堂　o. 北安普敦郡韦灵伯勒教堂　p. 北安普敦郡阿克切奇教堂塔尖台座样式

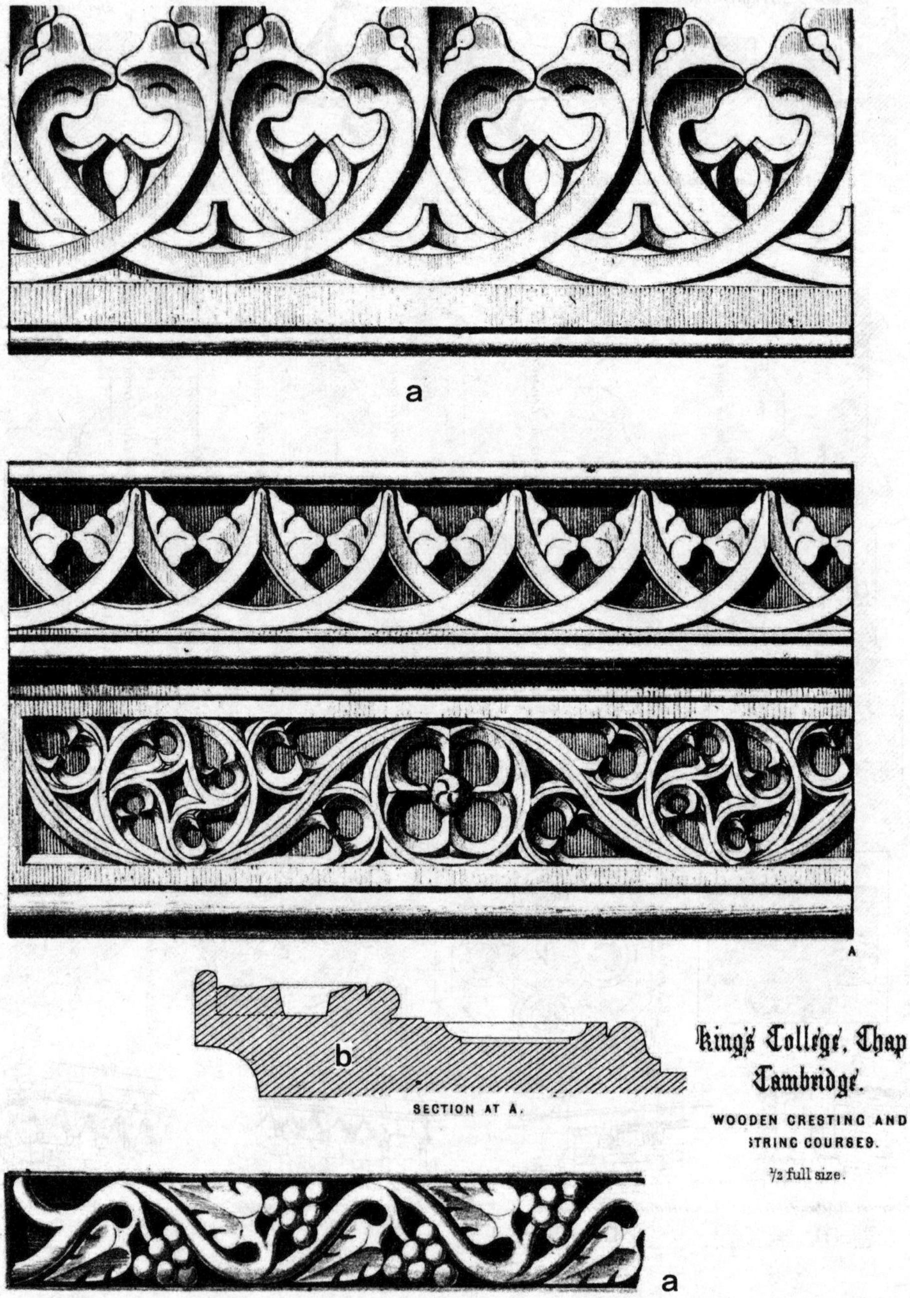

剑桥郡国王学院礼拜堂：a. 木制缘饰及束带层　b.A 处的剖面

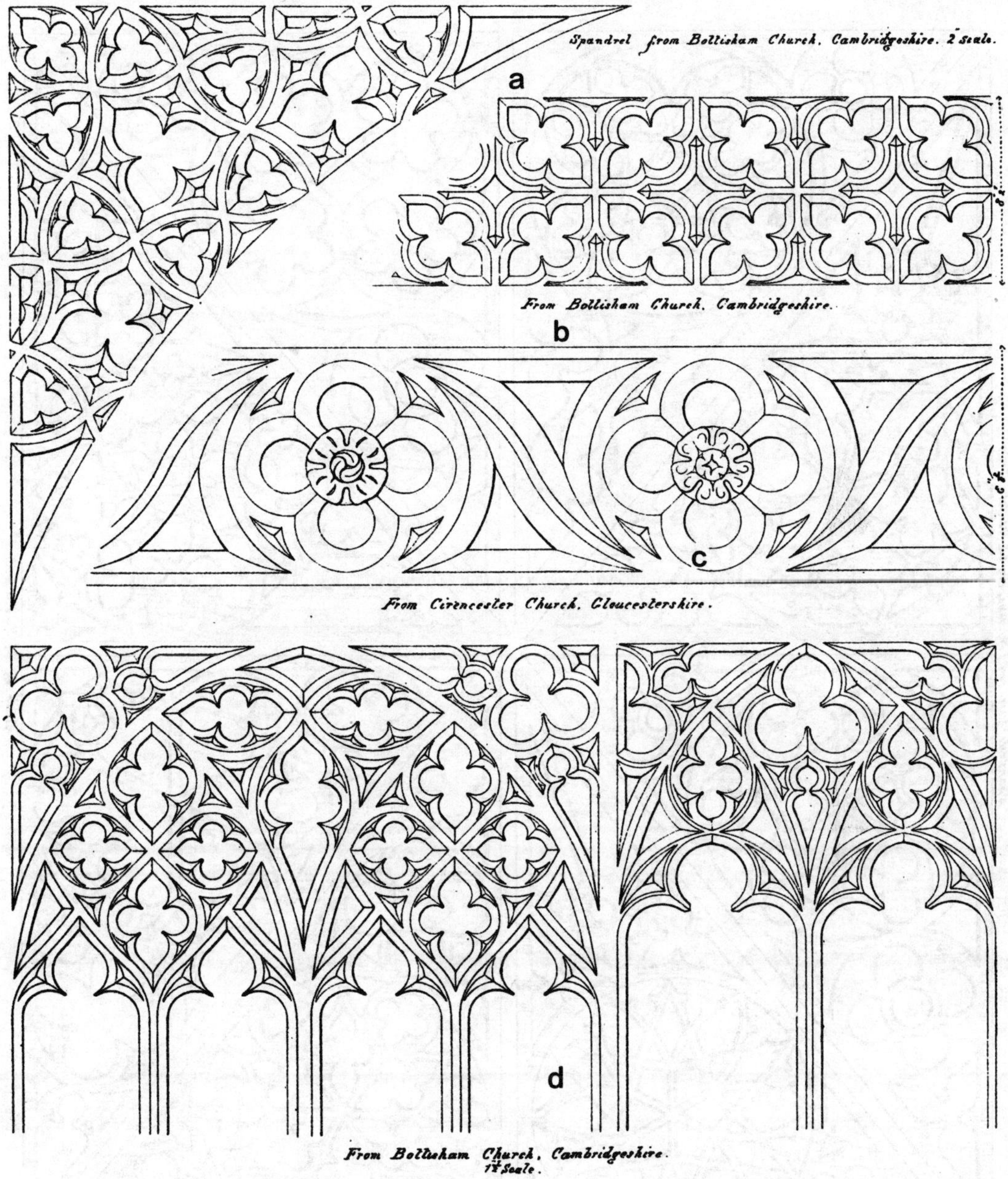

a. 剑桥郡博蒂舍姆教堂拱肩　b. 来自剑桥郡博蒂舍姆教堂　c. 格洛斯特郡赛伦塞斯特教堂　d. 来自剑桥郡博蒂舍姆教堂

诺福克郡怀蒙德汉姆教堂屋顶上拱肩

Spandrels from Doncaster Church, Yorkshire.
a
Spandrel from
Stoke Church,
Suffolk
b
Spandrel from
Stoke Church,
Suffolk.
b
Spandrels from Doncaster Church, Yorkshire.
a
Spandrel from
Stoke Church,
Suffolk.
b
Spandrel from
Stoke Church.
Suffolk.
b
Scale
0
1
2
3
4
5
6
Inches

a. 约克郡唐卡斯特教堂拱肩　b. 萨福克郡斯托克教堂拱肩

肯特郡石之教堂：a. 圣坛拱廊内石制拱肩　b.A 处的拱肩局部　c. 诺福克西沃尔顿教堂墓地内石制装饰

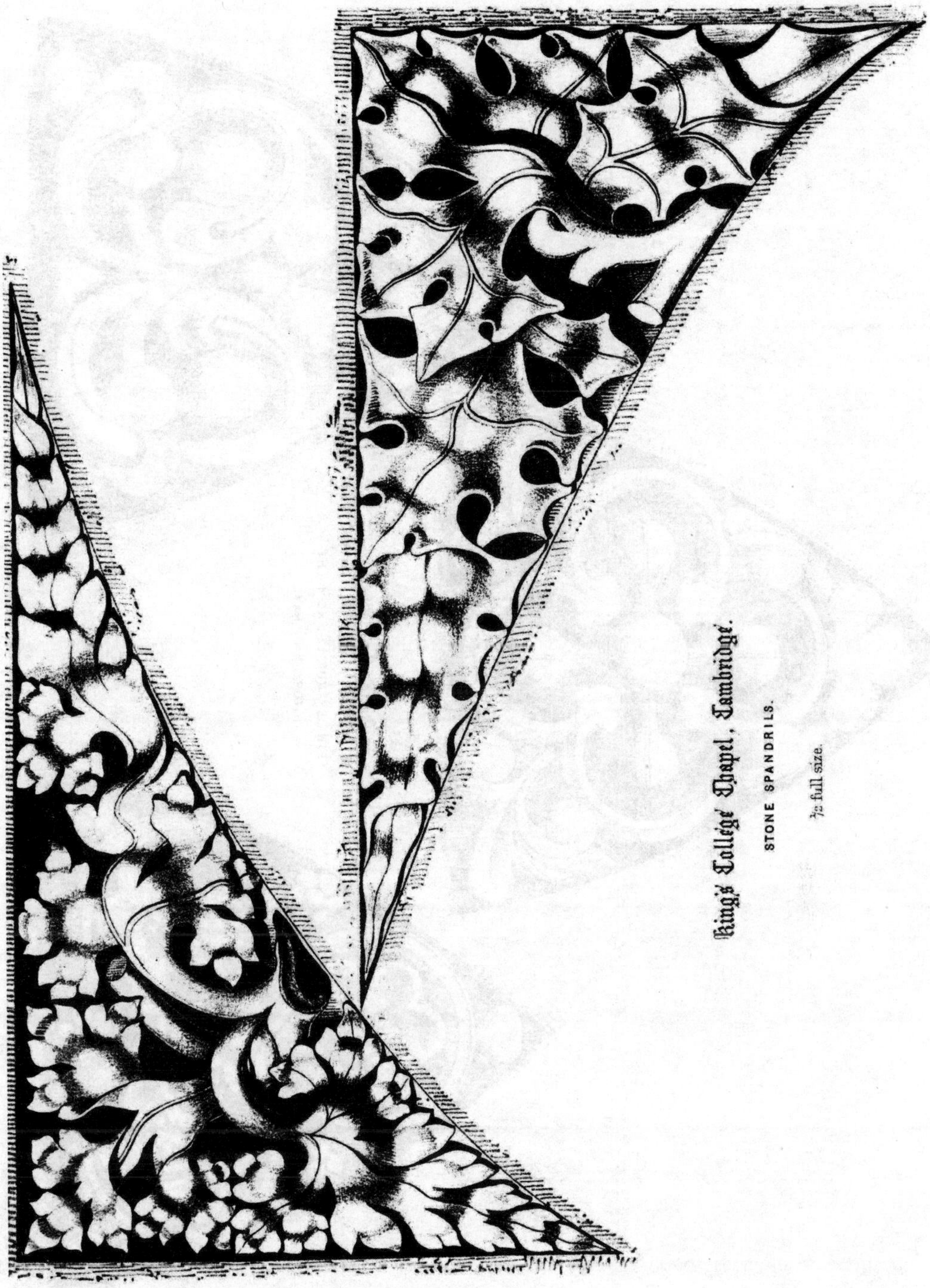

剑桥郡国王学院礼拜堂石制拱肩

威尔斯教堂中殿拱门上拱廊石制拱肩

Wells Cathedral.
UPPER CORNICE FROM BISHOP BECKINGTON'S SHRINE, ½ full size.
For Section see next Plate.

威尔斯教堂贝金顿主教圣堂上部檐口

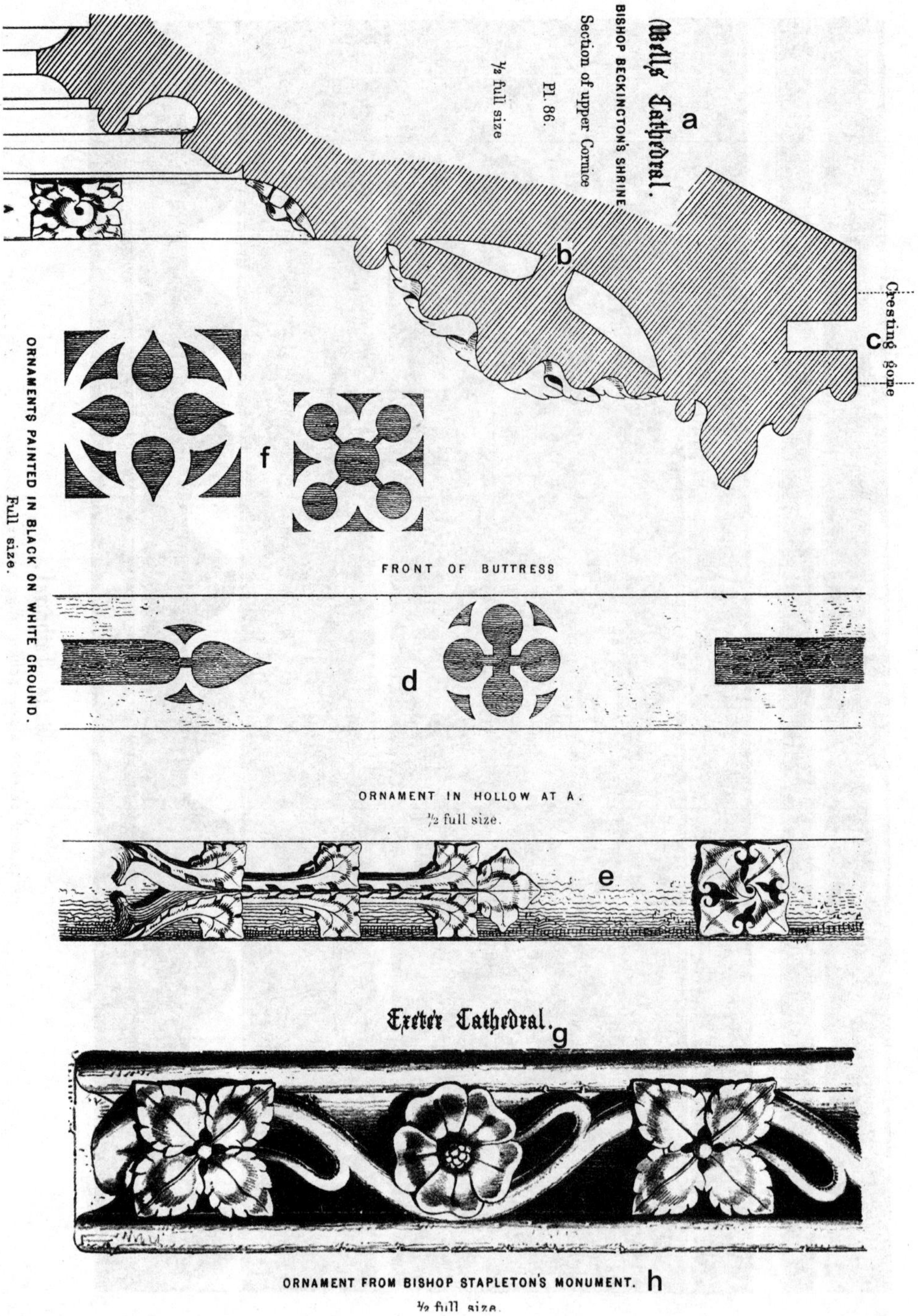

a. 威尔斯教堂贝金顿主教圣堂　b. 上部檐口剖面　c. 丢失的缘饰　d. 扶壁正面　e.A 处的凹槽内装饰　f. 画在白底上的黑色装饰　g. 埃克赛特教堂　h. 斯泰普尔顿主教纪念碑上装饰

第7章

托架

贝弗利修道院珀西神殿内的石制托架

威尔斯教堂：a. 中殿拱门上的拱廊石制托架　b. 拱线脚

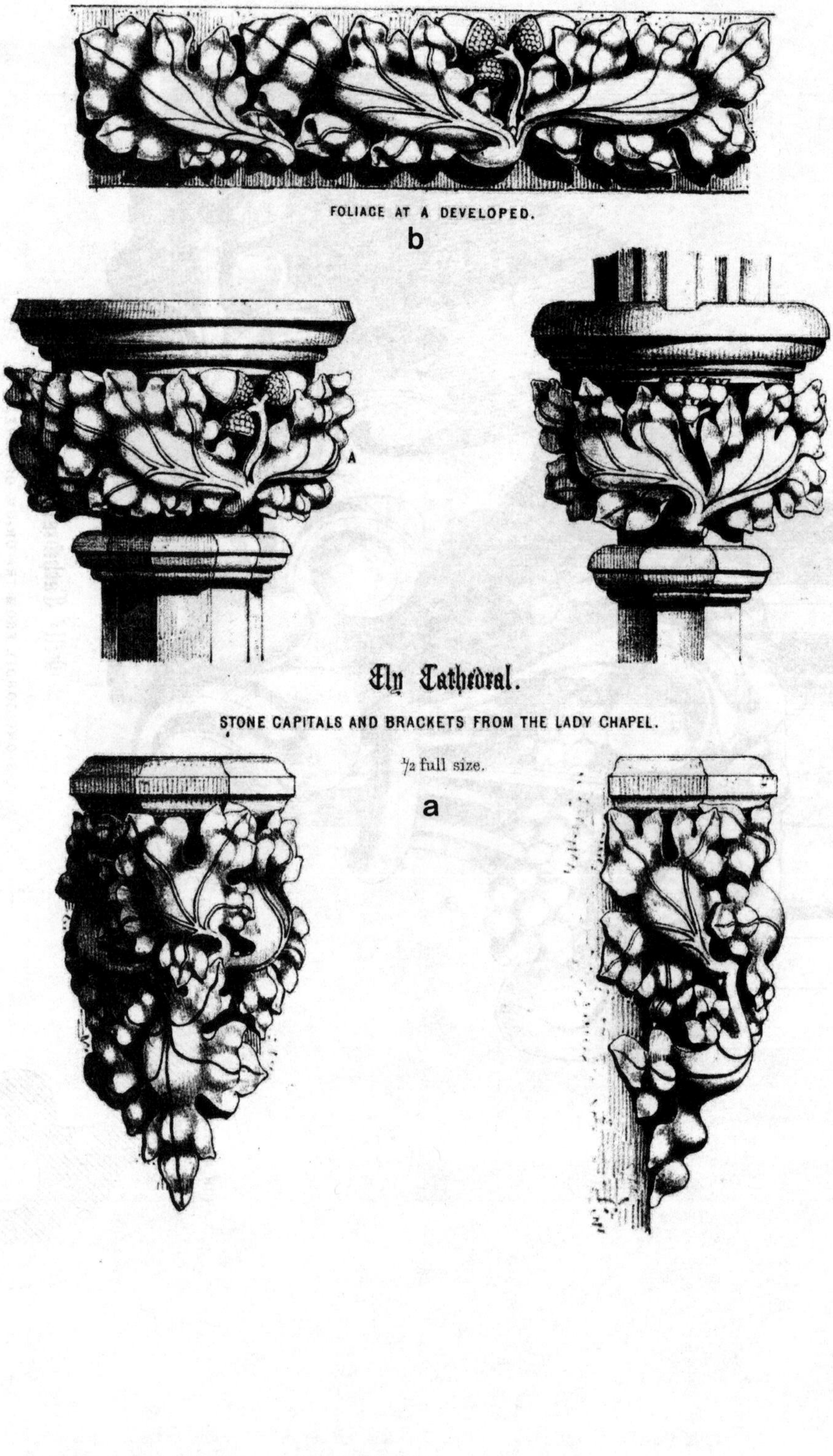

以利教堂：a. 圣母堂内石制柱头及托架　b.A 处的叶饰展开

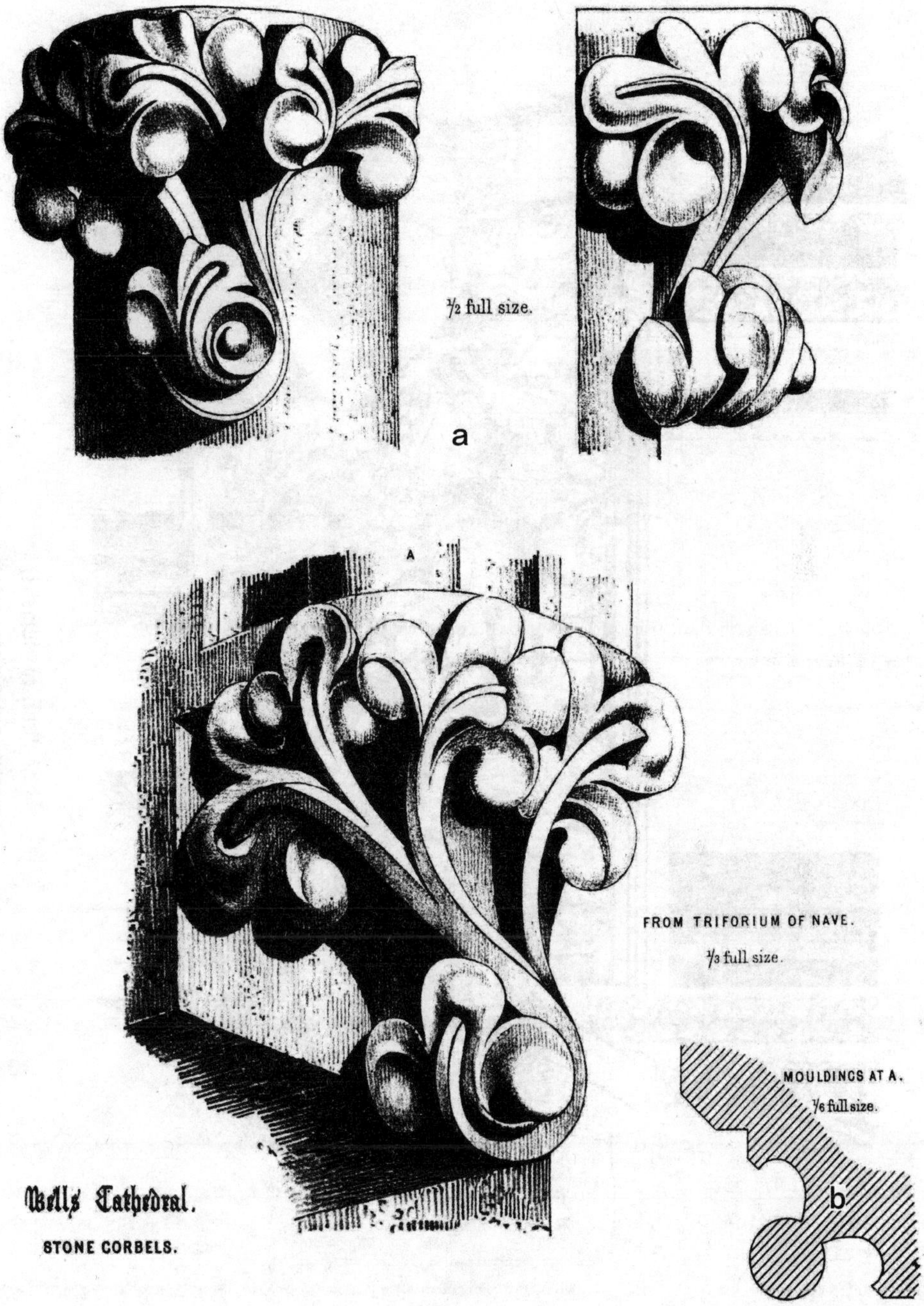

威尔斯教堂：a. 中殿拱门上拱廊石制托架　b.A 处的线脚

贝弗利修道院圣坛屏风背部壁龛石制托架

第8章

屋顶及顶棚

a. 来自格洛斯特郡布罗克沃思教堂北侧廊屋顶　b. 来自格洛斯特郡布罗克沃的思教堂南礼拜堂　c. 来自格洛斯特郡赛伦塞斯特教堂　d. 来自格洛斯特郡赛伦塞斯特教堂

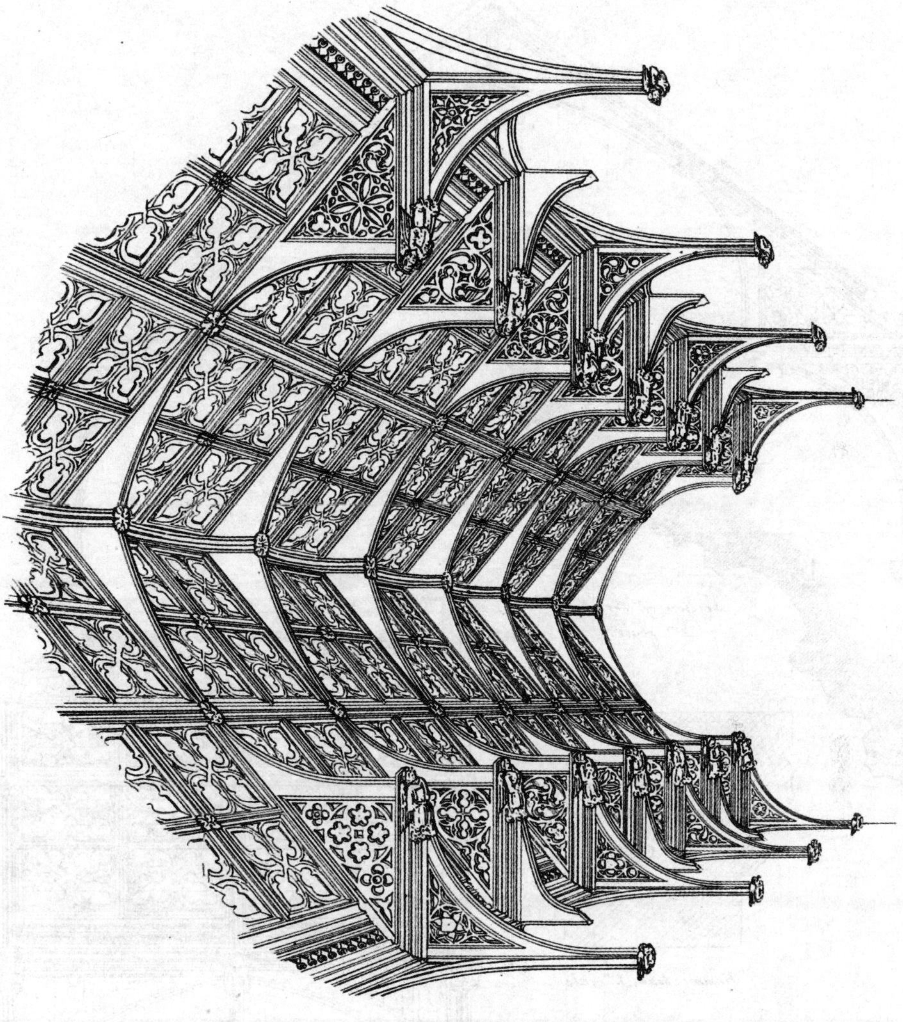

Roof over North Aisle, Wymondham Church, Norfolk.

诺福克郡怀蒙德汉姆教堂北侧廊屋顶

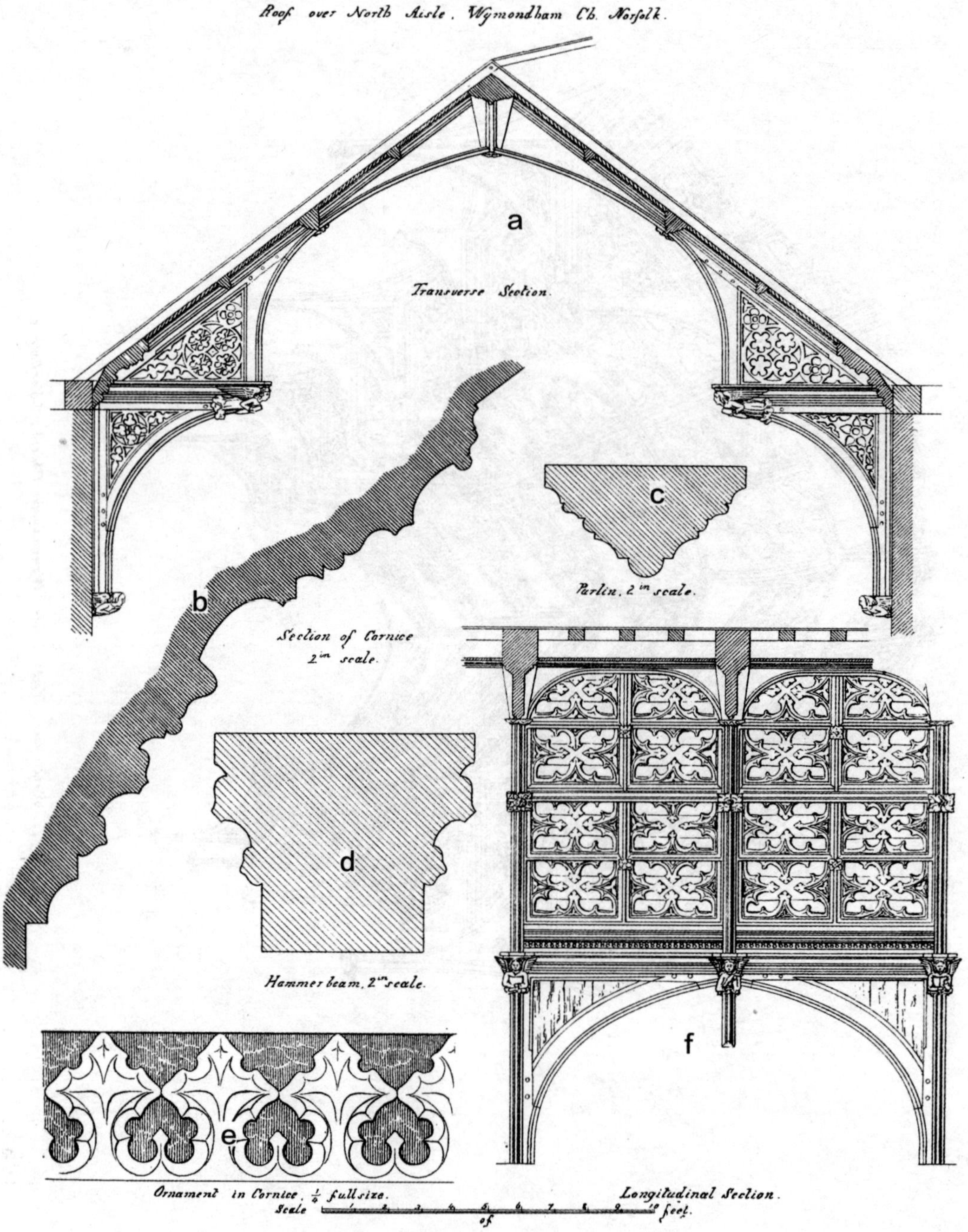

诺福克郡怀蒙德汉姆教堂北侧廊屋顶：a. 横剖面　b. 檐口剖面　c. 檩条　d. 托臂梁　e. 檐口装饰　f. 纵剖面

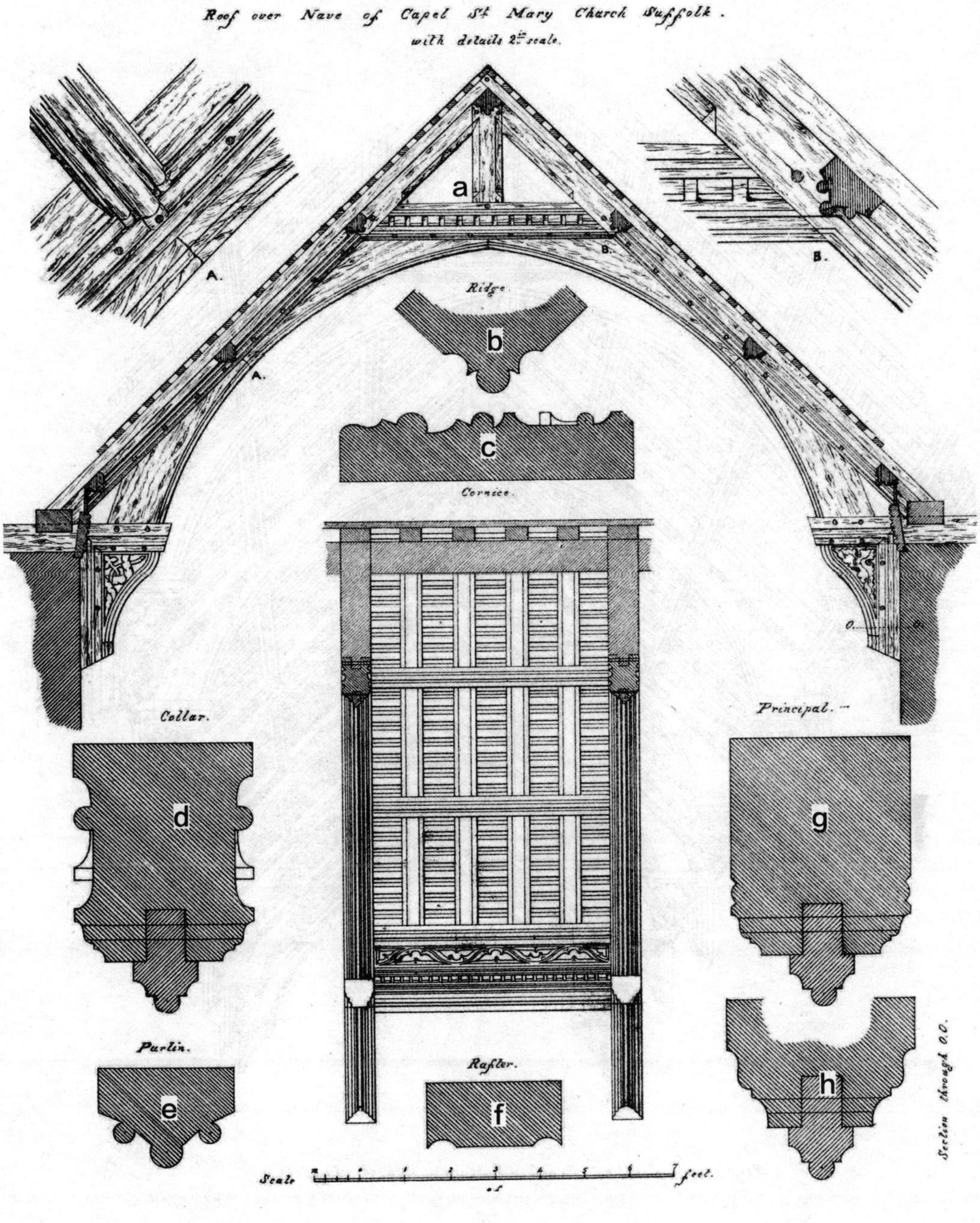

a. 萨福克郡卡佩尔圣玛丽教堂中殿屋顶　b. 屋脊　c. 檐口　d. 接头　e. 檩条　f. 椽子　g. 主要屋架　h.O-O 处的剖面

Roof over Nave of Capel St. Mary's Church Suffolk.

萨福克郡卡佩尔圣玛丽教堂中殿屋顶

a. 萨福克郡斯图斯顿教堂门廊屋顶　b. 纵剖面

Roof over the nave of Grundisburgh Ch. Suffolk.

萨福克郡格朗第斯伯格教堂中殿屋顶

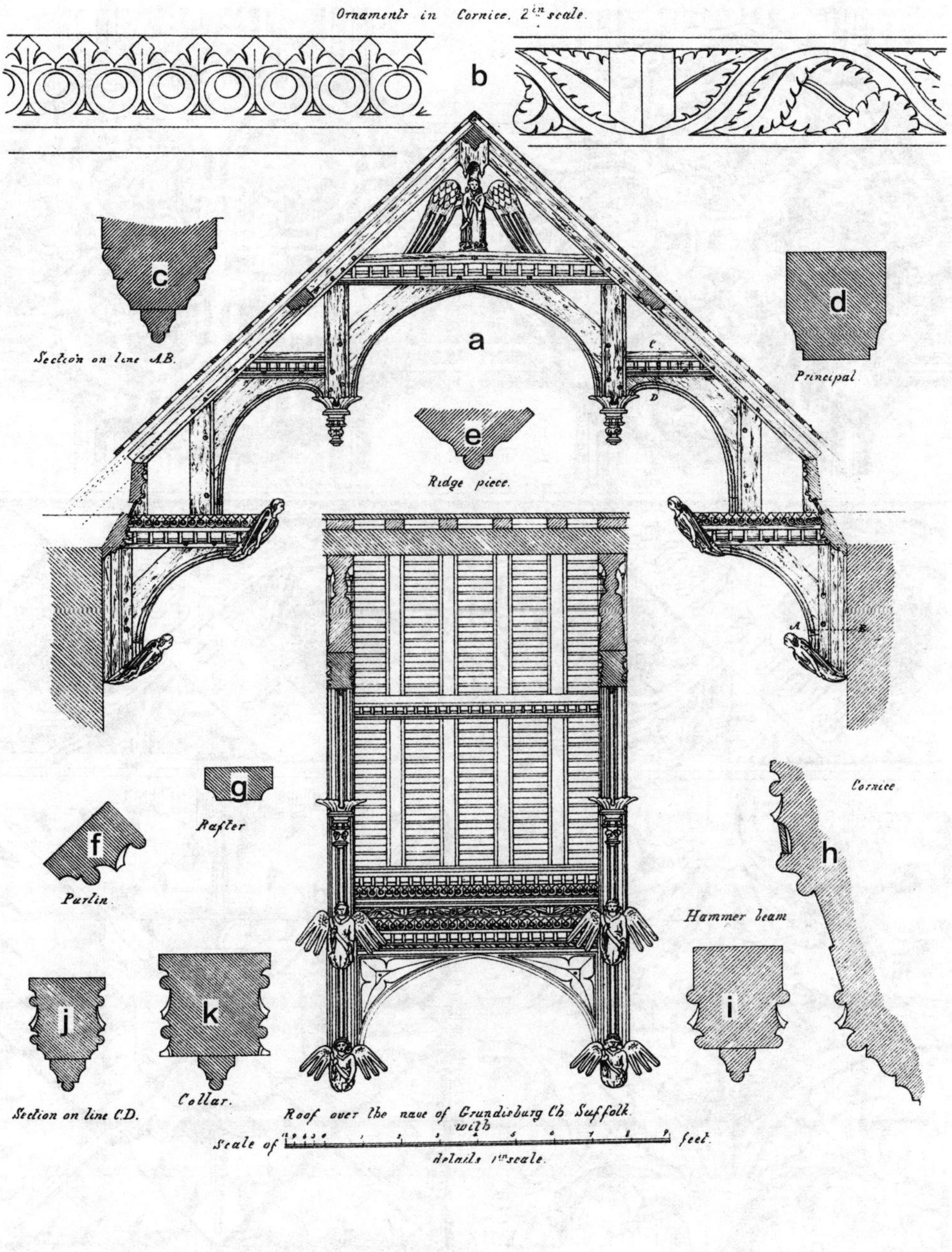

a. 萨福克郡格朗第斯伯格教堂中殿屋顶 b. 檐口上装饰 c.A–B 处的剖面 d. 主要屋架 e. 脊枋 f. 檩条 g. 椽子 h. 檐口 i. 托臂梁 j.C–D 处的剖面 k. 接头

a. 来自剑桥郡哈斯灵菲尔德教堂侧廊屋顶　b. 来自诺福克郡创奇教堂圣坛屏　c. 来自诺福克郡迪斯教堂北门廊屋顶　d. 来自格洛斯特郡布罗克沃斯教堂北侧廊屋顶

贝弗利修道院圣坛屏风背部壁龛石制顶棚

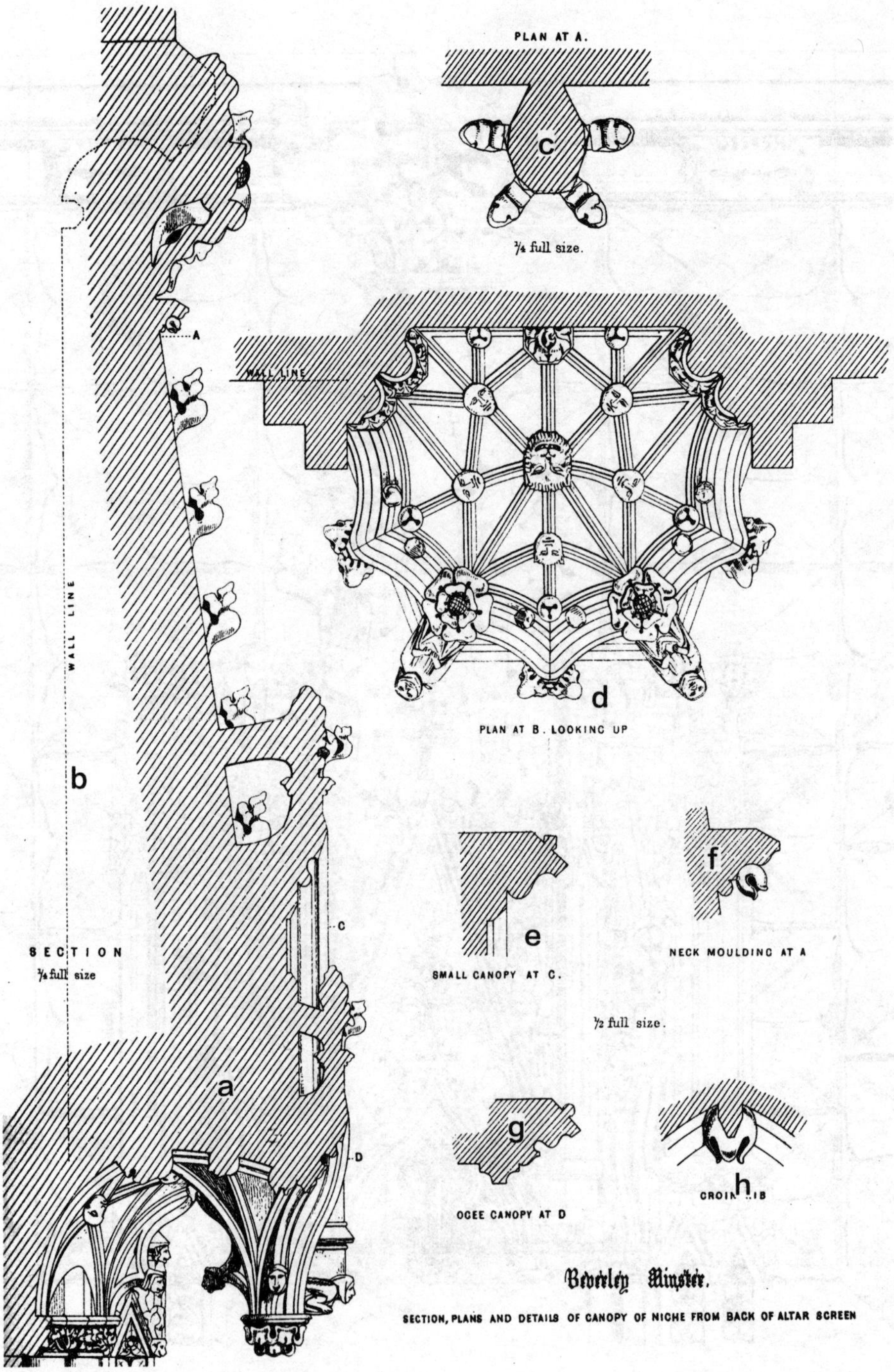

贝弗利修道院圣坛屏风背部壁龛石制顶棚剖面、平面及细部：a. 剖面　b. 墙线　c.A 处的平面　d.B 处的仰视平面　e.C 处的小顶棚　f.A 处的柱颈线脚　g.D 处的二心内心桃尖拱顶棚　h. 穹棱

第9章

圣坛

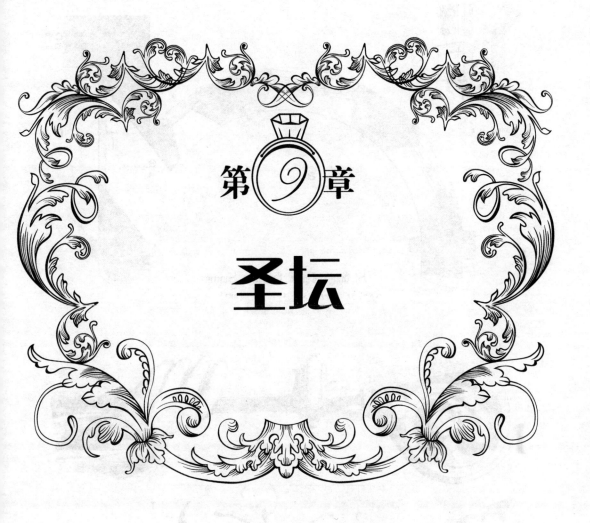

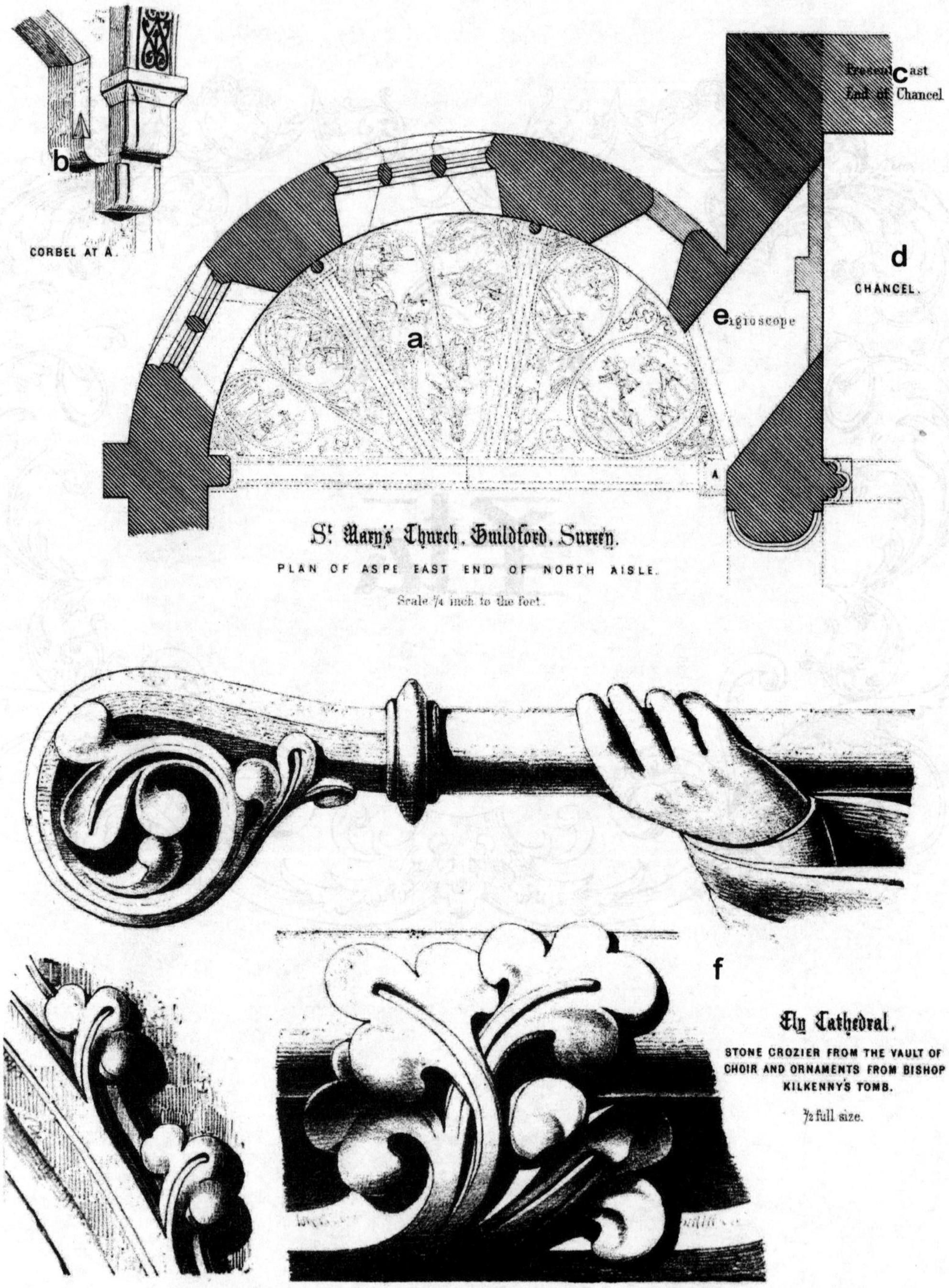

萨里郡吉尔福德圣玛丽教堂：a. 北侧廊东尽端半圆形后殿平面　b.A 处的托架　c. 现有圣坛东侧尽端　d. 圣坛
e. 内壁小窗　f. 以利教堂唱诗室拱顶上石制权杖及基尔肯尼主教墓穴装饰

Beverley Minster.

STONE ORNAMENTS IN CORNICE, BACK OF ALTAR.

½ full size.

贝弗利修道院圣坛背部檐口的石制装饰

贝弗利修道院圣坛穹棱上石制凸饰

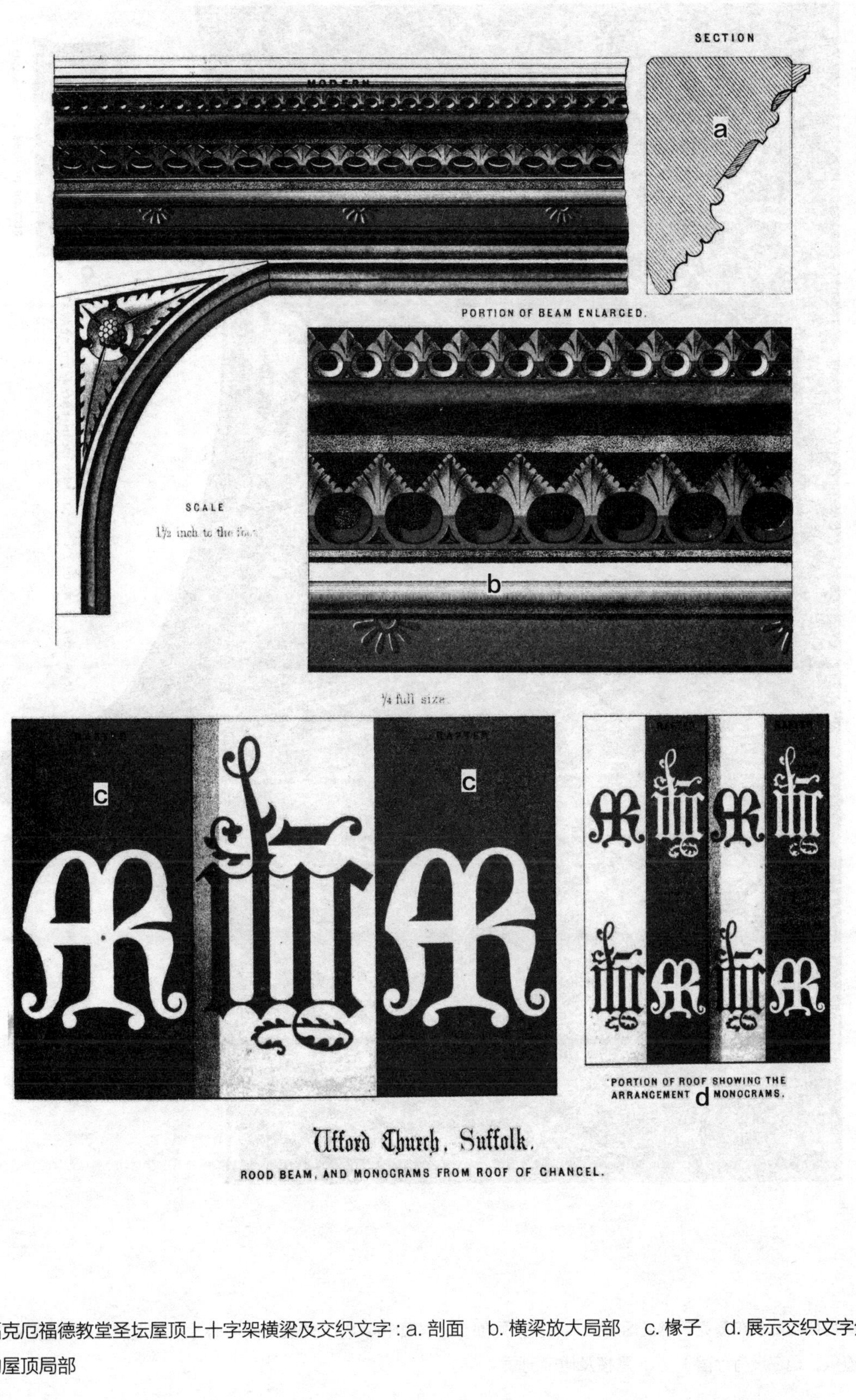

萨福克厄福德教堂圣坛屋顶上十字架横梁及交织文字：a. 剖面　b. 横梁放大局部　c. 椽子　d. 展示交织文字分布的屋顶局部

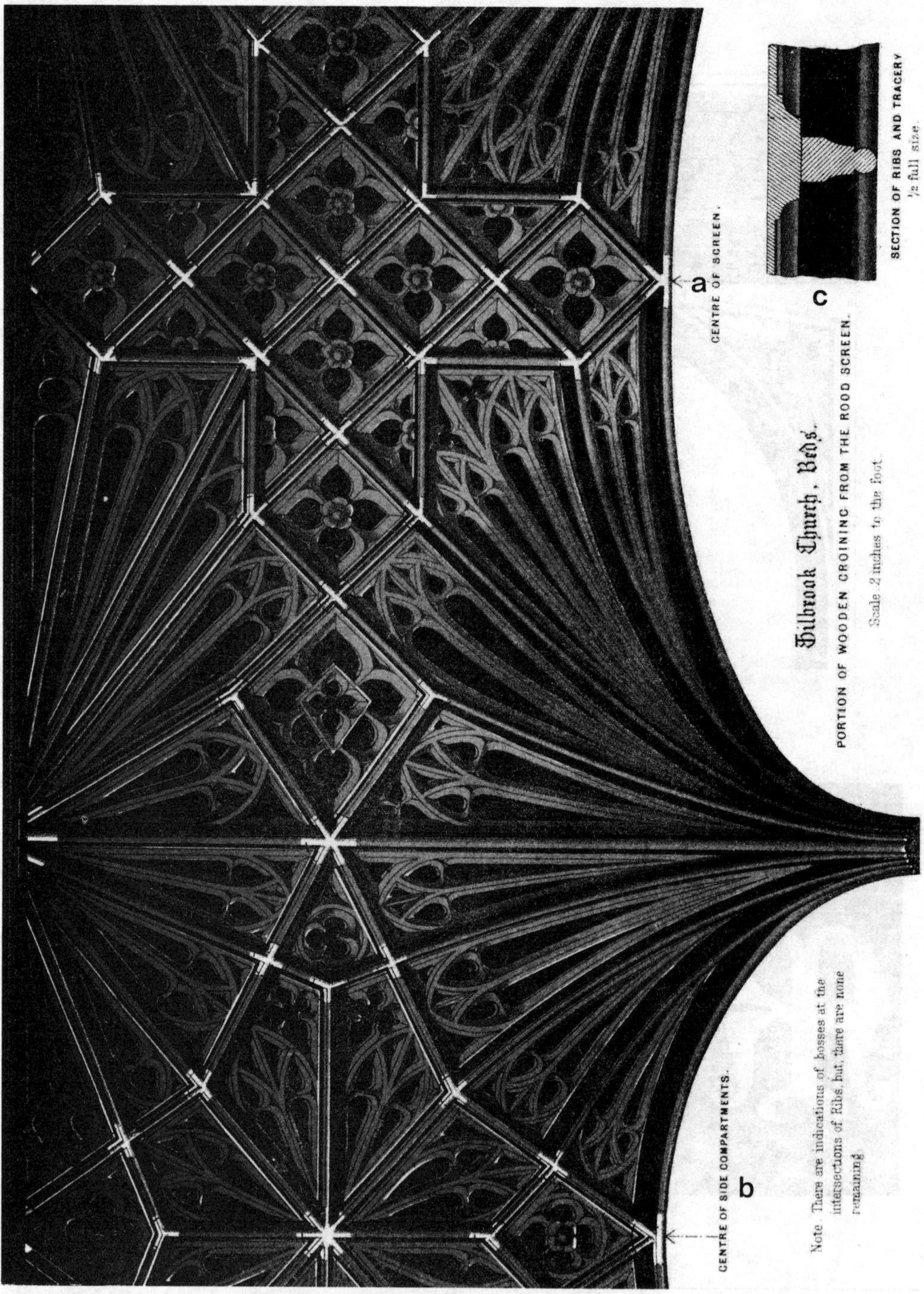

贝德福德郡蒂尔布鲁克教堂圣坛屏上木制穹顶局部：a. 屏风中心　b. 侧面隔间中心（在穹棱交汇处有凸饰存在过的迹象，现在没有存留）　c. 穹棱及檐口剖面

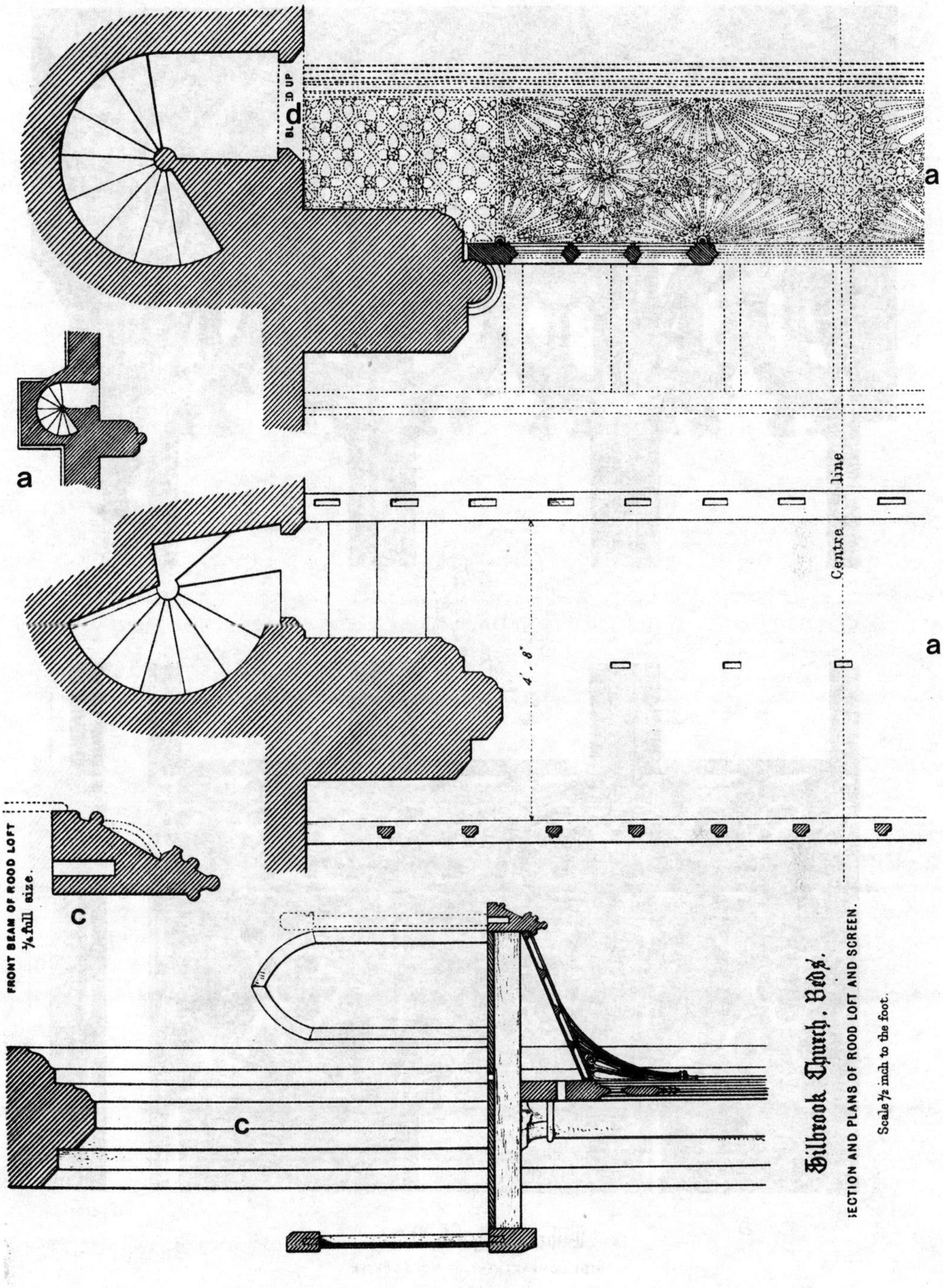

贝德福德郡蒂尔布鲁克教堂：a. 十字架龛及圣坛屏剖面和平面　b. 圣坛拱　c. 十字架龛前横梁　d. 已封堵

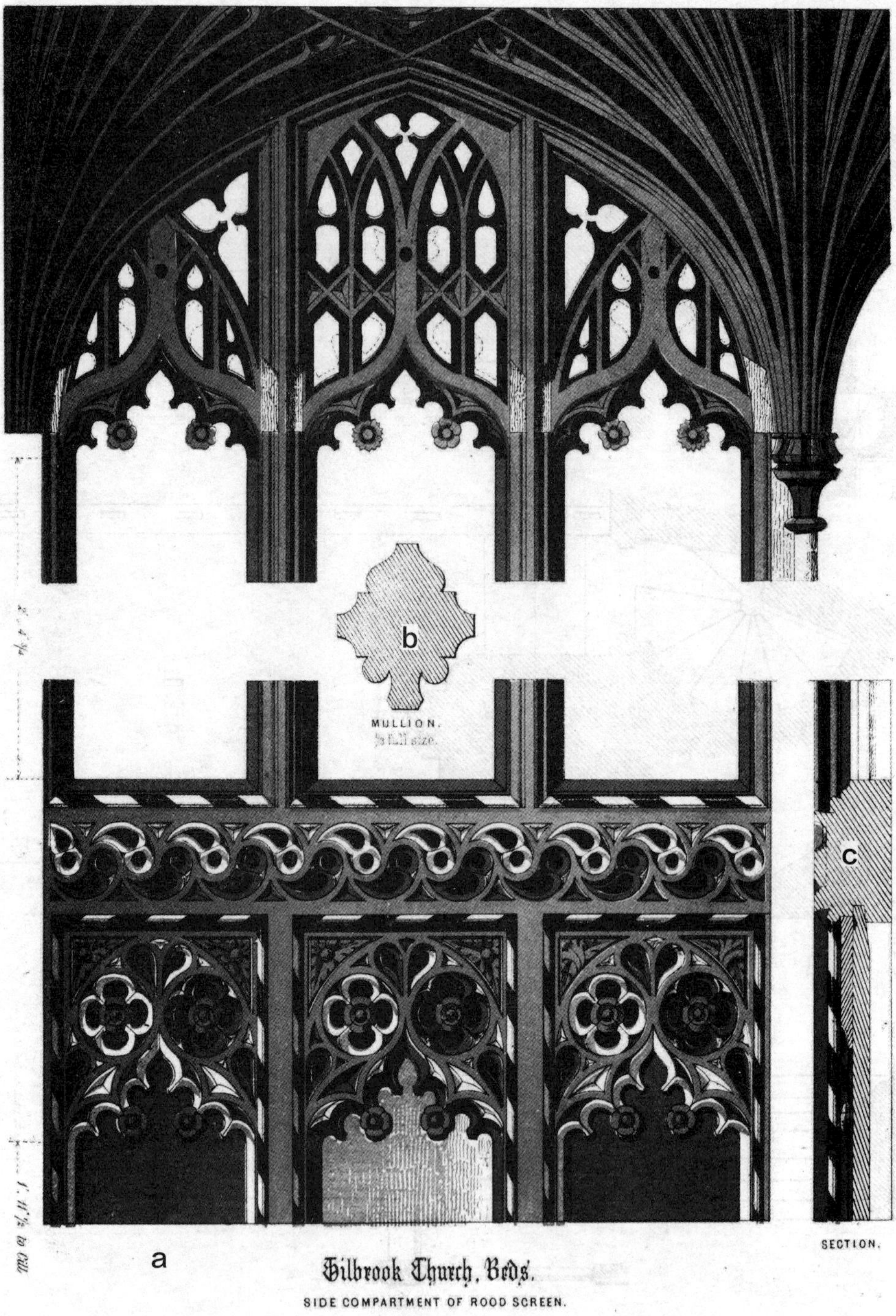

贝德福德郡蒂尔布鲁克教堂：a. 圣坛屏侧面隔间　b. 竖框　c. 剖面

贝弗利修道院：a. 修复过的圣坛屏风背部壁龛　b. 平面

第10章

顶饰及十字架

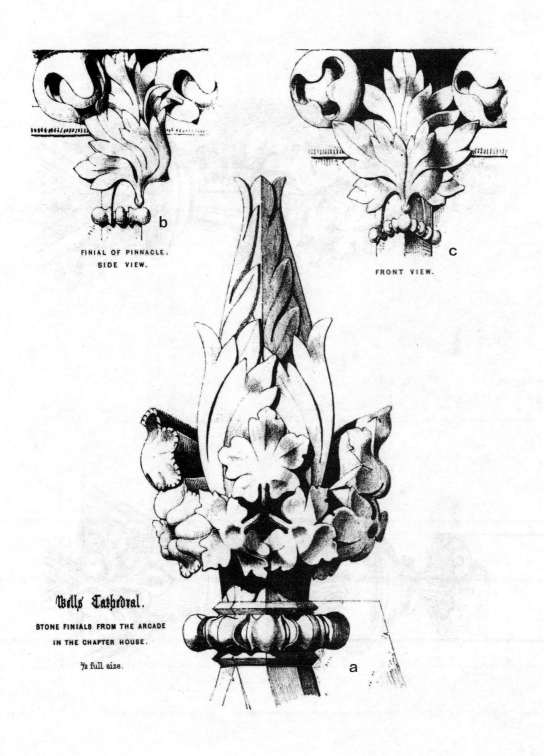

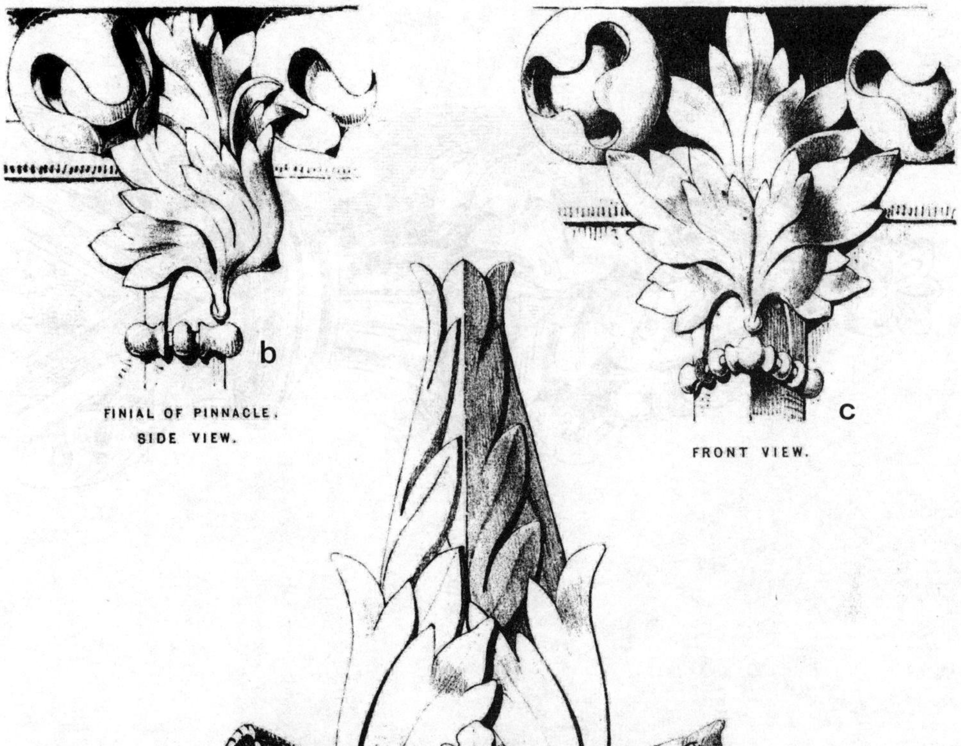

威尔斯教堂：a. 牧师会礼堂拱廊上的石制尖顶饰　b. 小尖塔尖顶饰侧面　c. 正面

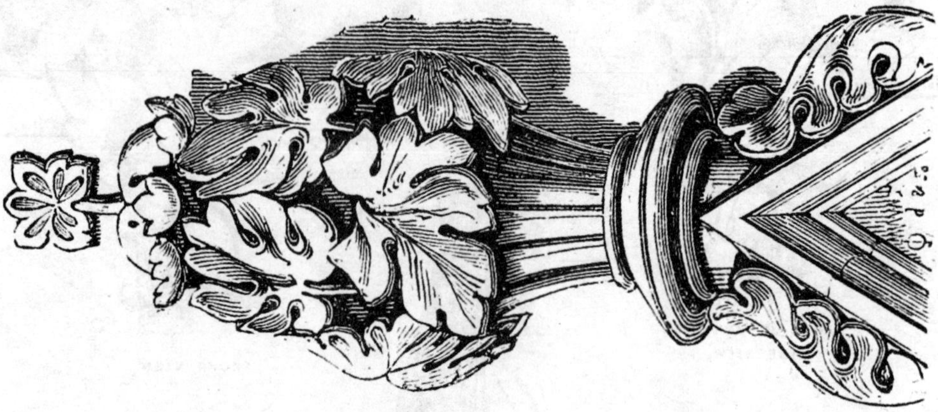

a

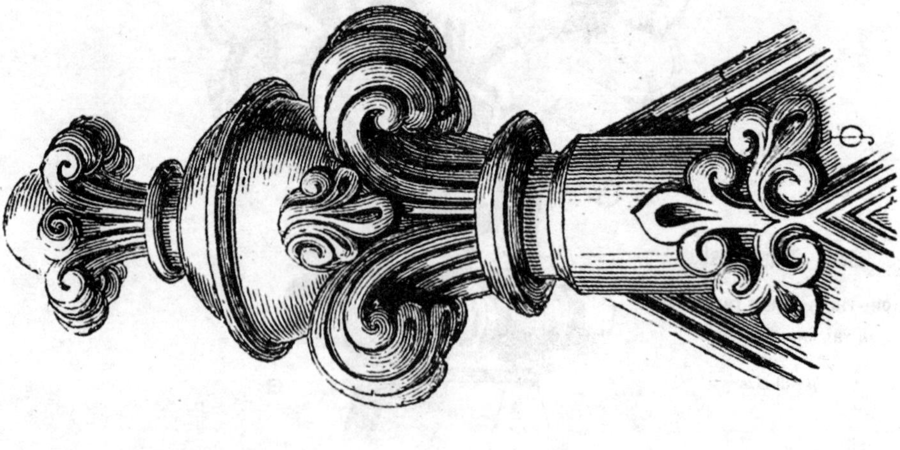

b

a. 牛津大学默顿教堂　b. 林肯教堂

诺丁汉郡伊顿教堂

a

b

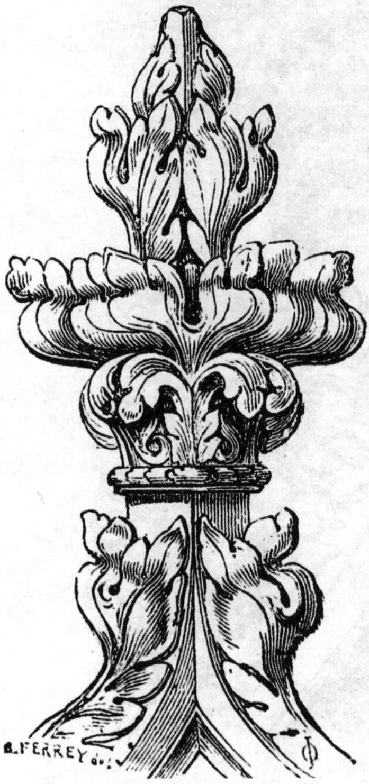

c

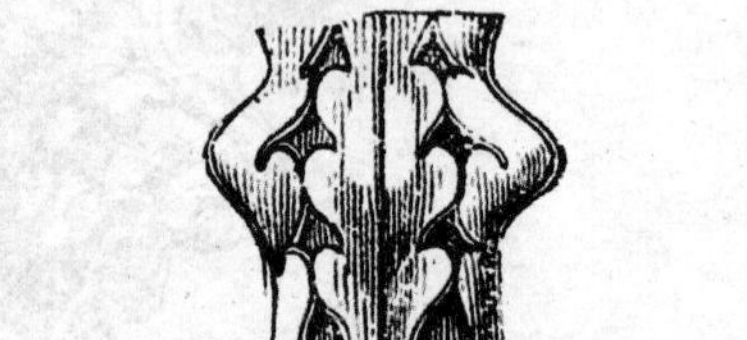

d

a. 剑桥郡国王学院礼拜堂　b. 德朗菲尔德教堂　c. 温伯恩大教堂　d. 德文郡齐托汉普

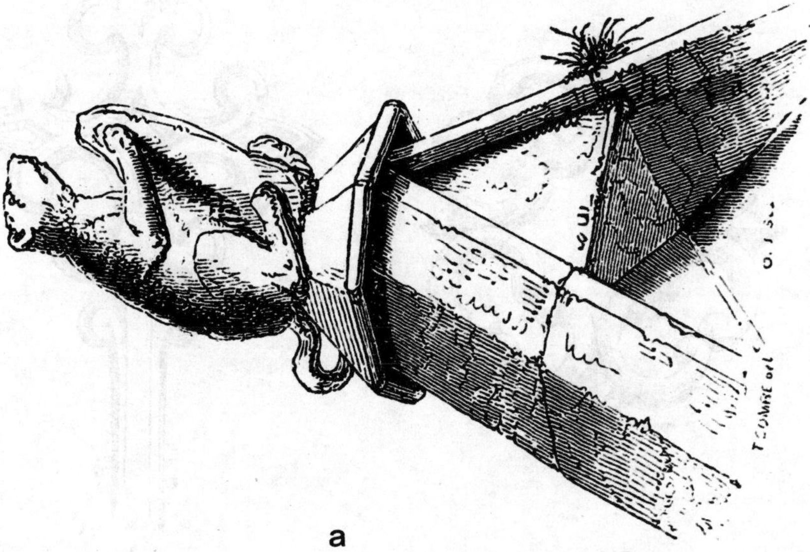

a

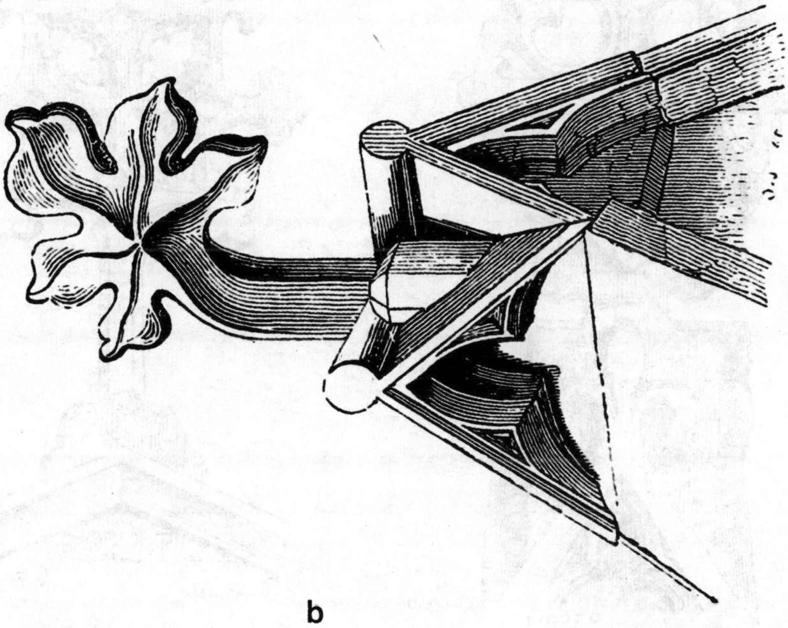

b

a. 巴斯汉普顿谷仓　b. 多塞特郡沃尔弗顿礼堂

a

b

c

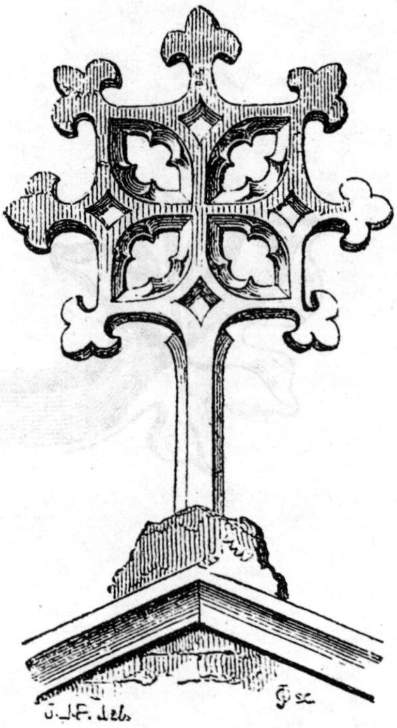

d

a. 林肯郡默顿教堂　b. 彼得伯勒教堂　c. 北安普敦郡瓦克顿　d. 林肯郡默顿教堂

温彻斯特大教堂

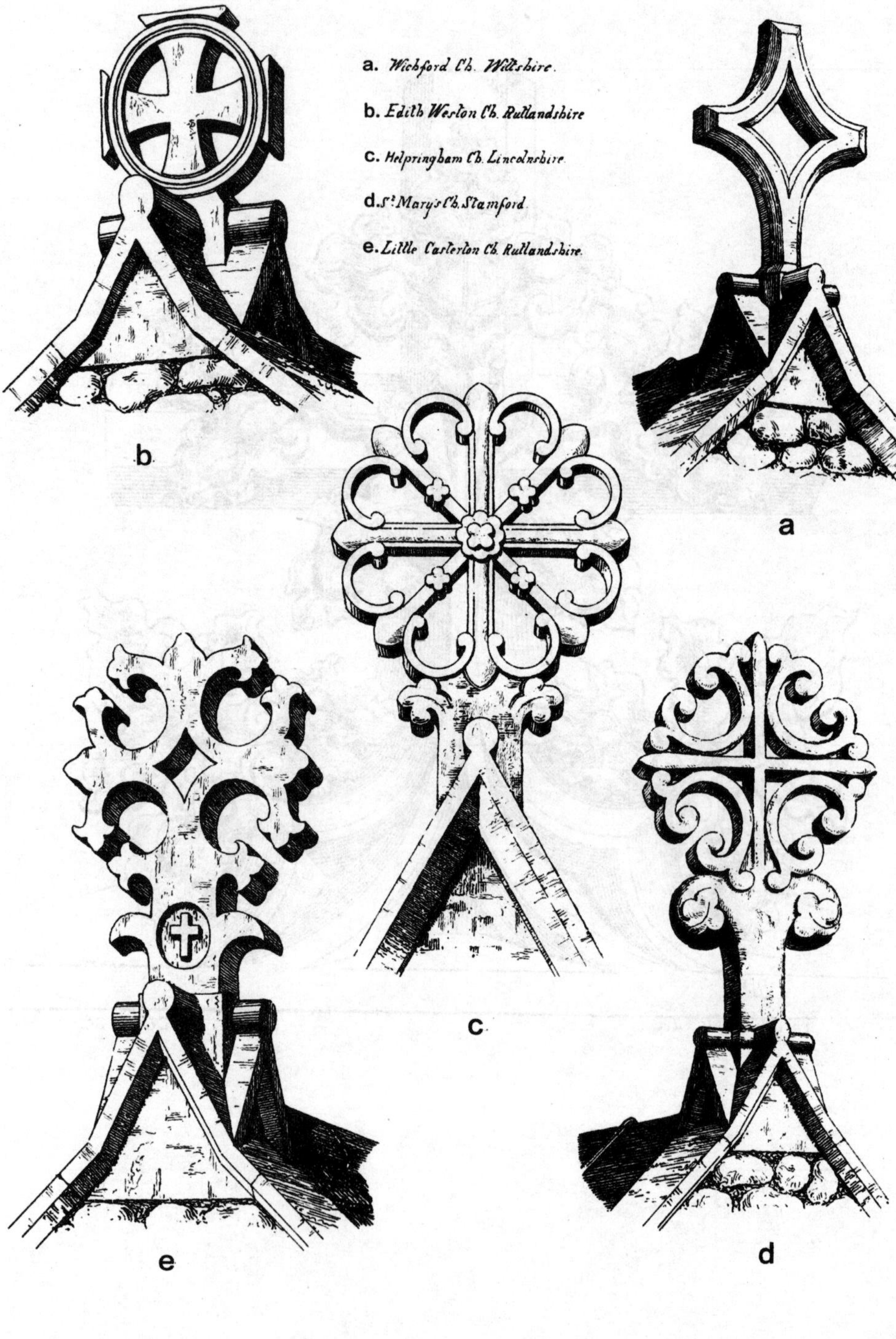

a. 威尔特郡威奇福德教堂 b. 拉特兰郡伊迪丝韦斯顿教堂 c. 林肯郡赫尔普灵厄姆教堂 d. 斯坦福德圣玛丽教堂 e. 拉特兰郡小卡斯特顿教堂

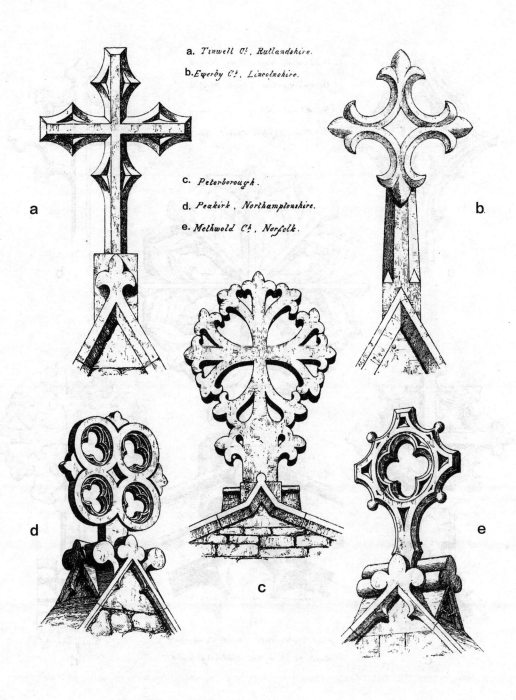

a. 拉特兰郡廷韦尔教堂　b. 林肯郡尤尔比教堂　c. 彼得伯勒　d. 北安普敦郡皮克尔克　e. 诺福克郡梅斯沃尔德教堂

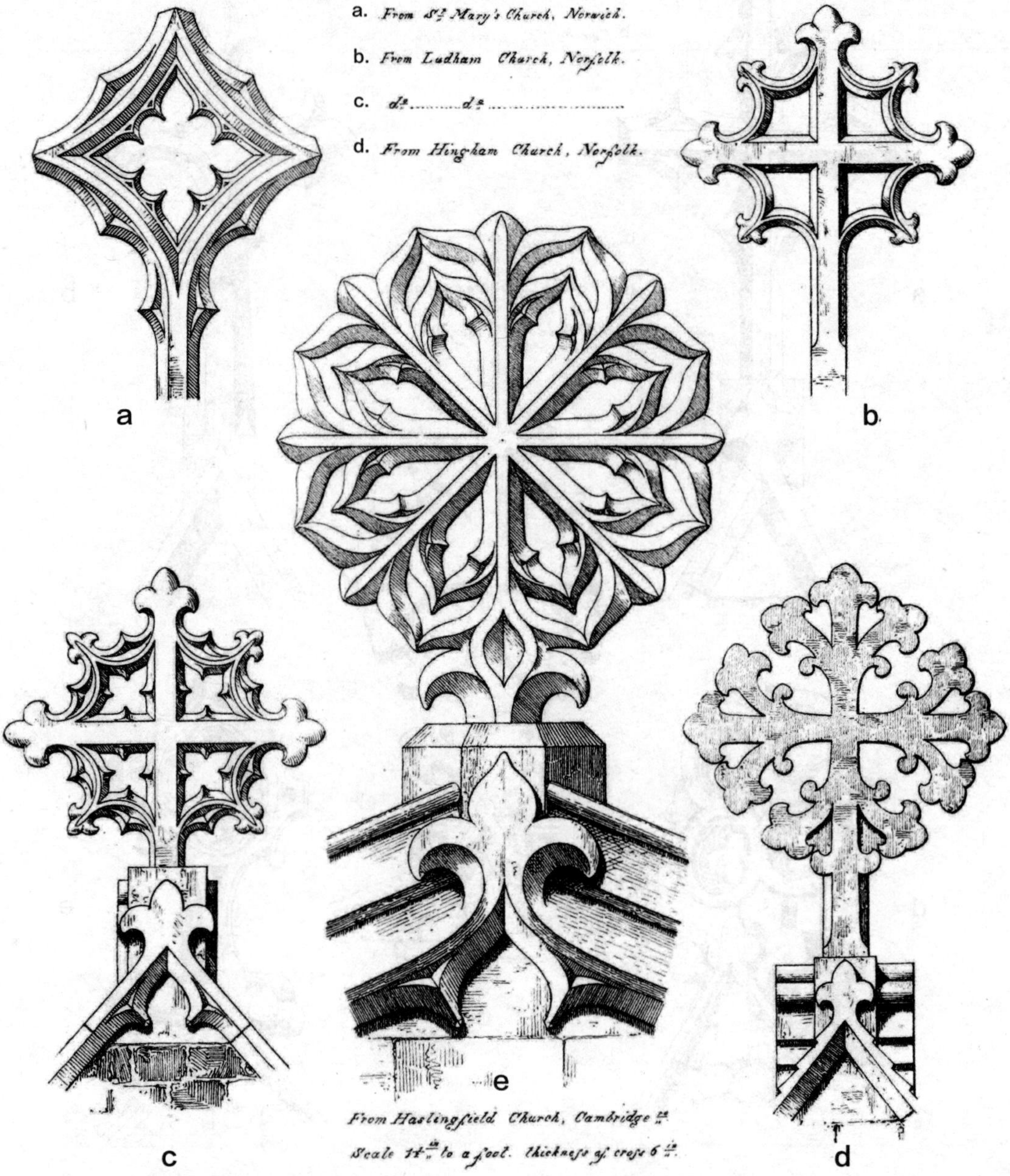

a. 来自诺威奇圣玛丽教堂　b. 来自诺福克郡路德姆教堂　c. 来自诺福克郡路德姆教堂　d. 来自诺福克郡欣厄姆教堂　e. 来自剑桥郡哈斯灵菲尔德教堂

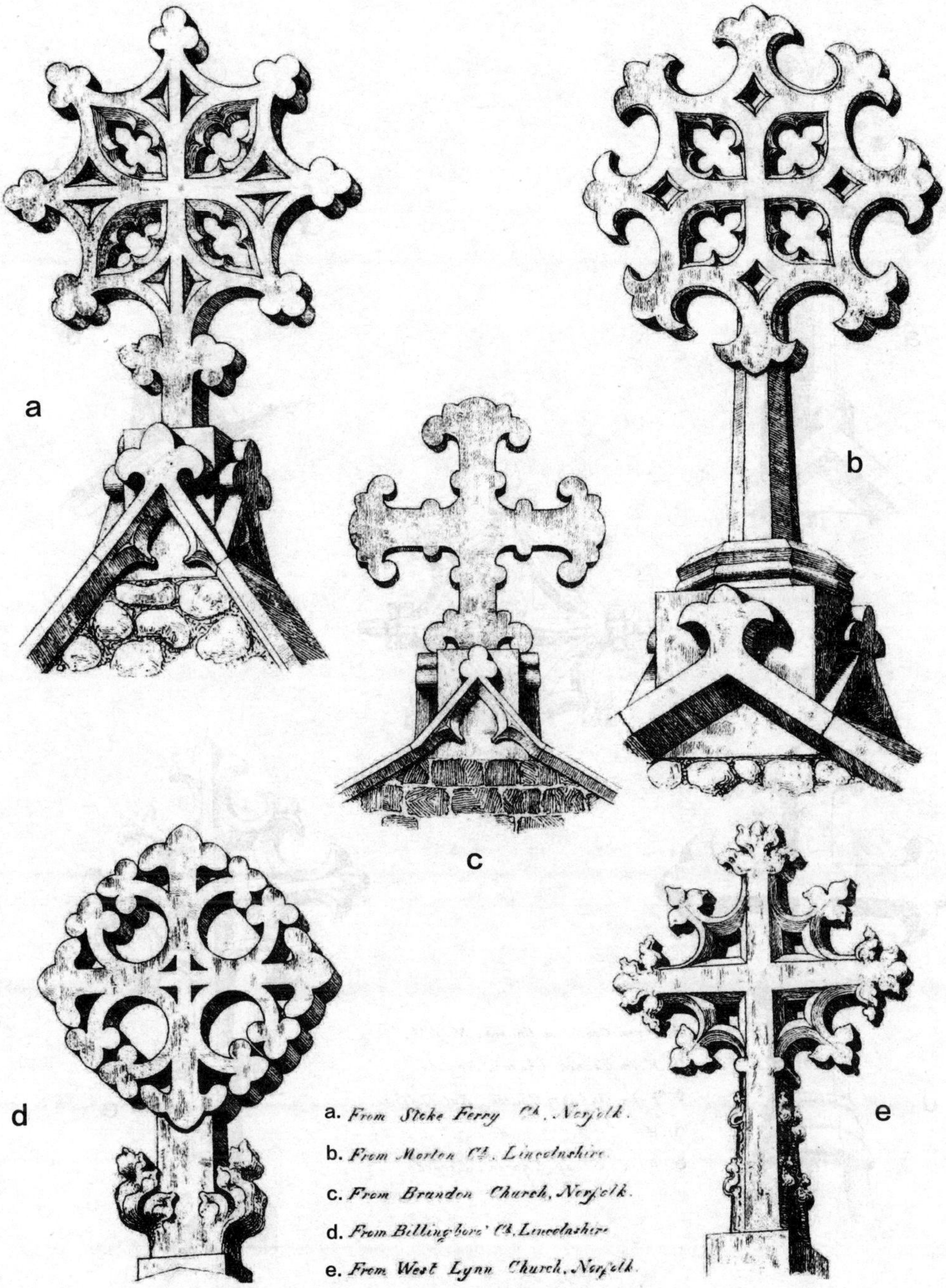

a. 来自诺福克郡斯托克费里教堂　b. 来自林肯郡默顿教堂　c. 来自诺福克郡布兰登教堂　d. 来自林肯郡比灵波洛教堂　e. 来自诺福克郡西林恩教堂

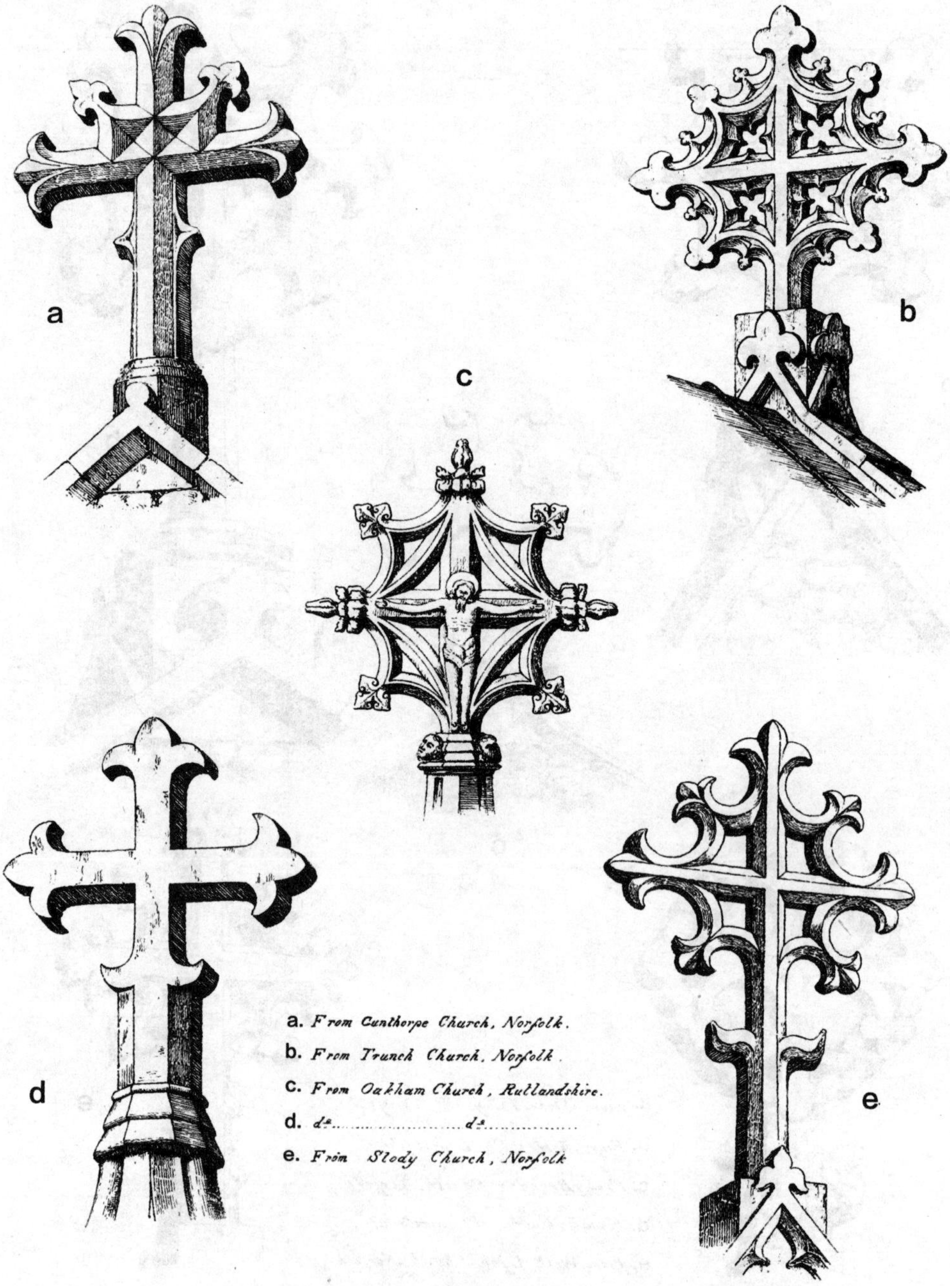

a. 诺福克郡冈索普教堂　b. 诺福克郡创奇教堂　c. 拉特兰郡奥克汉教堂　d. 拉特兰郡奥克汉教堂　e. 诺福克郡斯托地教堂

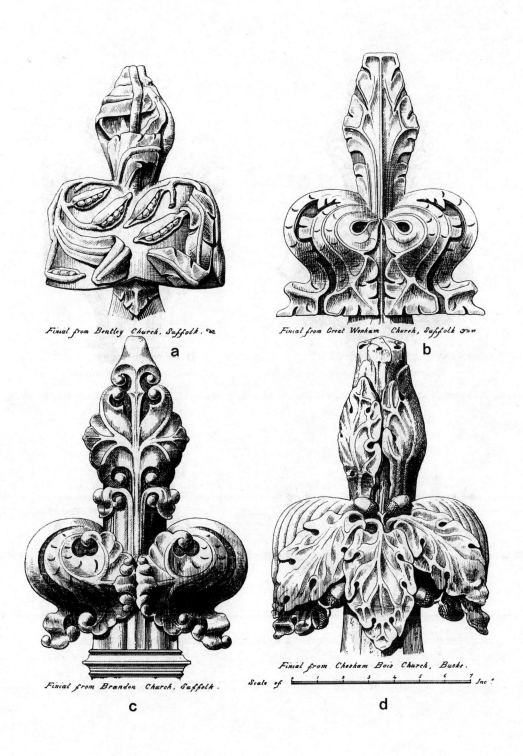

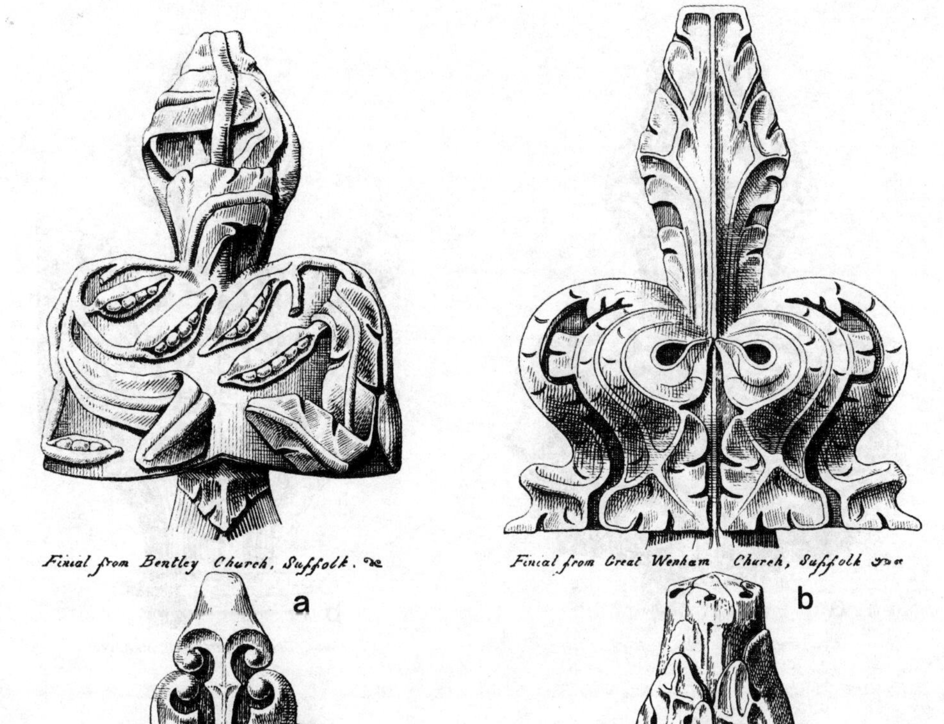

a. 萨福克郡本特利教堂尖顶饰 b. 萨福克郡韦纳姆教堂尖顶饰 c. 萨福克郡布兰登教堂尖顶饰 d. 巴克斯郡切舍姆博伊斯教堂尖顶饰

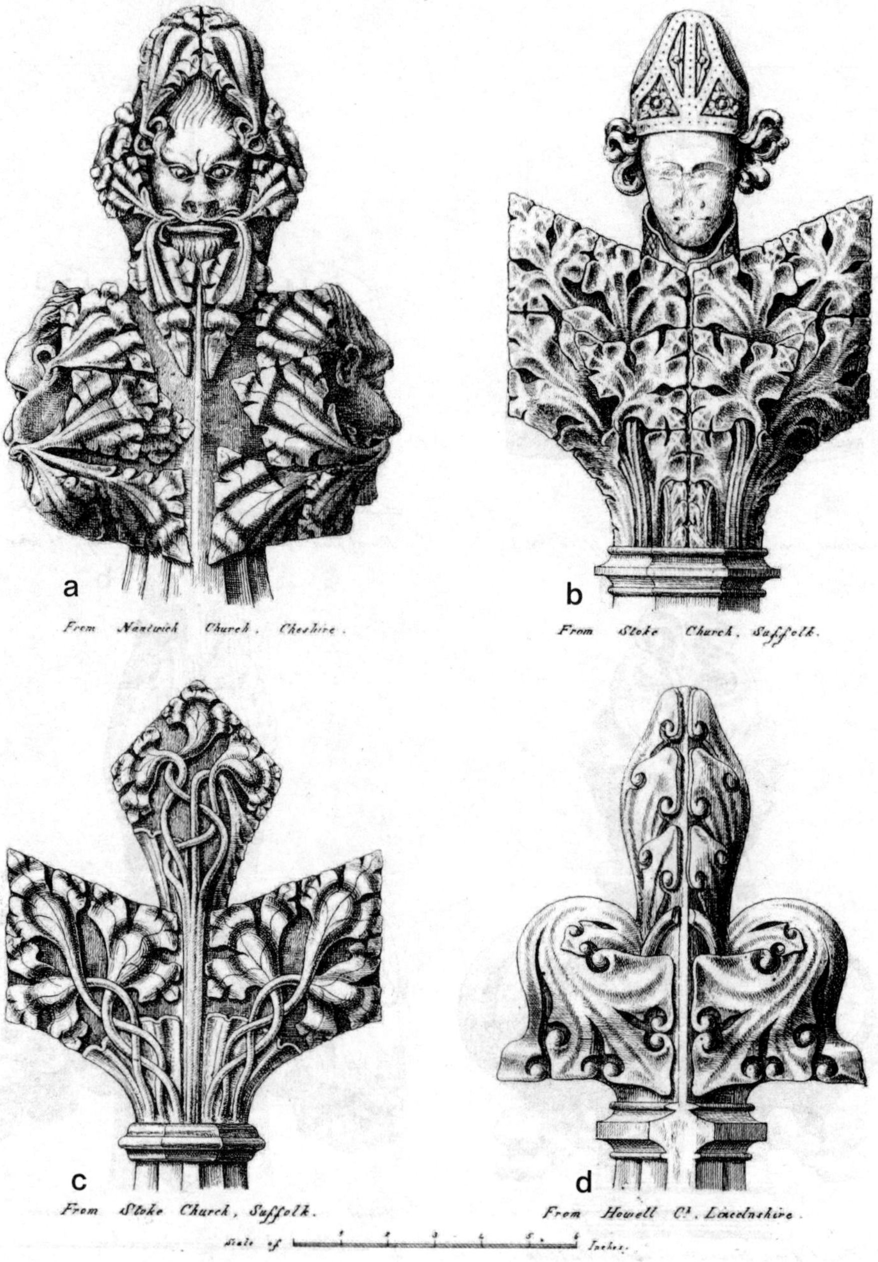

a. 来自柴郡南特威奇教堂 b. 来自萨福克郡斯托克教堂 c. 来自萨福克郡斯托克教堂 d. 来自林肯郡豪威尔教堂

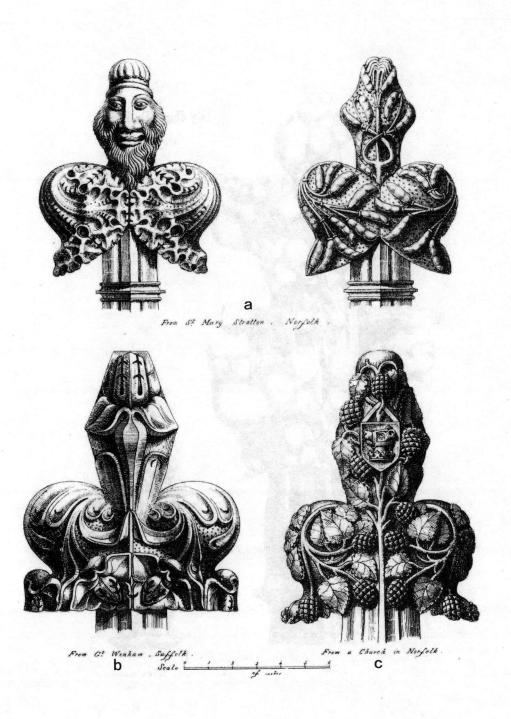

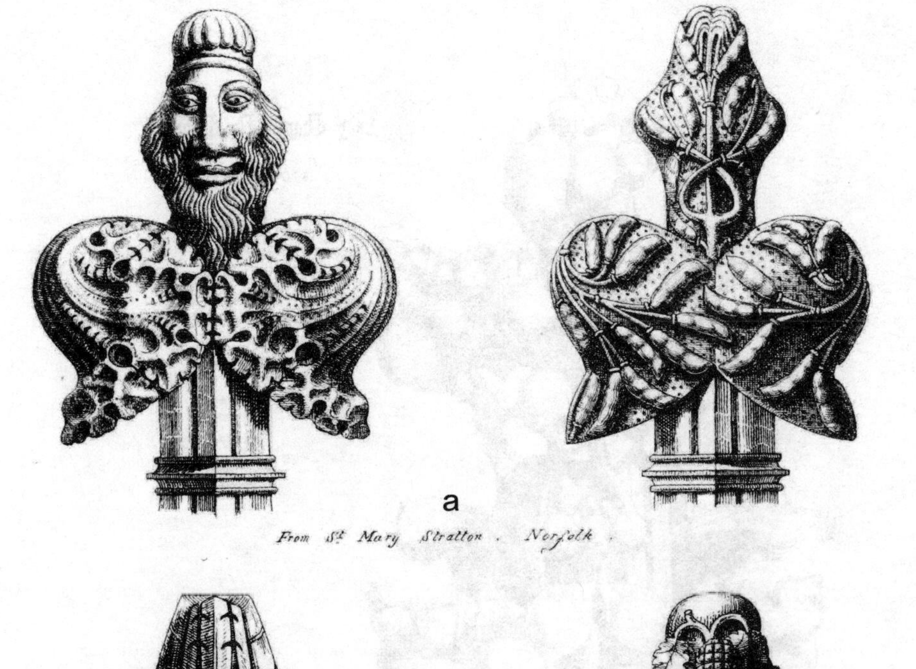

a. 来自诺福克郡斯特拉顿圣玛丽教堂　b. 萨福克郡韦纳姆教堂　c. 来自诺福克郡的某教堂

以利教堂圣母堂：a. 石制尖顶饰　b. 从斜视角看尖顶饰

山墙顶饰样式：a. 来自萨福克郡卡佩尔教堂　b. 来自诺福克郡斯瓦菲尔德教堂　c. 来自剑桥郡小谢尔福德教堂　d. 来自诺福克郡腾斯戴德教堂　e. 来自剑桥郡斯泰普尔福德教堂　f. 来自诺福克郡创奇教堂　g. 来自剑桥郡小谢尔福德教堂　h. 来自诺福克郡科尔蒂瑟尔教堂

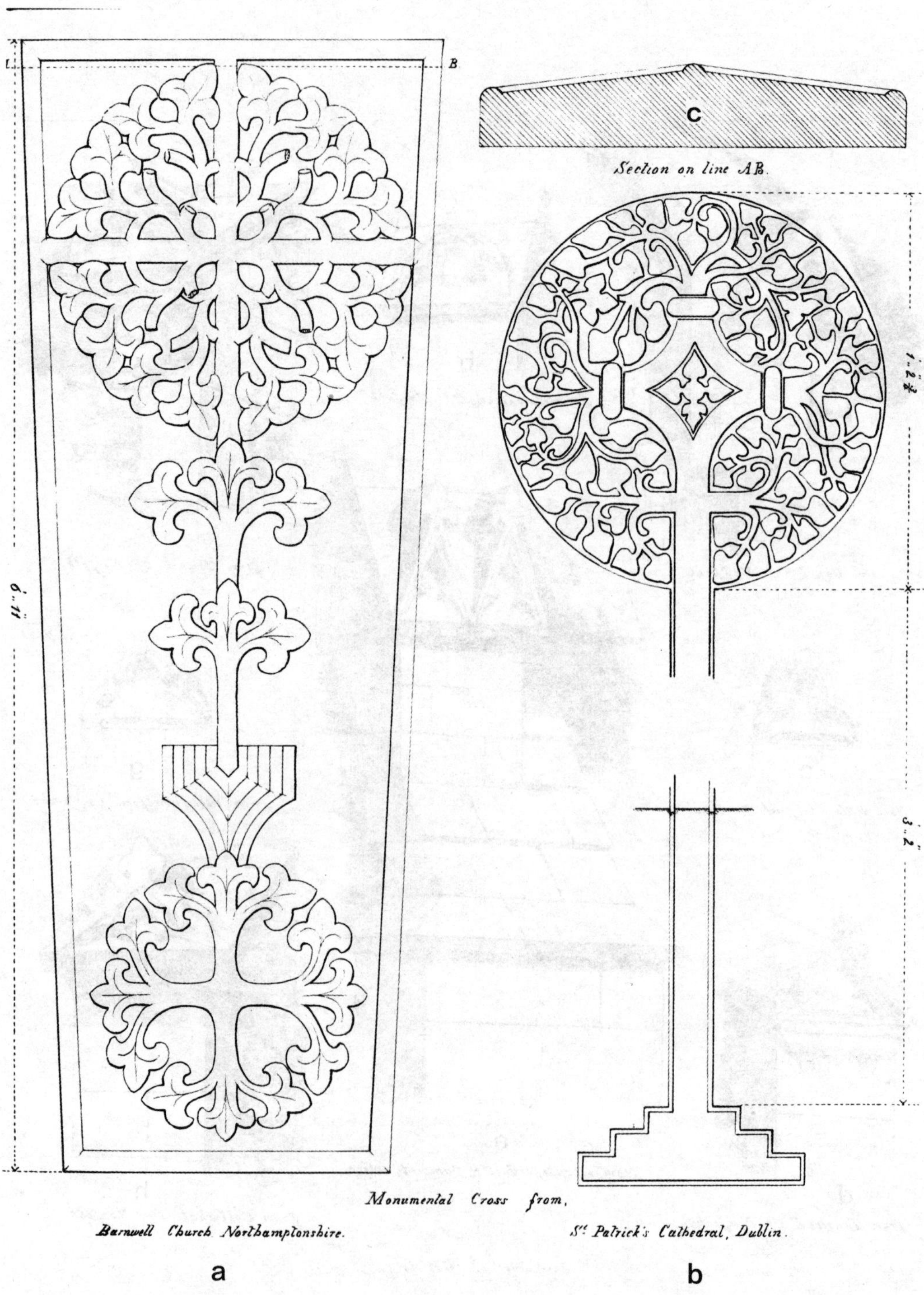

巨大的十字架：a. 北安普敦郡巴恩韦尔教堂　b. 都柏林圣帕特里克教堂　c.A–B 处的剖面

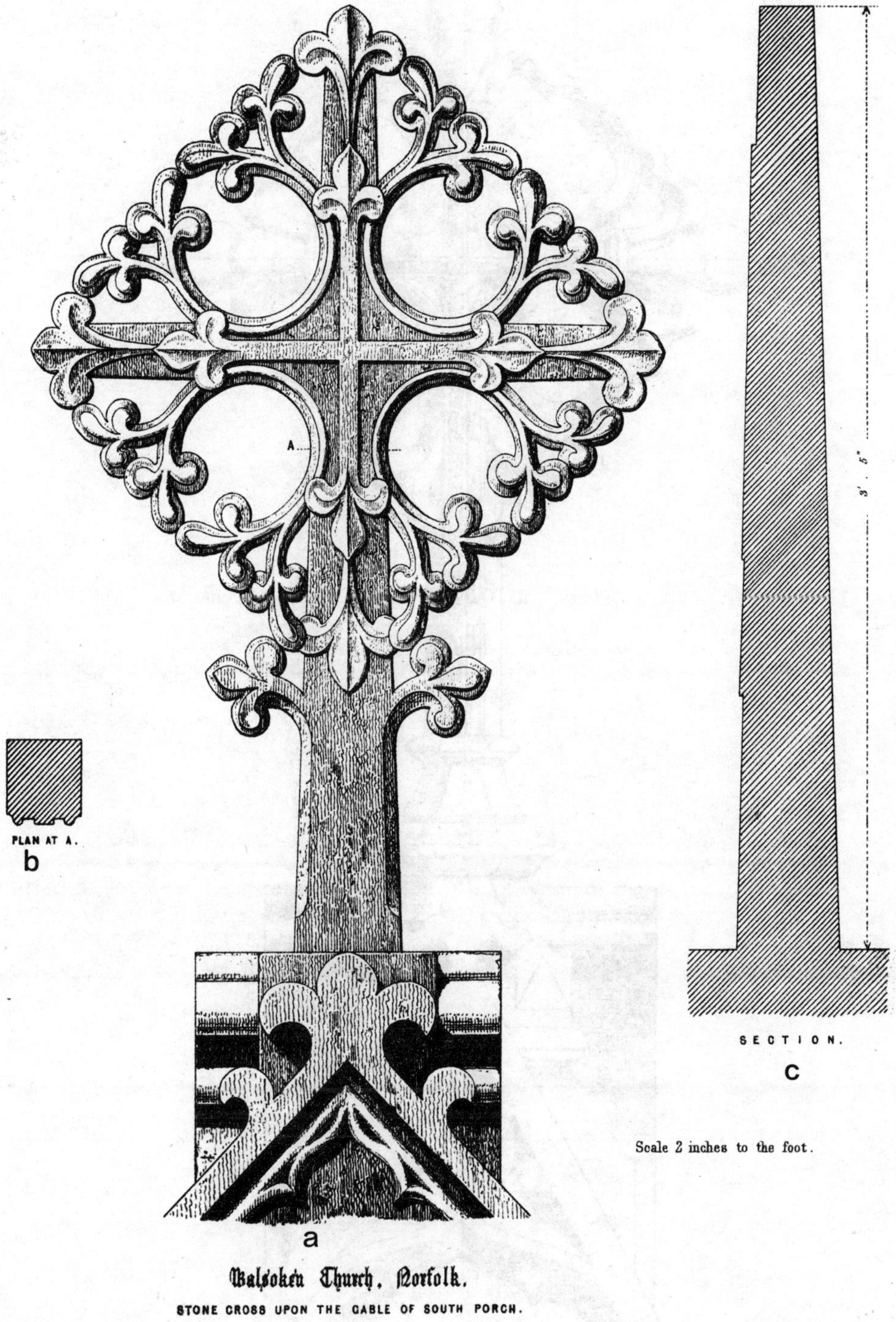

诺福克郡沃尔索肯教堂：a. 南门廊山墙上石制十字架　b.A 处的平面　c. 剖面

十字架

第11章

卷叶、花饰

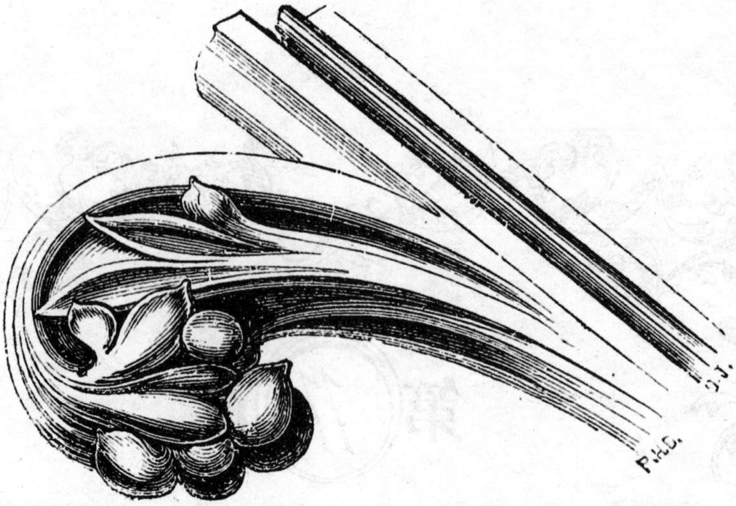

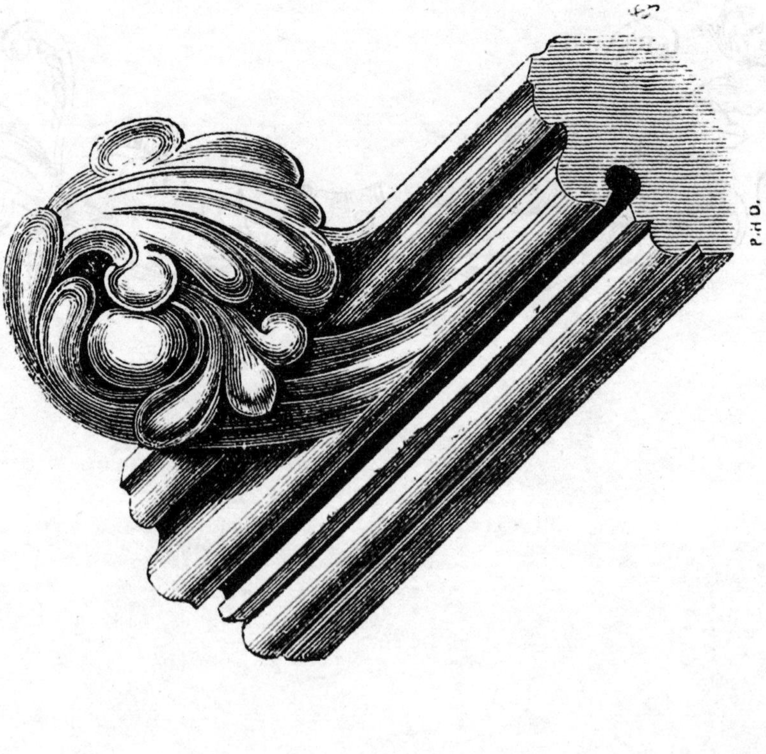

索尔兹伯里教堂

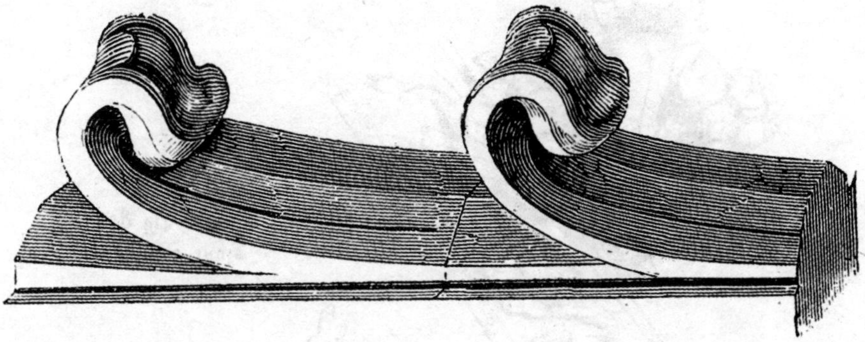

a

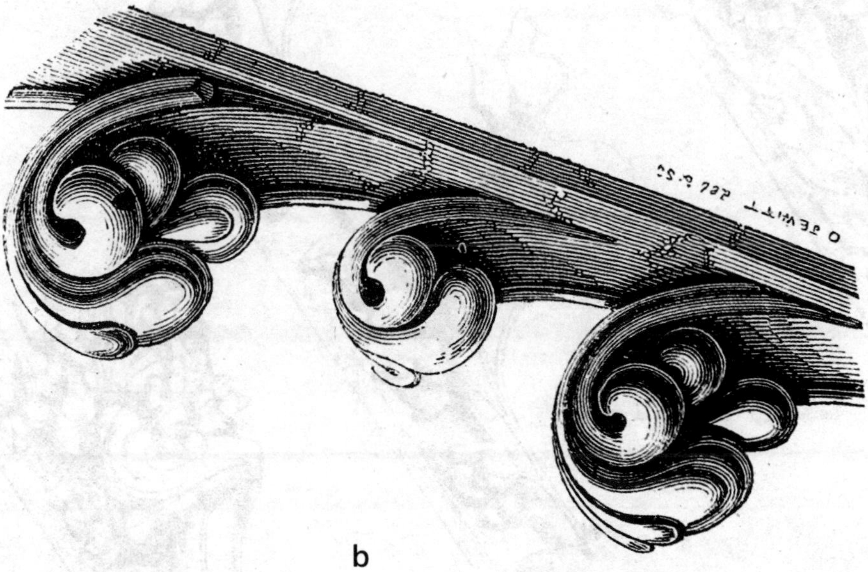

b

a. 沃尔特 · 格雷墓穴上卷叶形花饰　b. 林肯教堂唱诗室

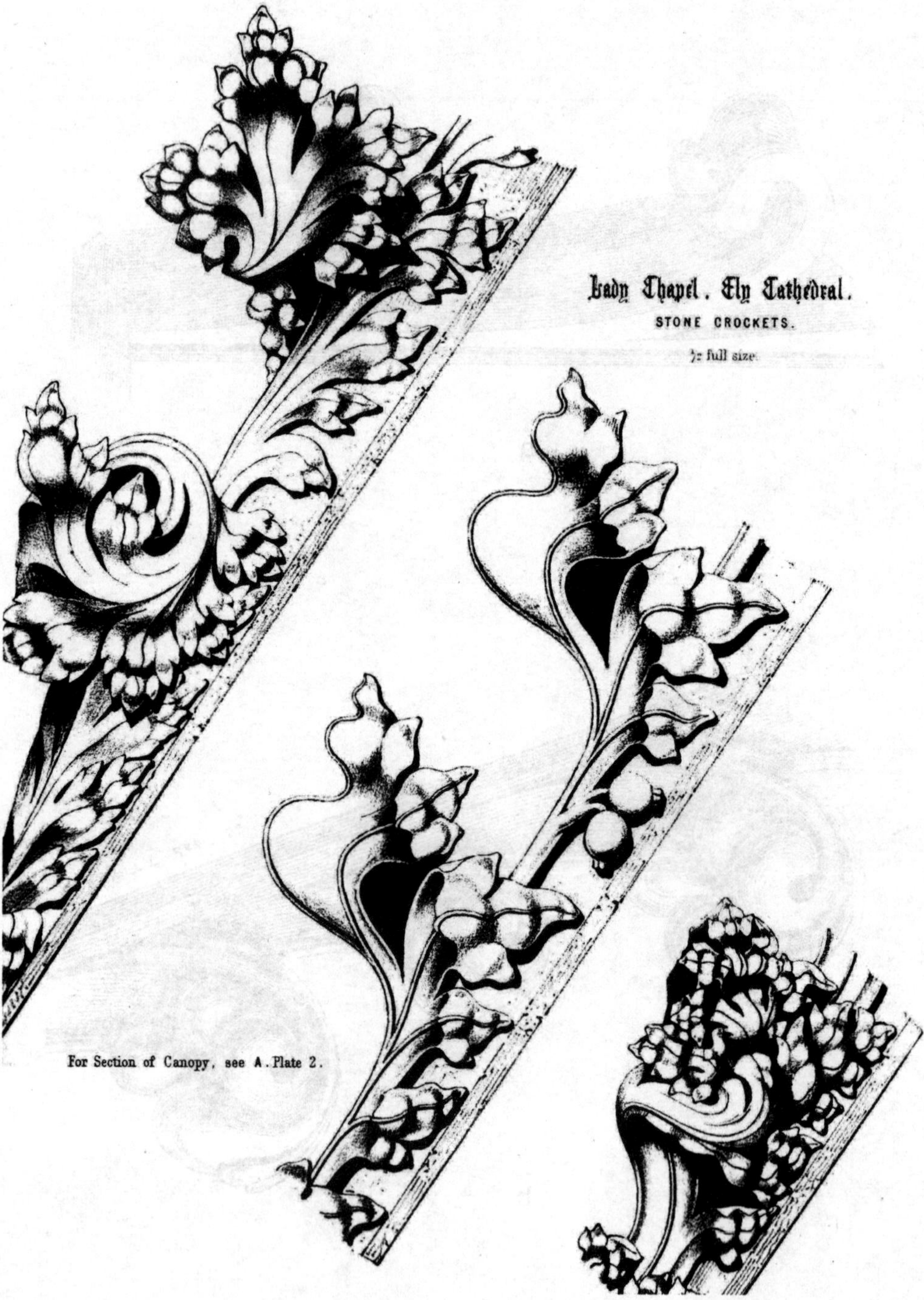

以利教堂圣母堂石制卷叶形花饰

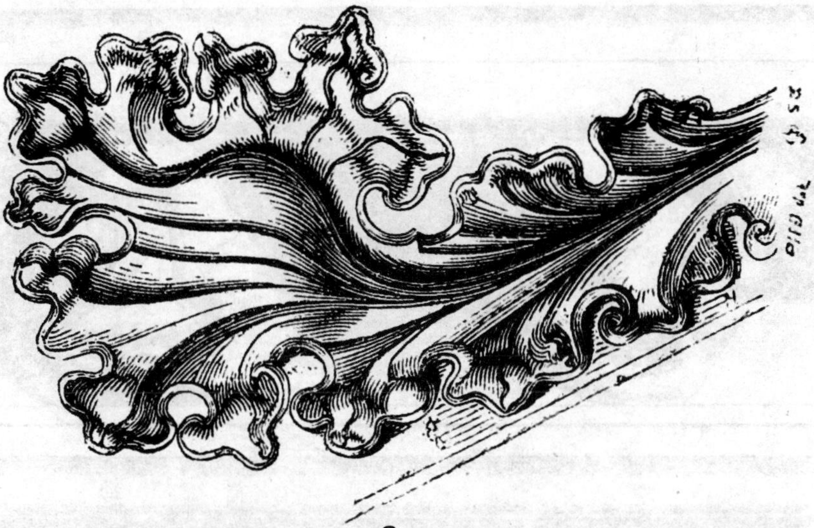

a

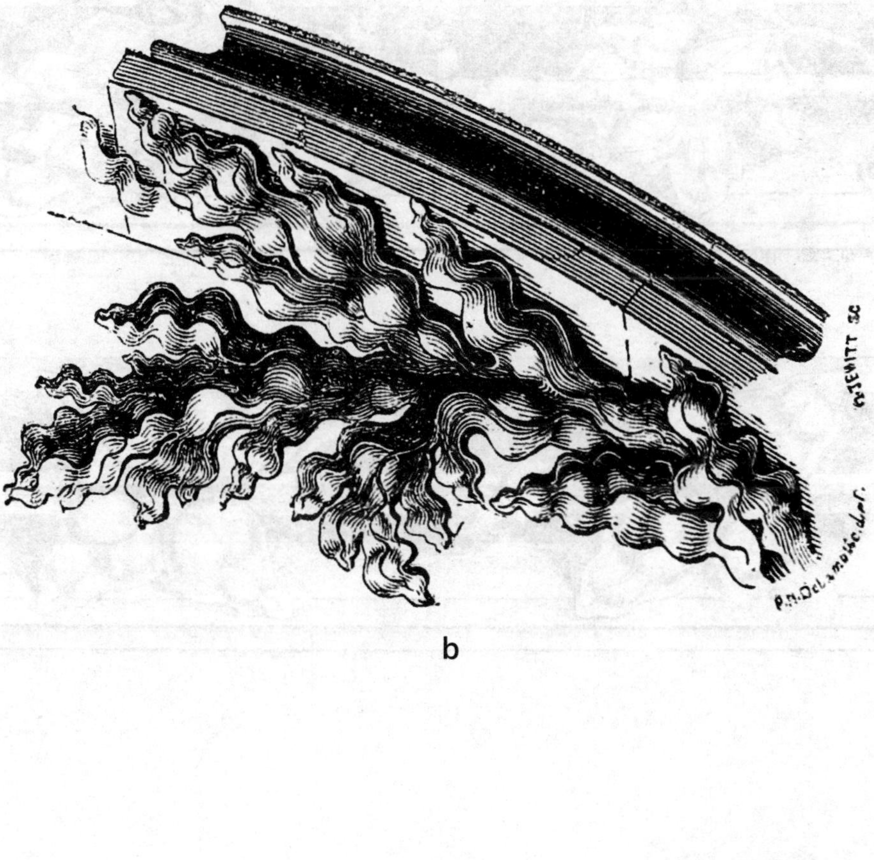

b

a. 诺丁汉郡霍顿教堂　b. 林肯教堂

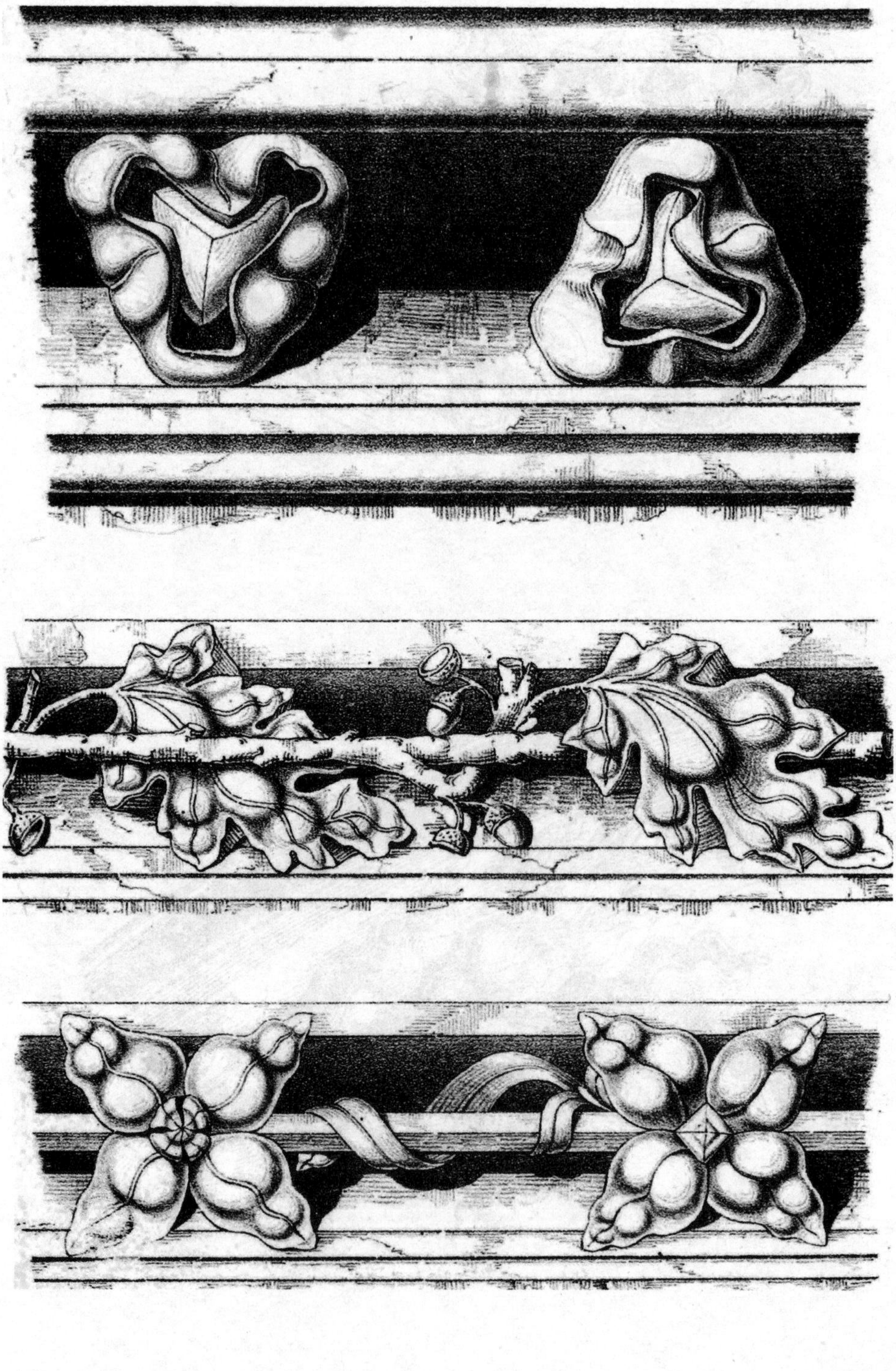

卷叶形花饰

卷叶形花饰

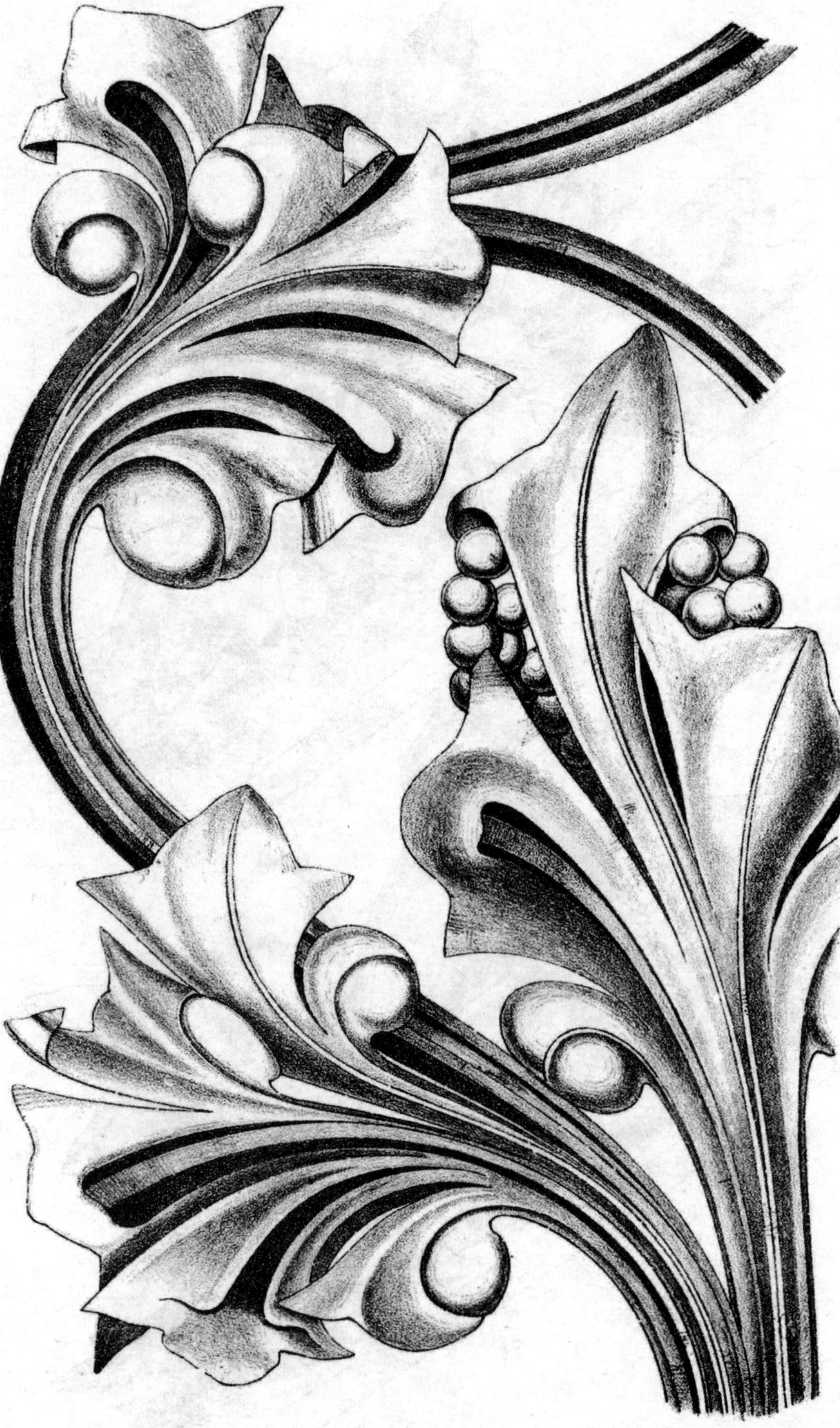

卷叶形花饰

卷叶形花饰

卷叶形花饰

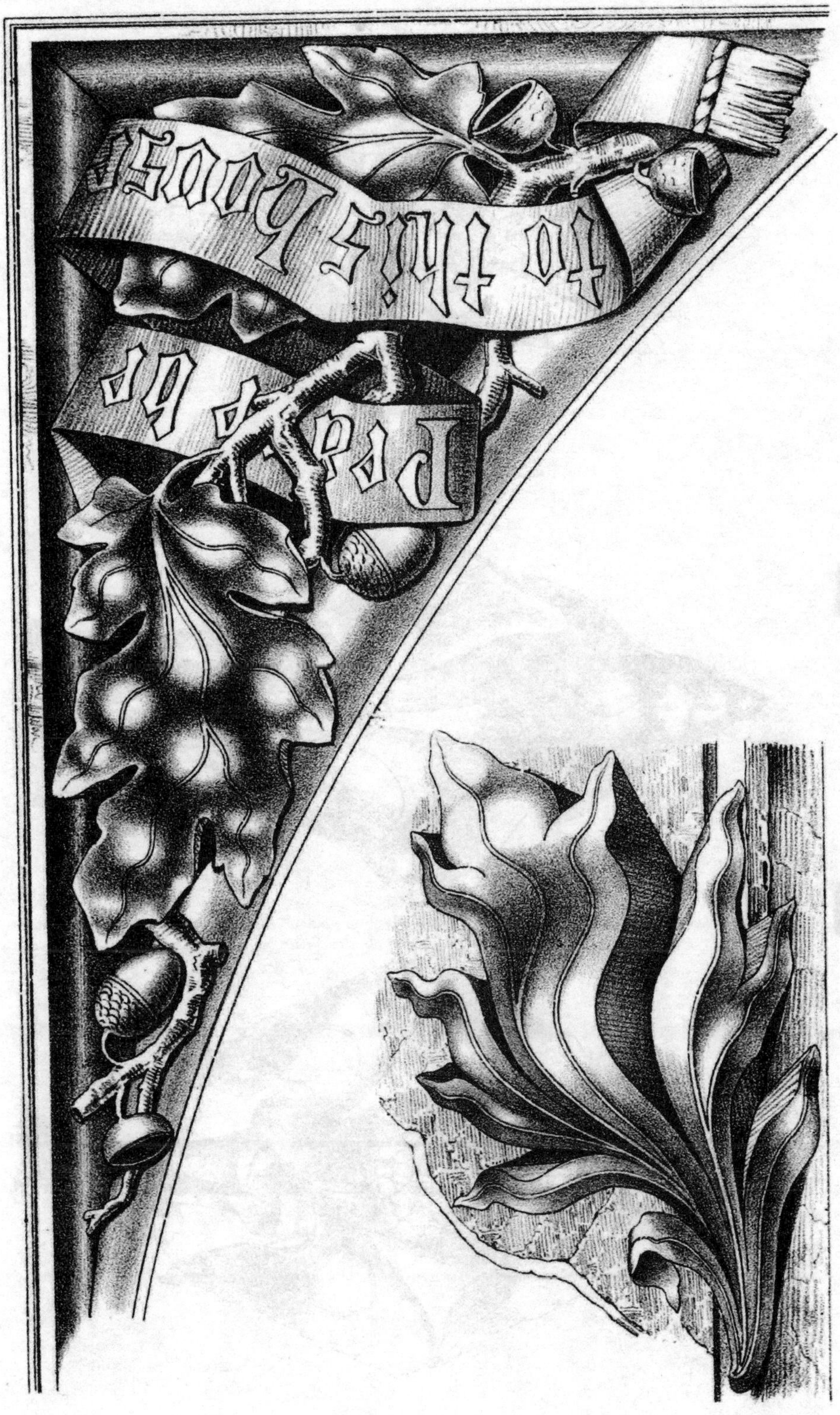
to this house
Peace be

卷叶形花饰

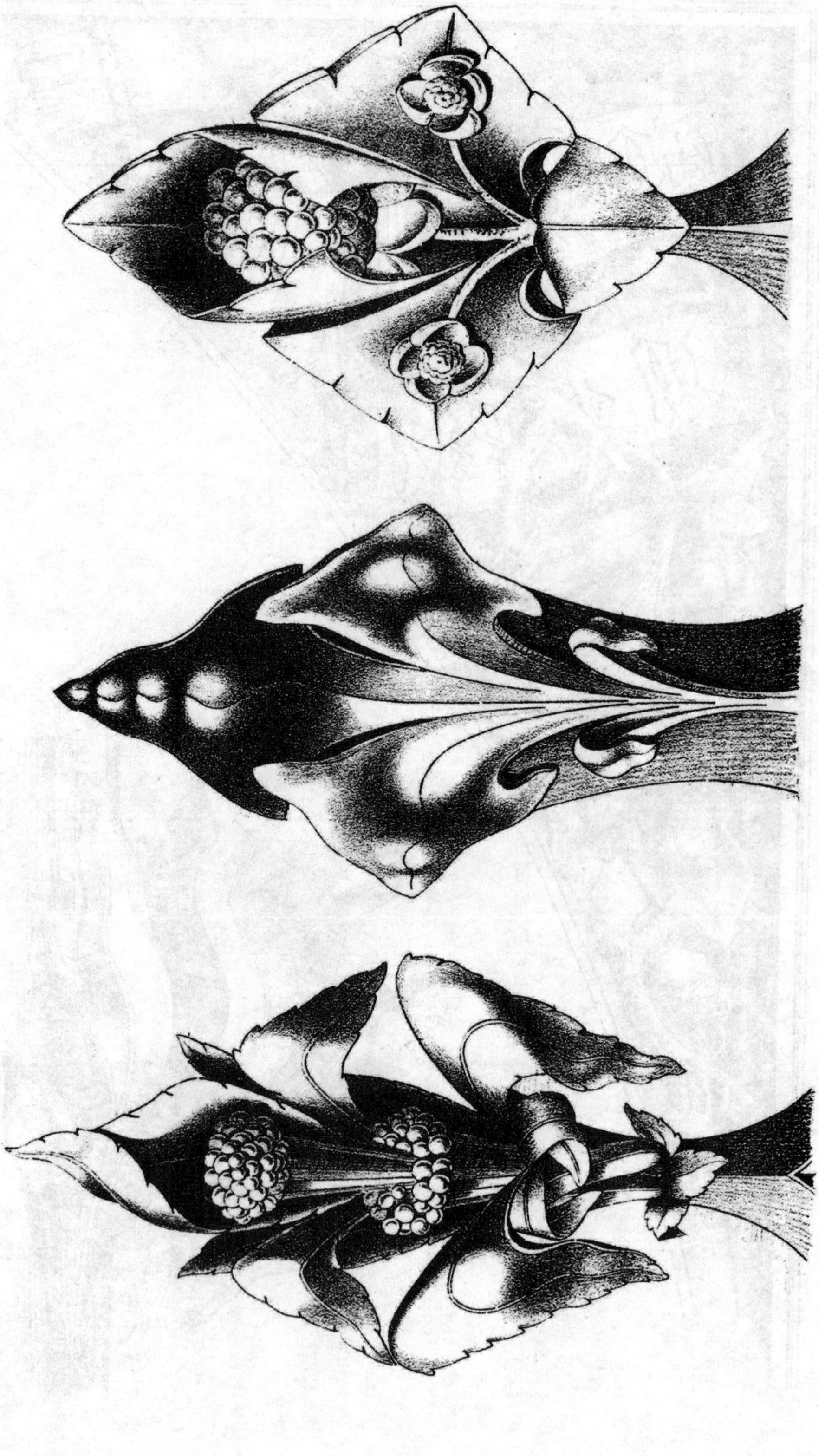

卷叶形花饰

卷叶形花饰

卷叶形花饰

卷叶形花饰

卷叶形花饰

卷叶形花饰

卷叶形花饰

卷叶形花饰

卷叶形花饰

卷叶形花饰

卷叶形花饰

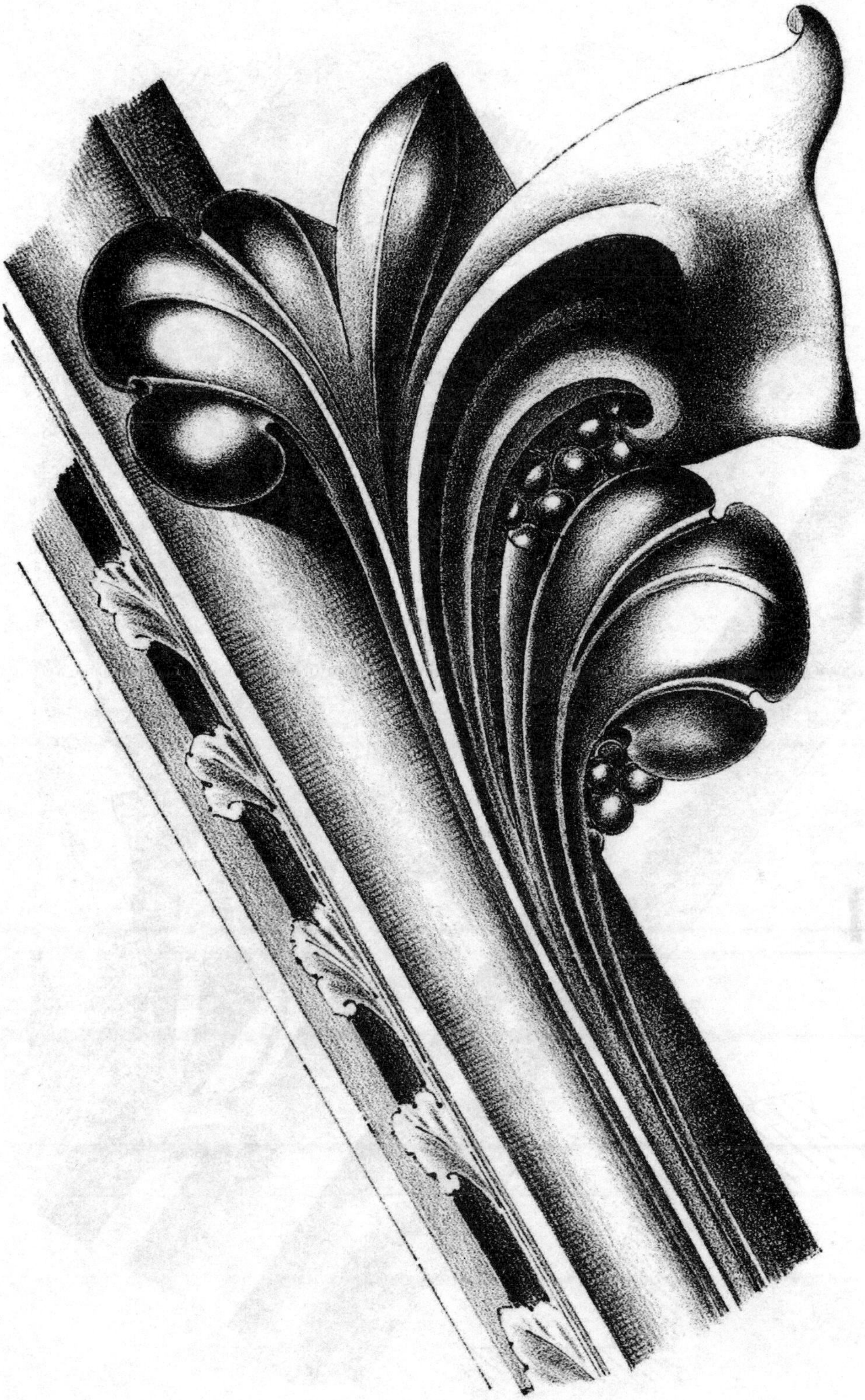

卷叶形花饰

卷叶形花饰

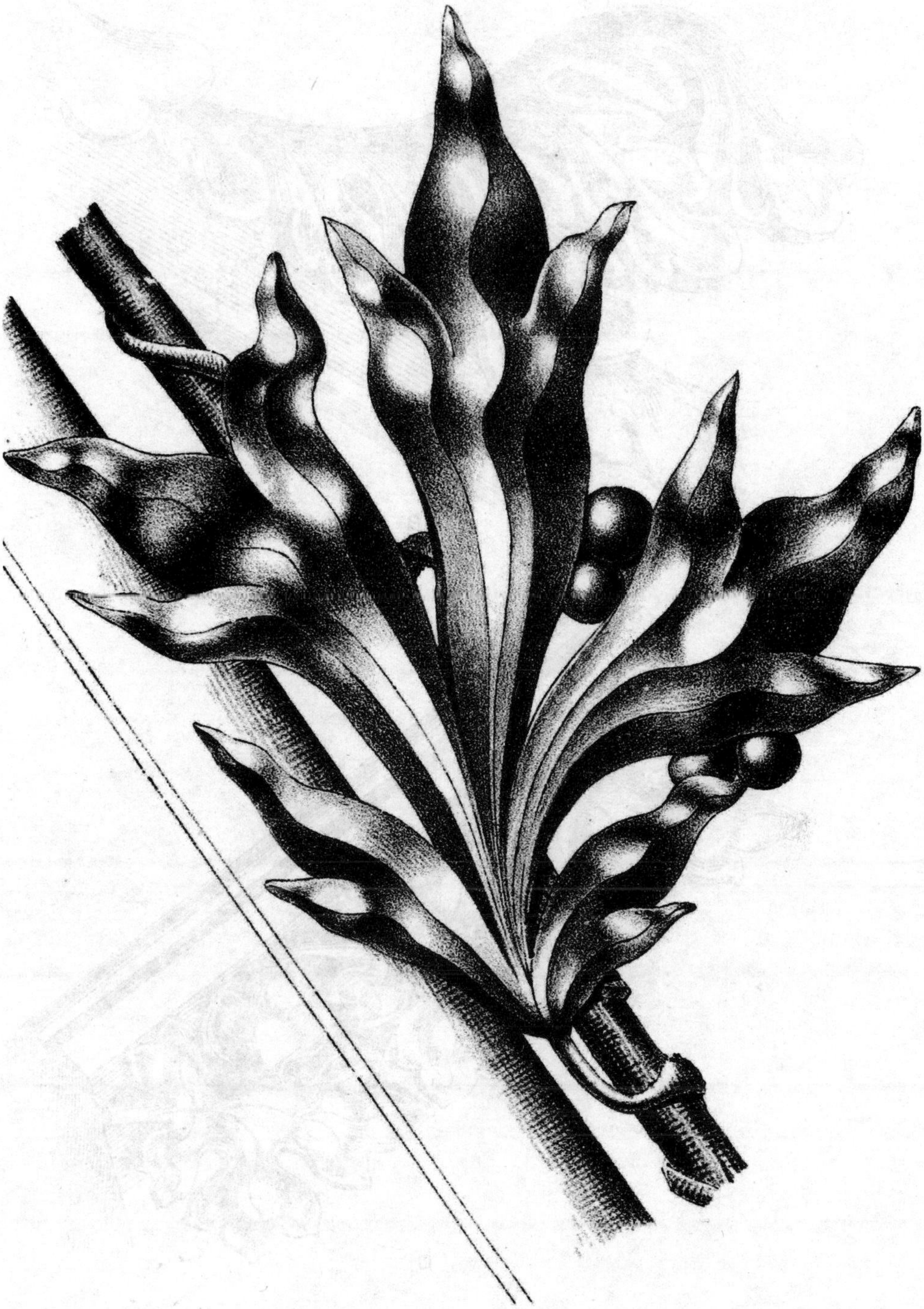

卷叶形花饰

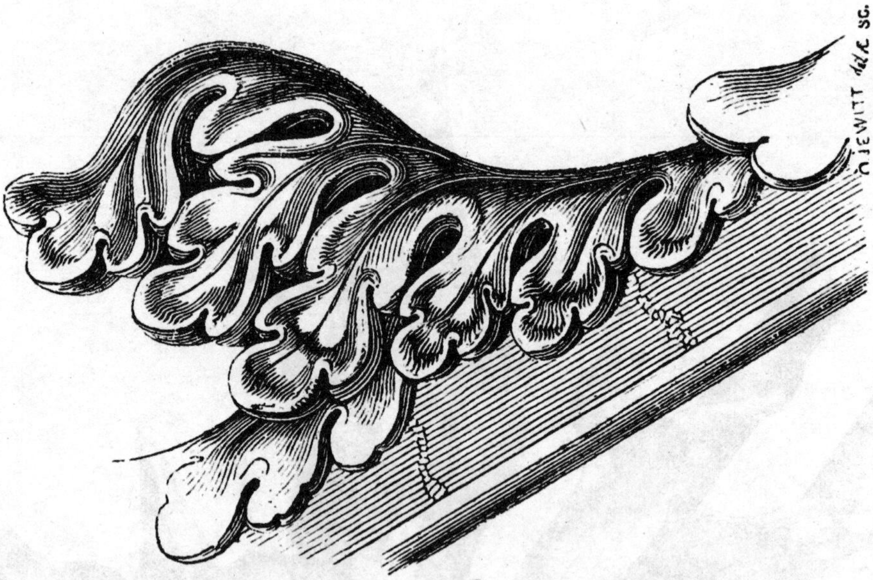

a

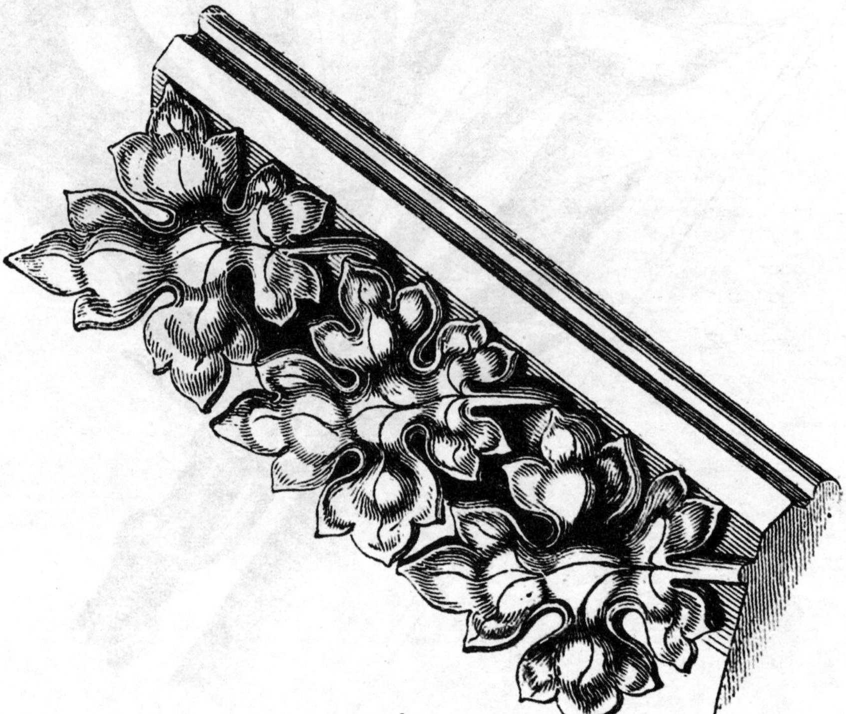

b

a. 约克郡吉斯伯勒修道院　b. 诺丁汉郡索斯韦尔教堂

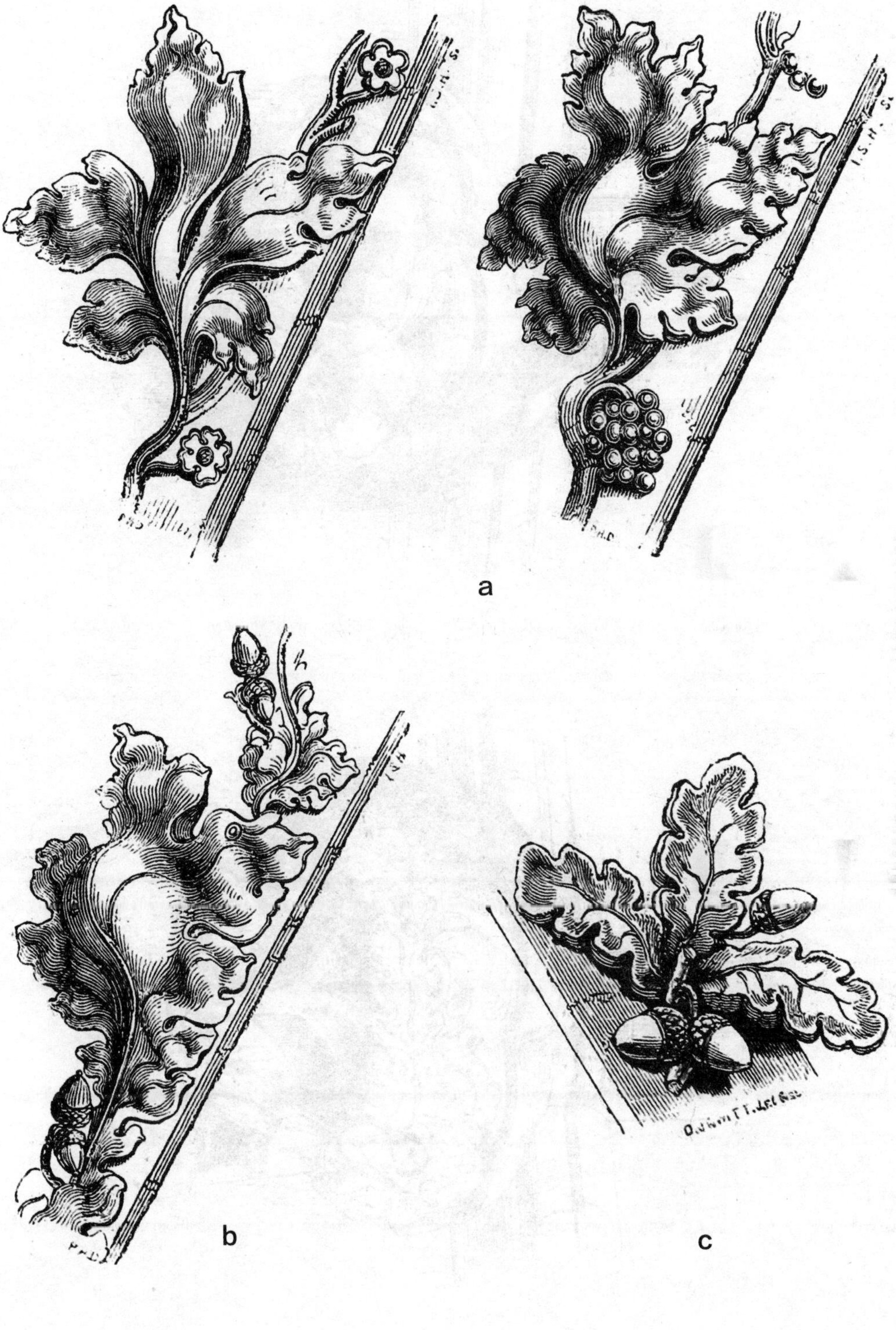

a. 温彻斯特大教堂　b. 温彻斯特大教堂　c. 埃克赛特大教堂

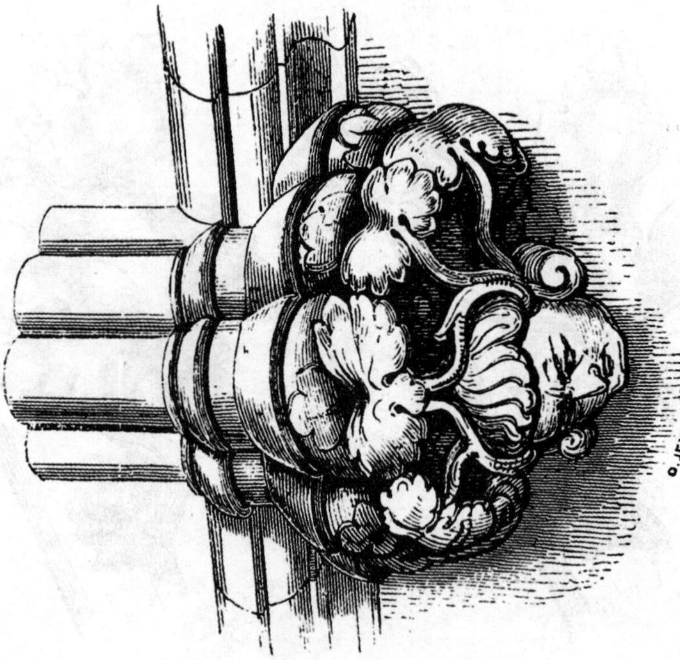

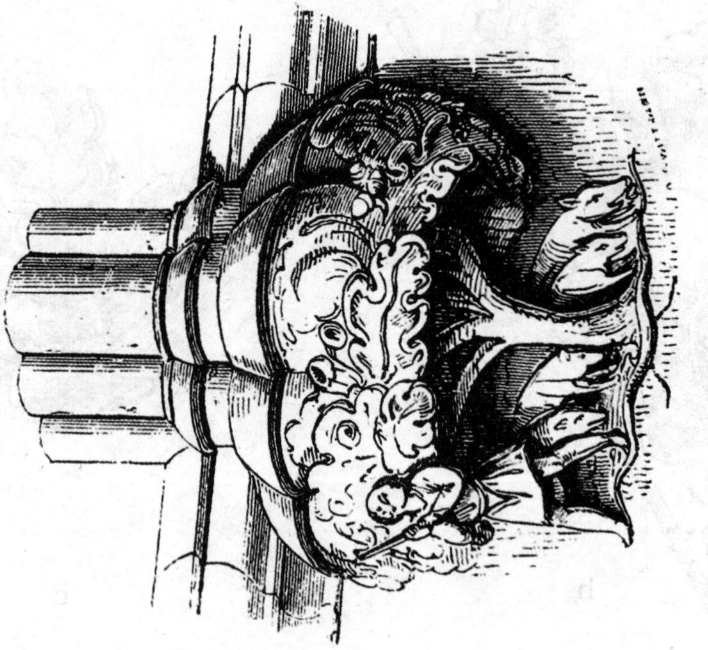

牛津郡默顿学院教堂

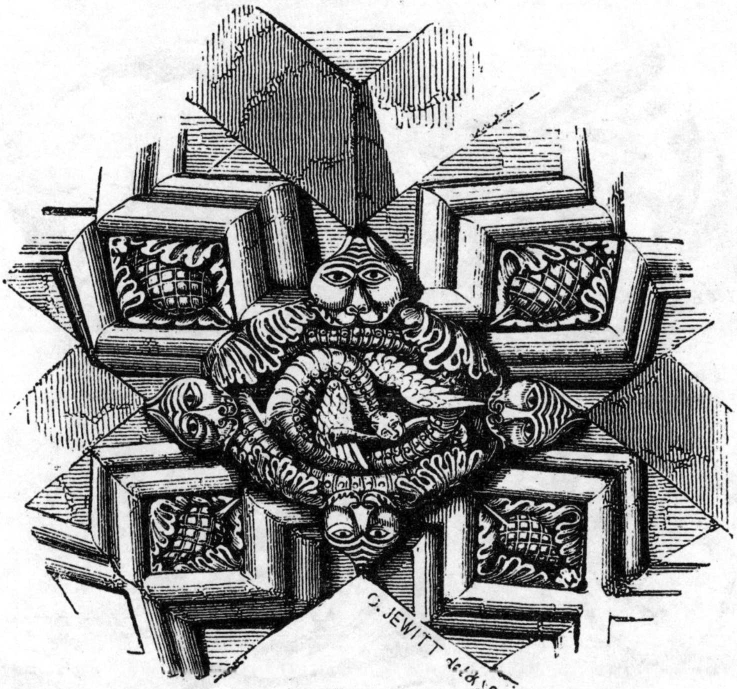

a
Iffley, Oxfordshire.

b
St. Alban's Abbey, Herts.

a. 剑桥郡伊夫雷教堂　b. 赫特福德郡圣奥尔本修道院

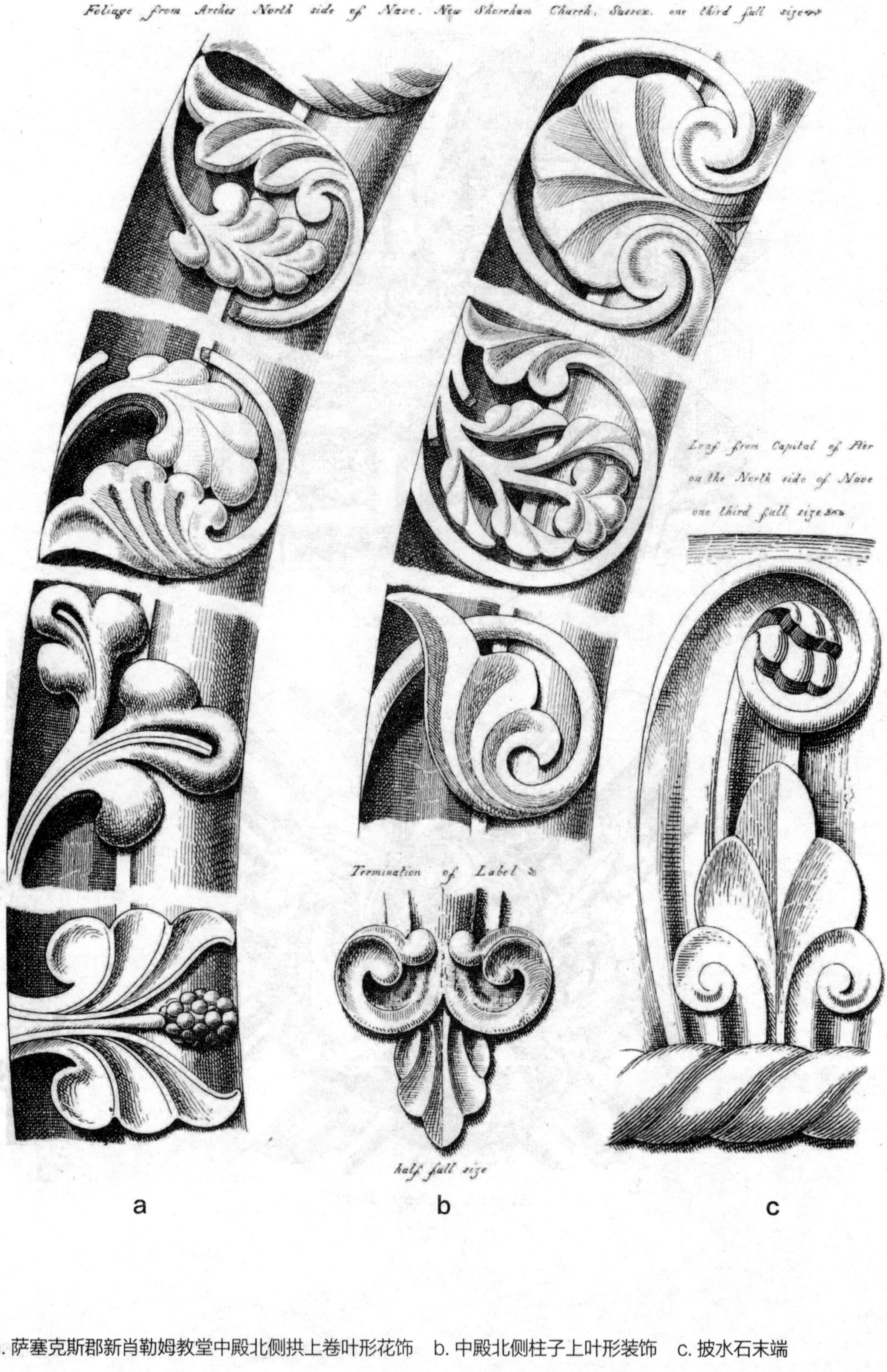

a　　b　　c

a. 萨塞克斯郡新肖勒姆教堂中殿北侧拱上卷叶形花饰　b. 中殿北侧柱子上叶形装饰　c. 披水石末端

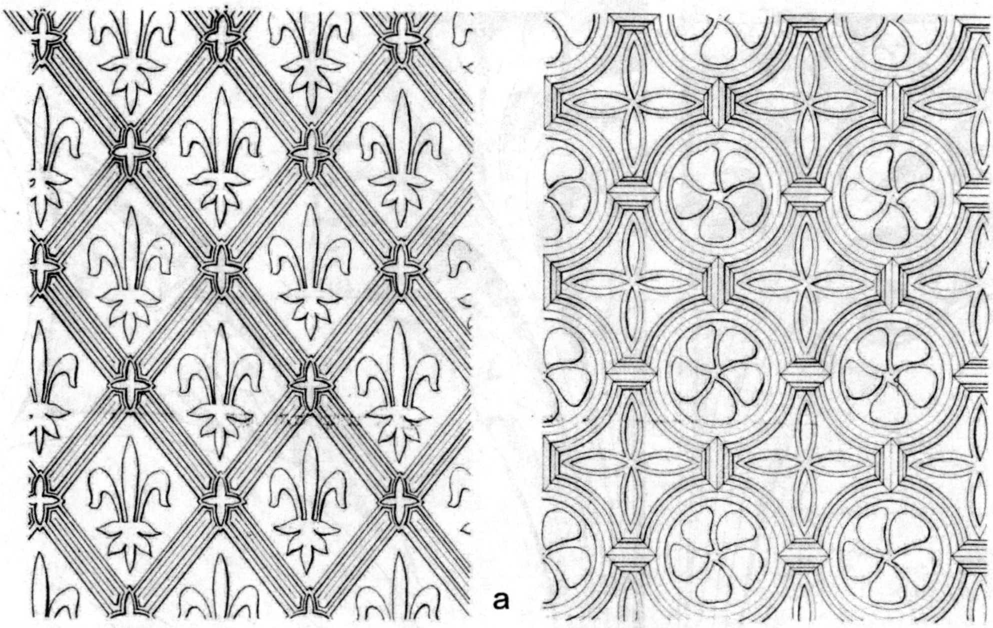

Diaper from Hatfield Broadoaks Ch. Essex.
1/2 full size.

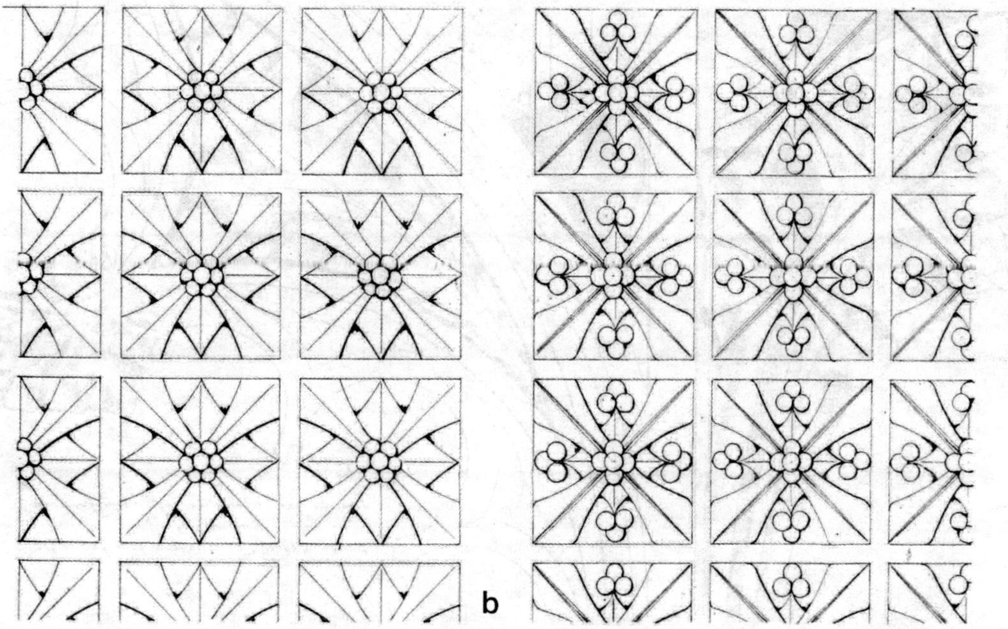

Diapers in Westminster Abbey Church.
1/6 full size.

a. 埃塞克斯郡福莱特菲尔德布罗德奥克斯教堂菱形花饰　b. 威斯敏斯特修道院菱形花饰

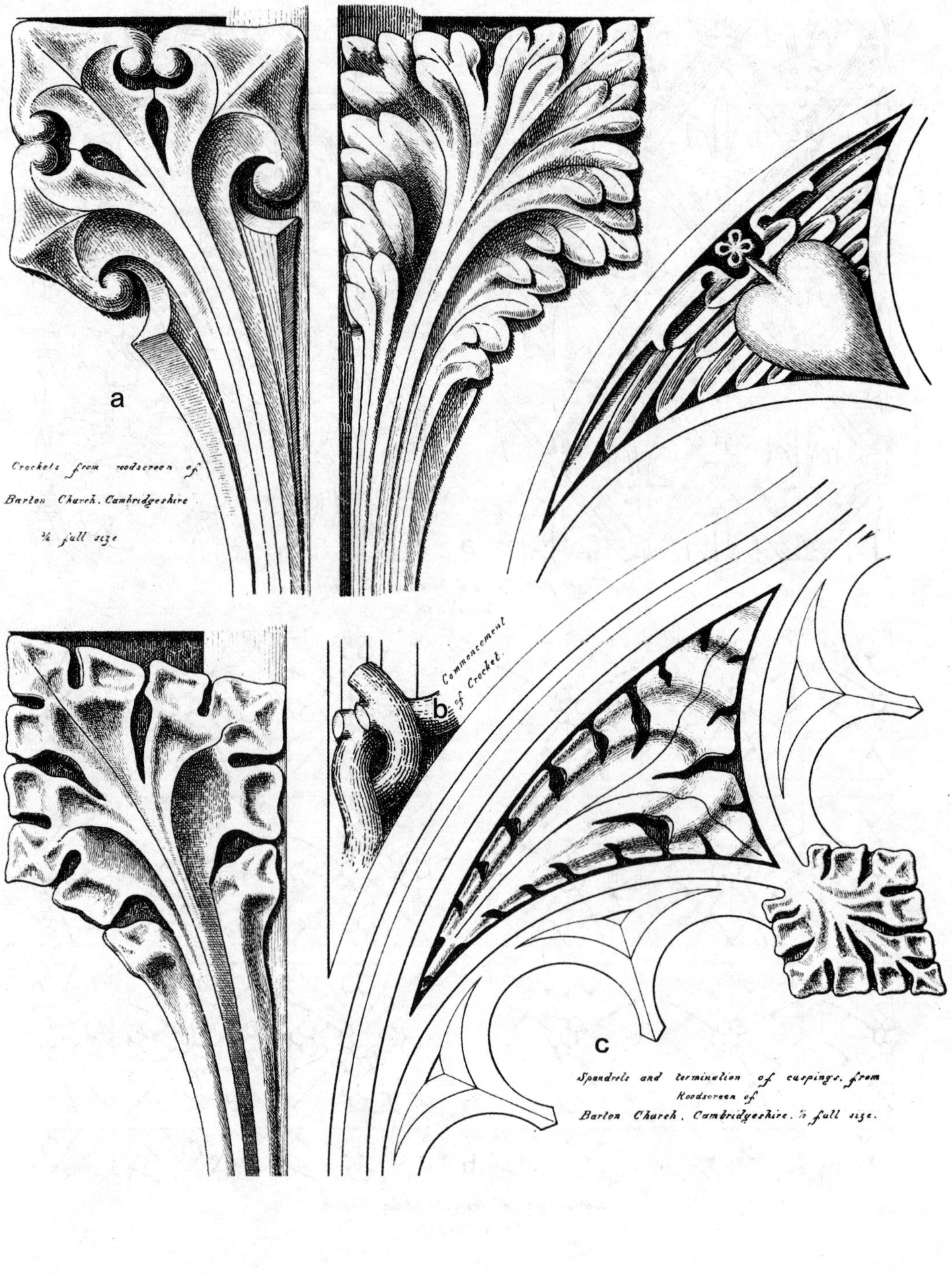

a. 剑桥郡巴顿教堂圣坛屏上卷叶形花饰　b. 卷叶形花饰开端　c. 剑桥郡巴顿教堂圣坛屏上拱肩及尖饰端部

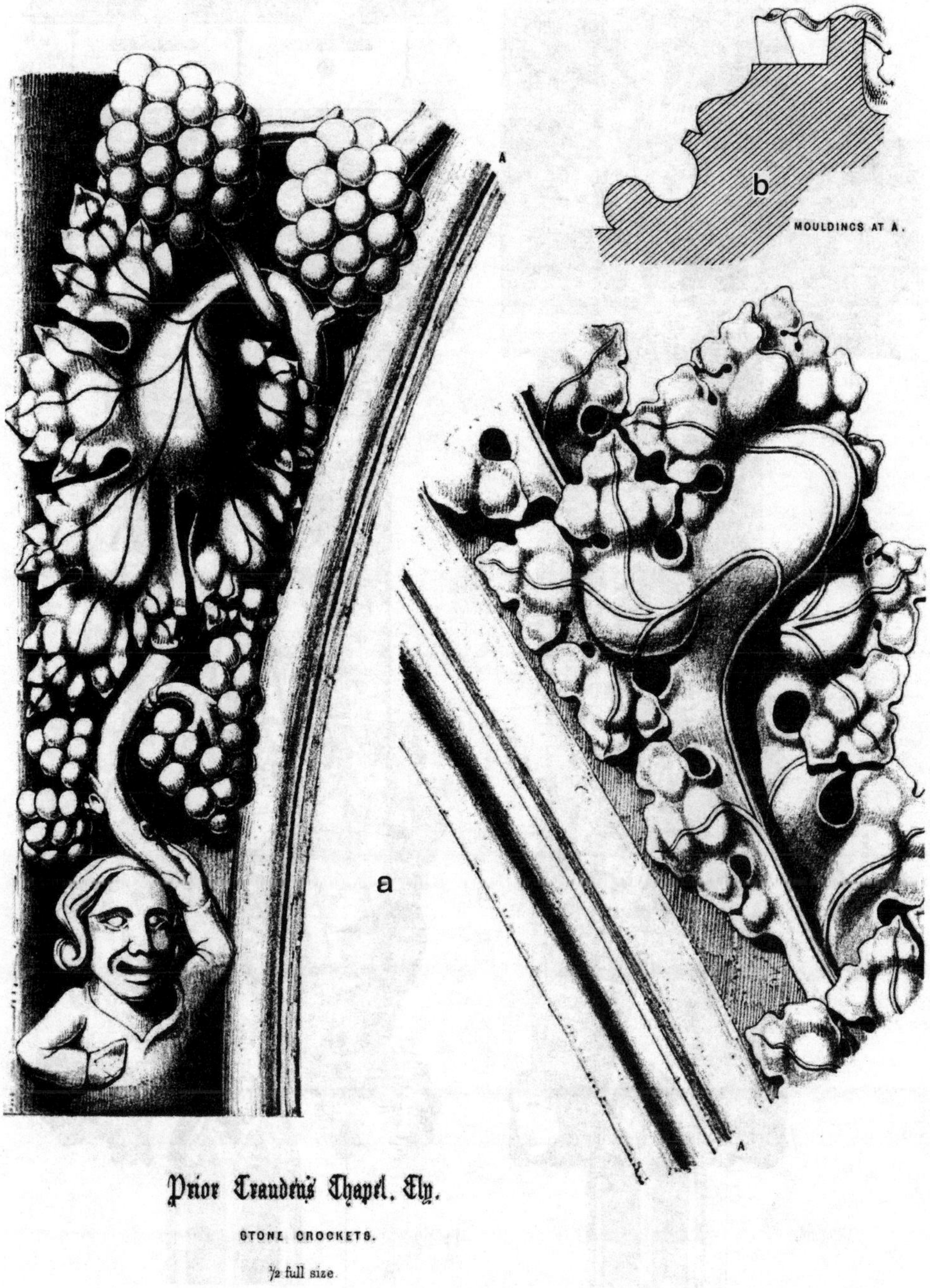

伊利普赖尔克兰登教堂：a. 石制卷叶形花饰　b.A 处的线脚

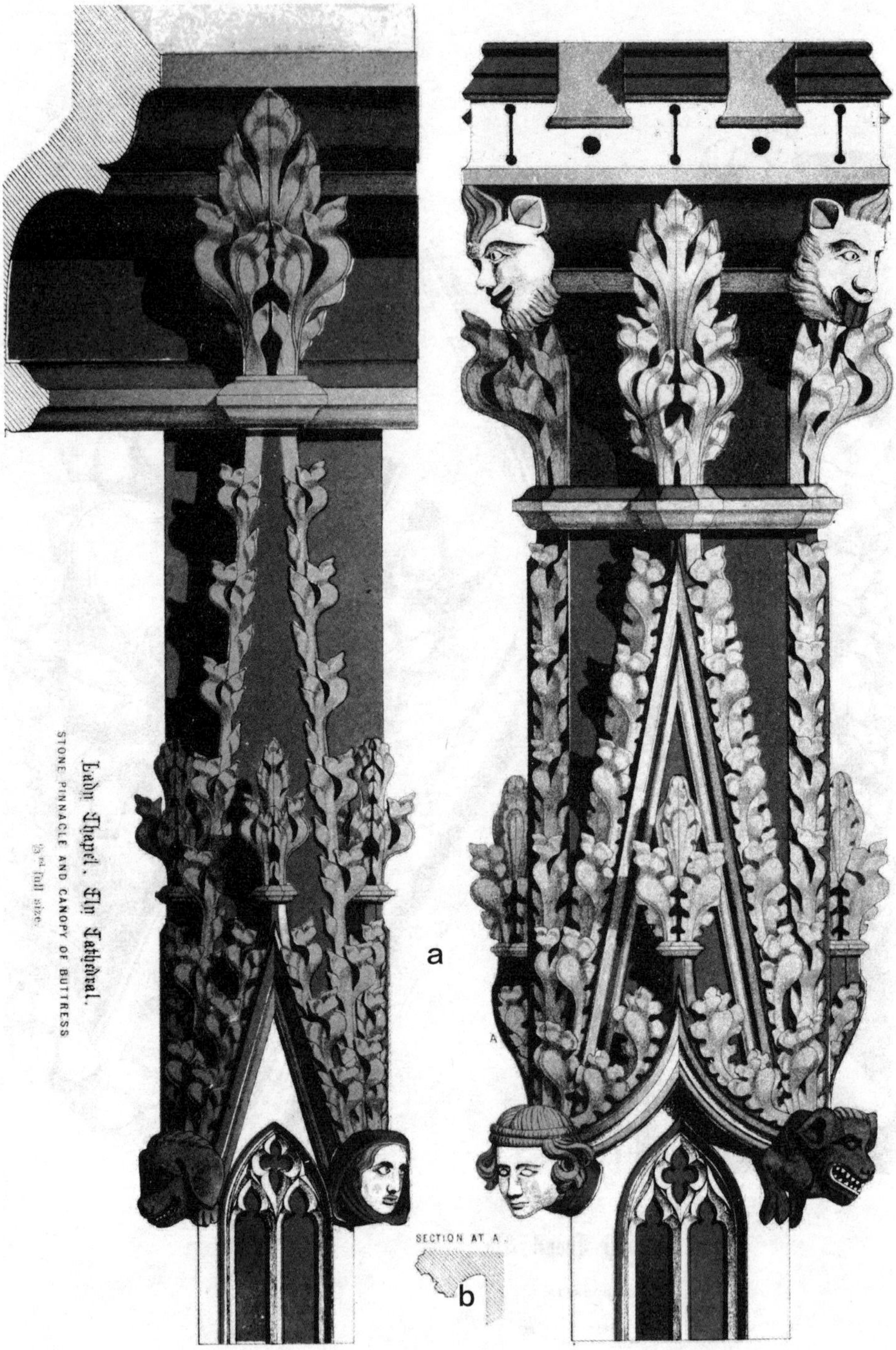

以利教堂圣母堂：a. 扶壁上石制小尖塔及顶棚　b.A 处的剖面

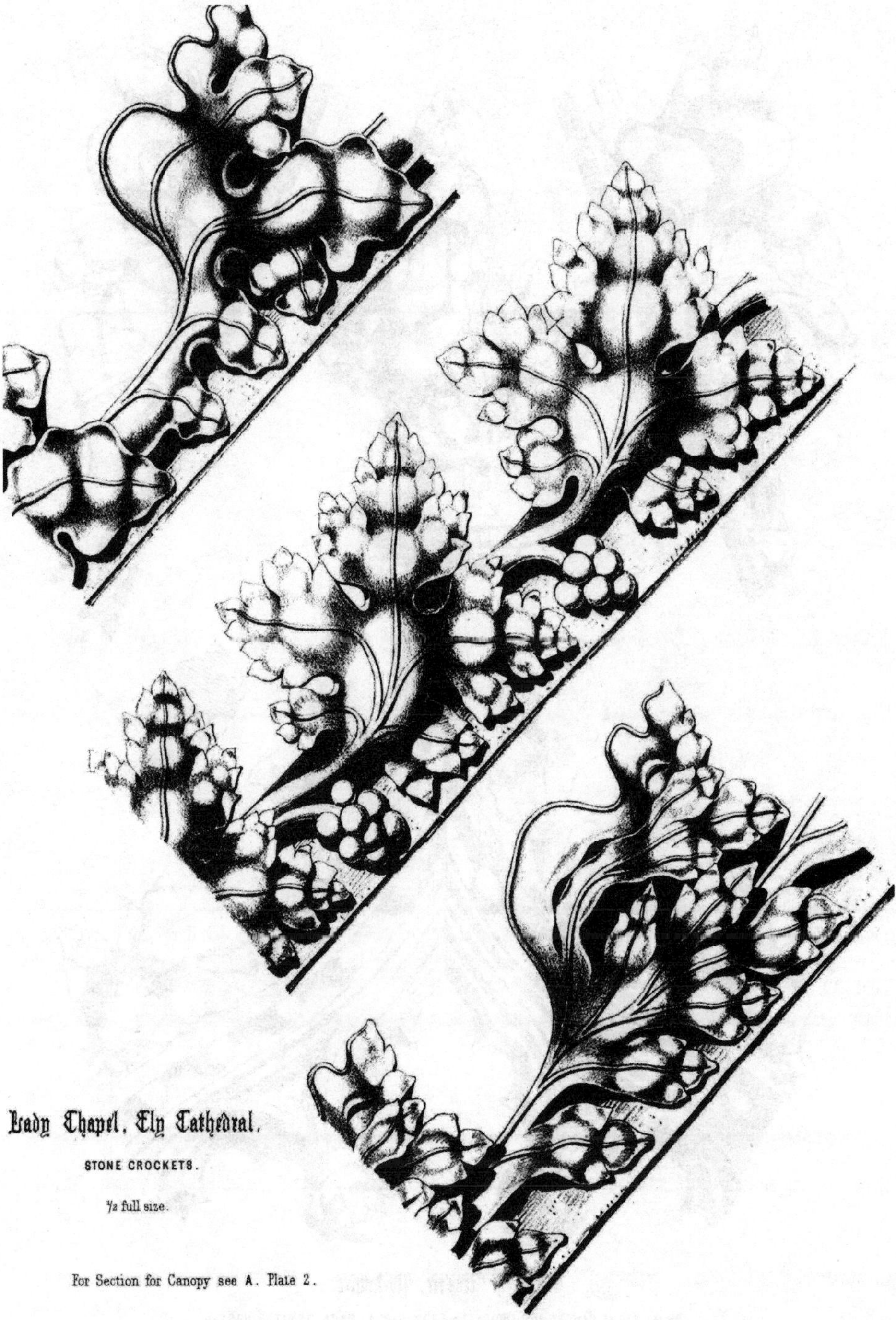

Lady Chapel, Ely Cathedral.

STONE CROCKETS.

½ full size.

For Section for Canopy see A. Plate 2.

以利教堂圣母堂石制卷叶形花饰

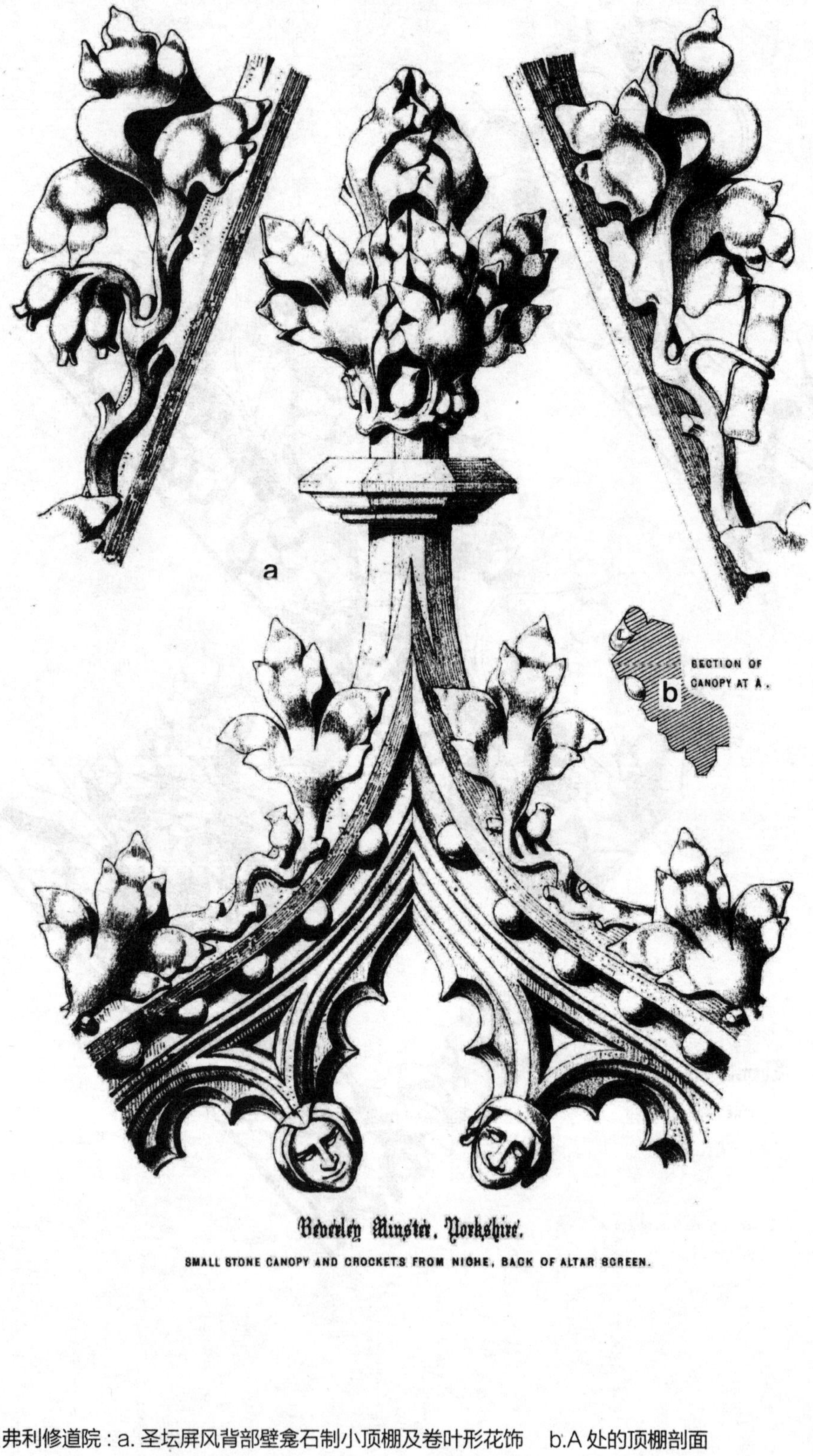

约克郡贝弗利修道院：a. 圣坛屏风背部壁龛石制小顶棚及卷叶形花饰　b.A 处的顶棚剖面

第12章

披水石、切角端部及垂饰

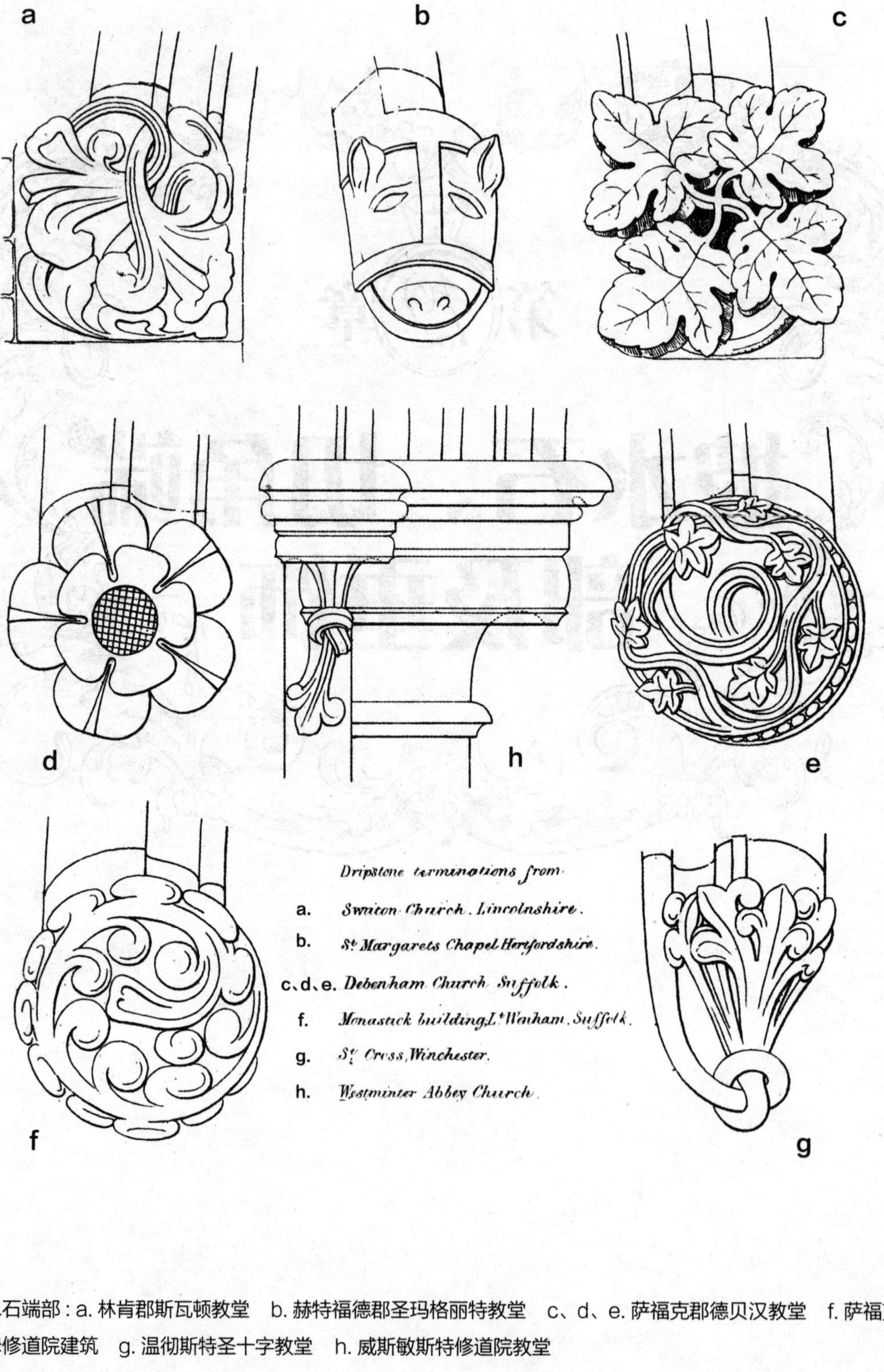

披水石端部：a. 林肯郡斯瓦顿教堂　b. 赫特福德郡圣玛格丽特教堂　c、d、e. 萨福克郡德贝汉教堂　f. 萨福克韦纳姆修道院建筑　g. 温彻斯特圣十字教堂　h. 威斯敏斯特修道院教堂

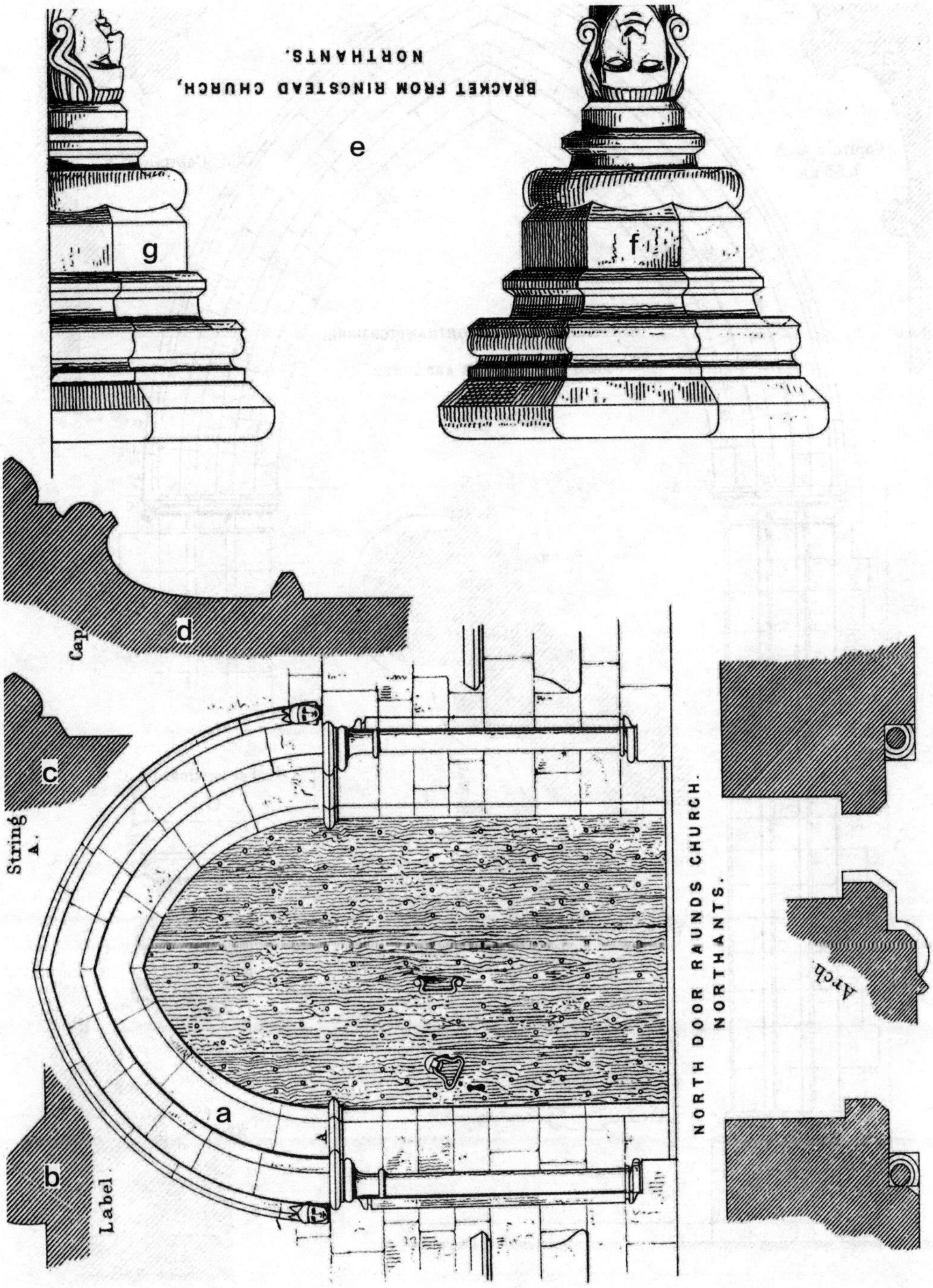

北安普敦郡朗兹教堂：a. 北门 b. 披水石 c. 束带层 d. 柱头 e. 北安普敦灵斯特德教堂牛腿 f. 正面 g. 侧面

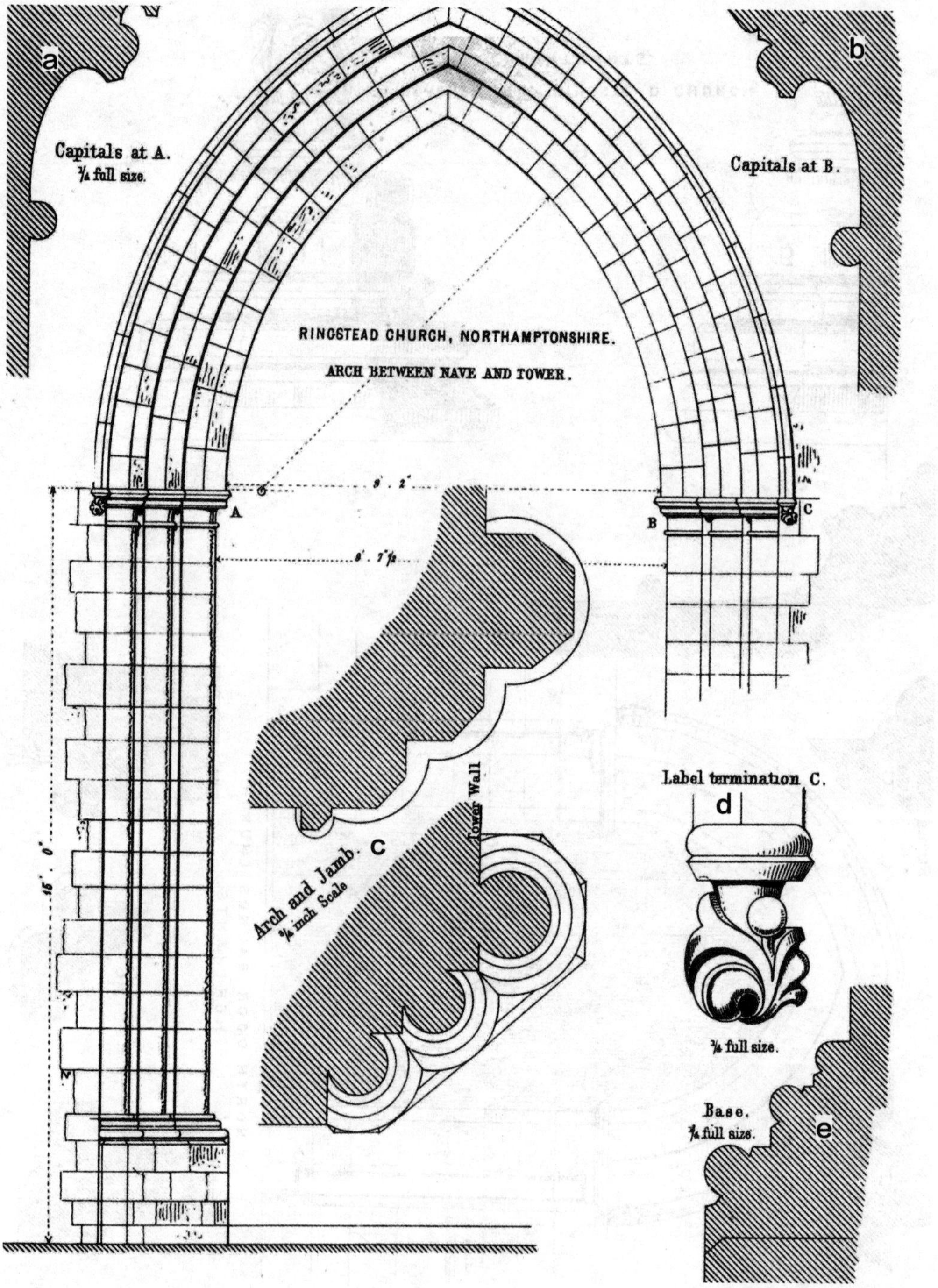

北安普敦灵斯特德教堂中殿和塔楼之间拱门：a.A 处的柱头　b.B 处的柱头　c. 拱和侧壁　d.C 处披水石端部　e. 柱础

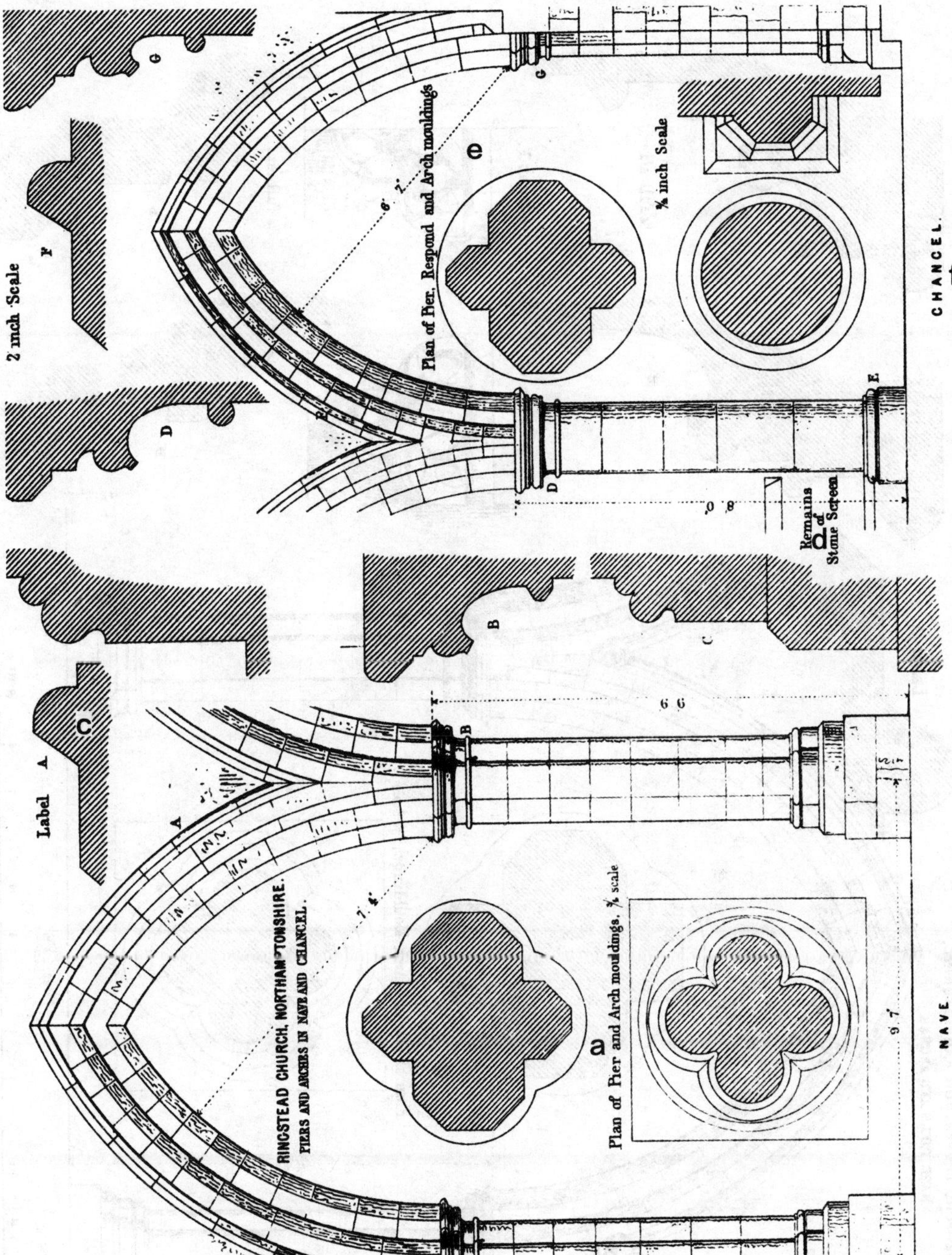

北安普敦灵斯特德教堂中殿和圣坛中集束柱和拱：a. 集束柱平面和拱线脚　b. 中殿　c.A 处披水石　d. 存留的石隔断　e. 柱子、壁联平面及拱线脚　f. 圣坛

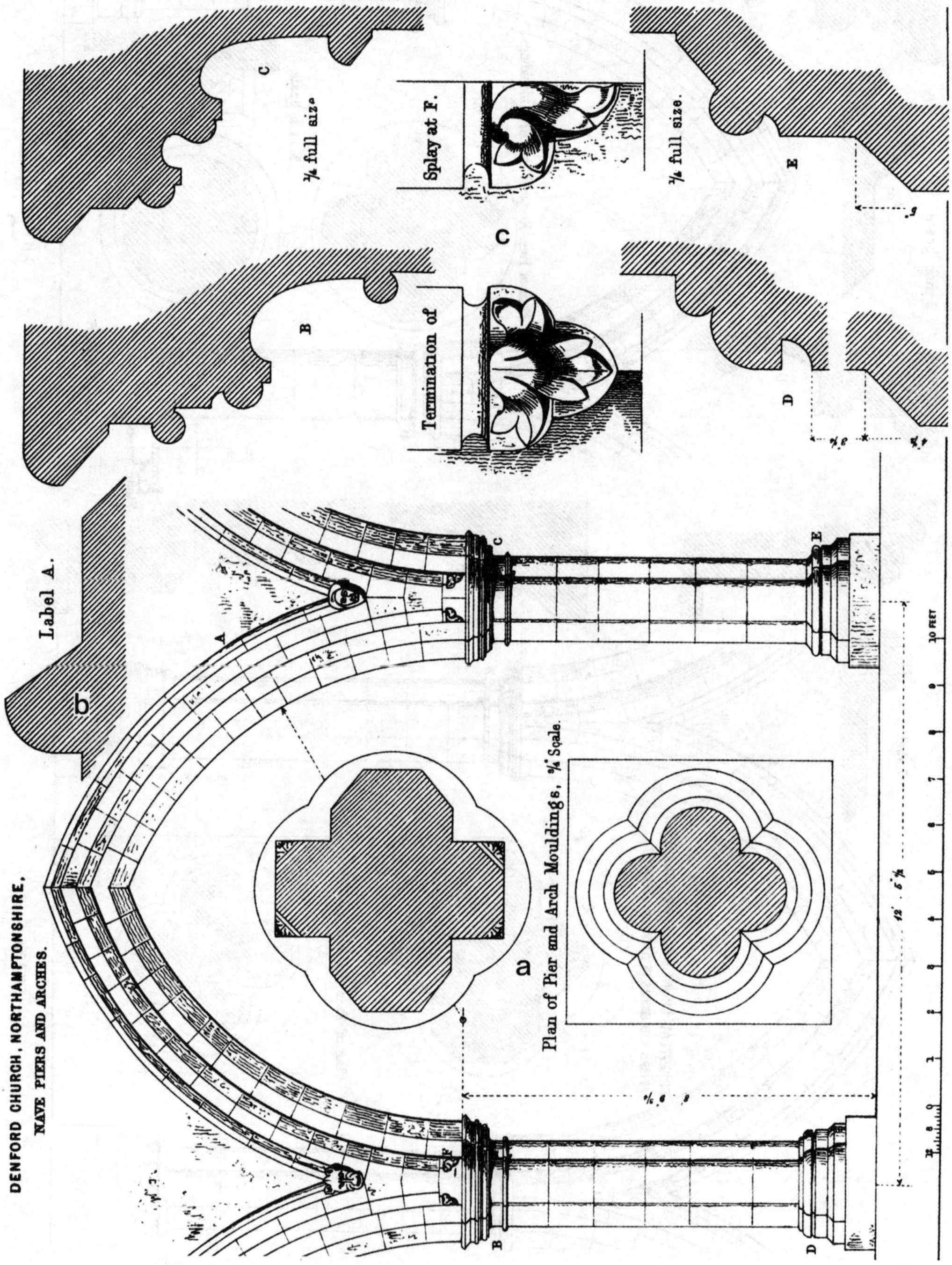

北安普敦郡邓福德教堂中殿集束柱及拱：a. 束柱及拱线脚平面　b.A 处的披水石　c. 斜面 F 端部

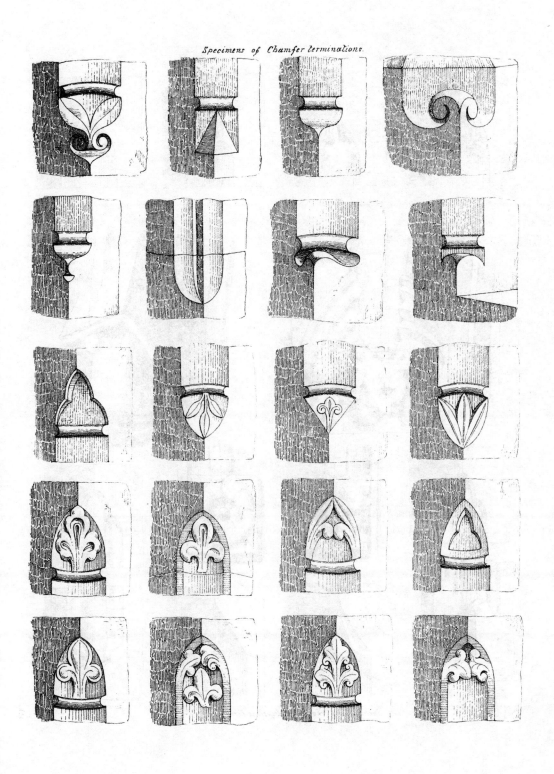

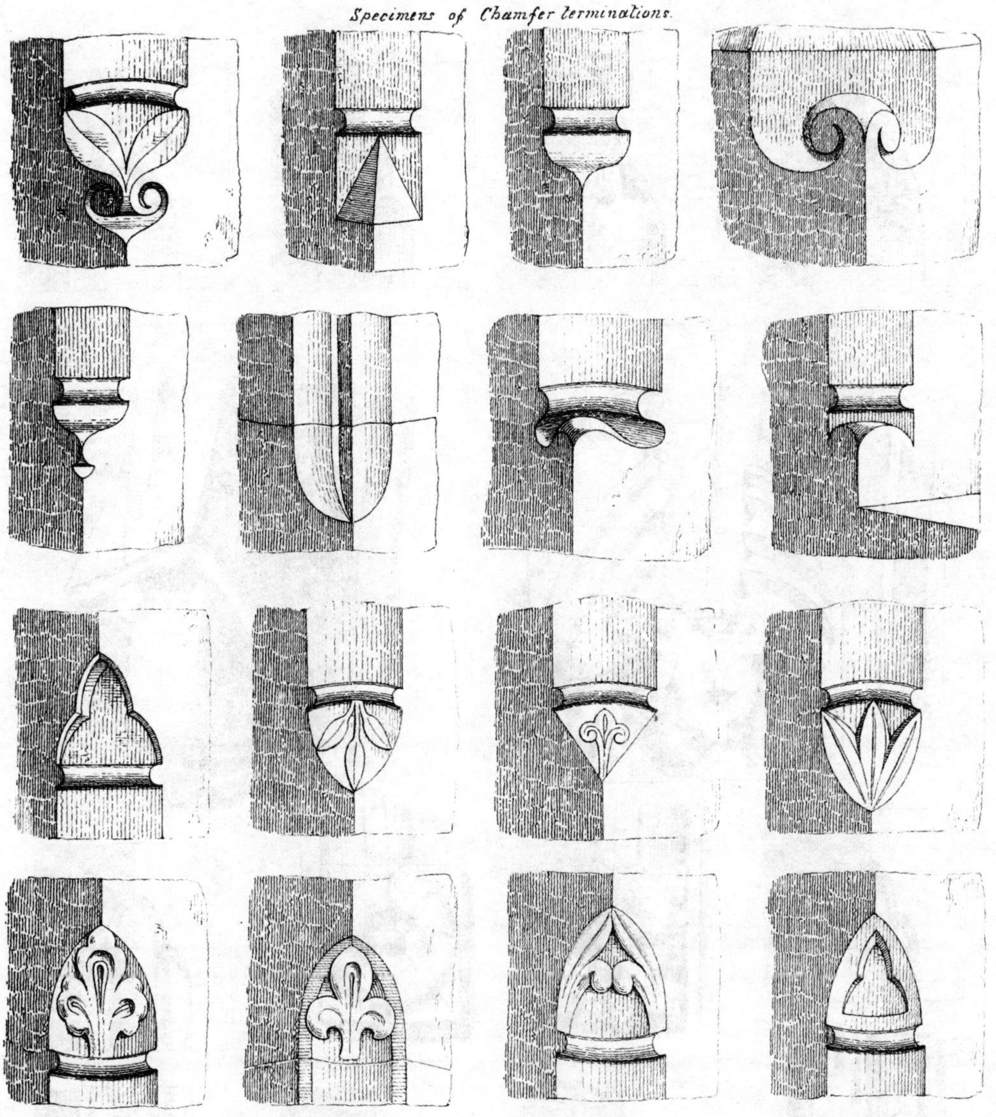
Specimens of Chamfer terminations.

切角端部式样

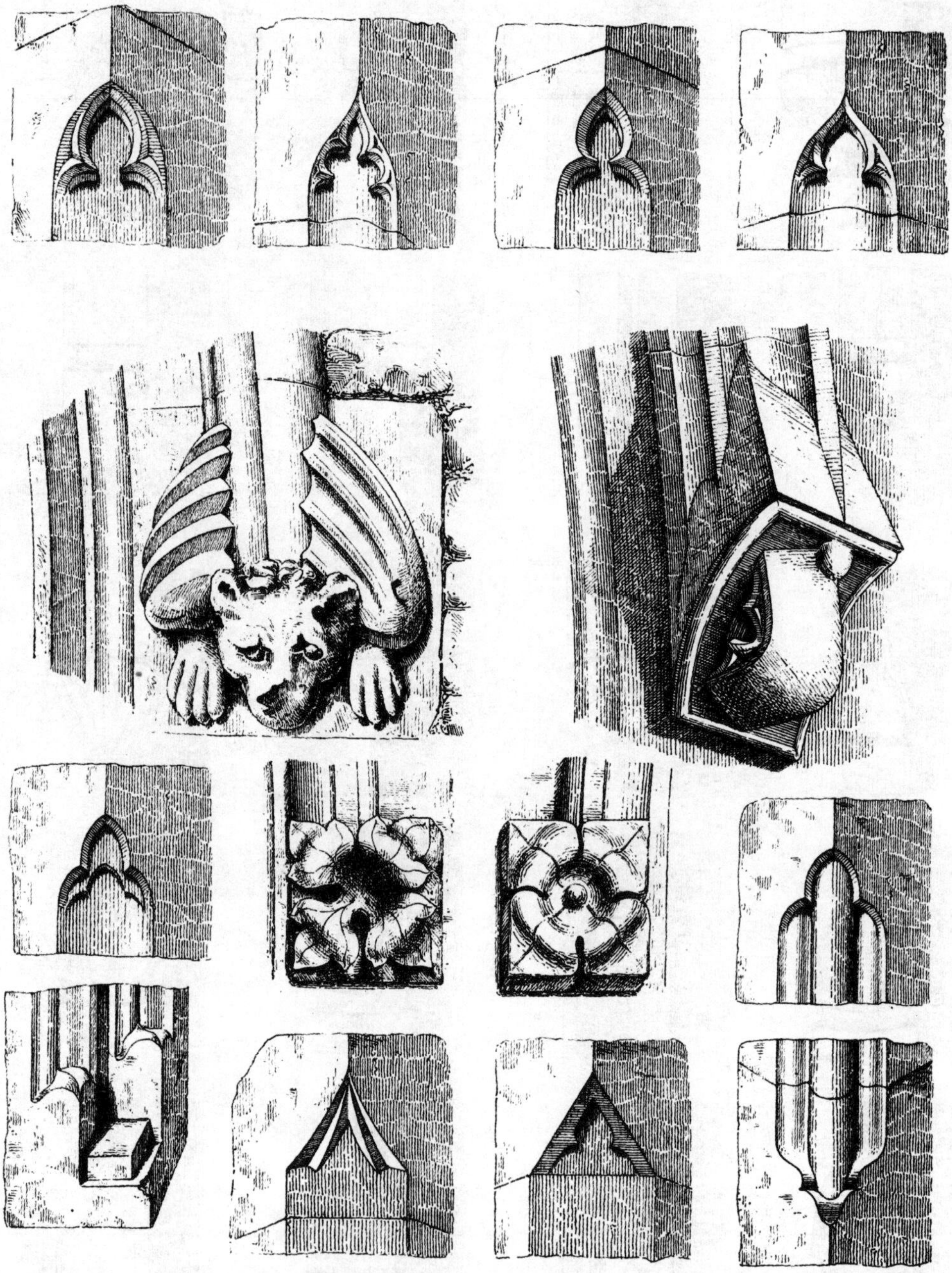

Specimens of chamfer & dripstone terminations.

切角及披水石端部式样

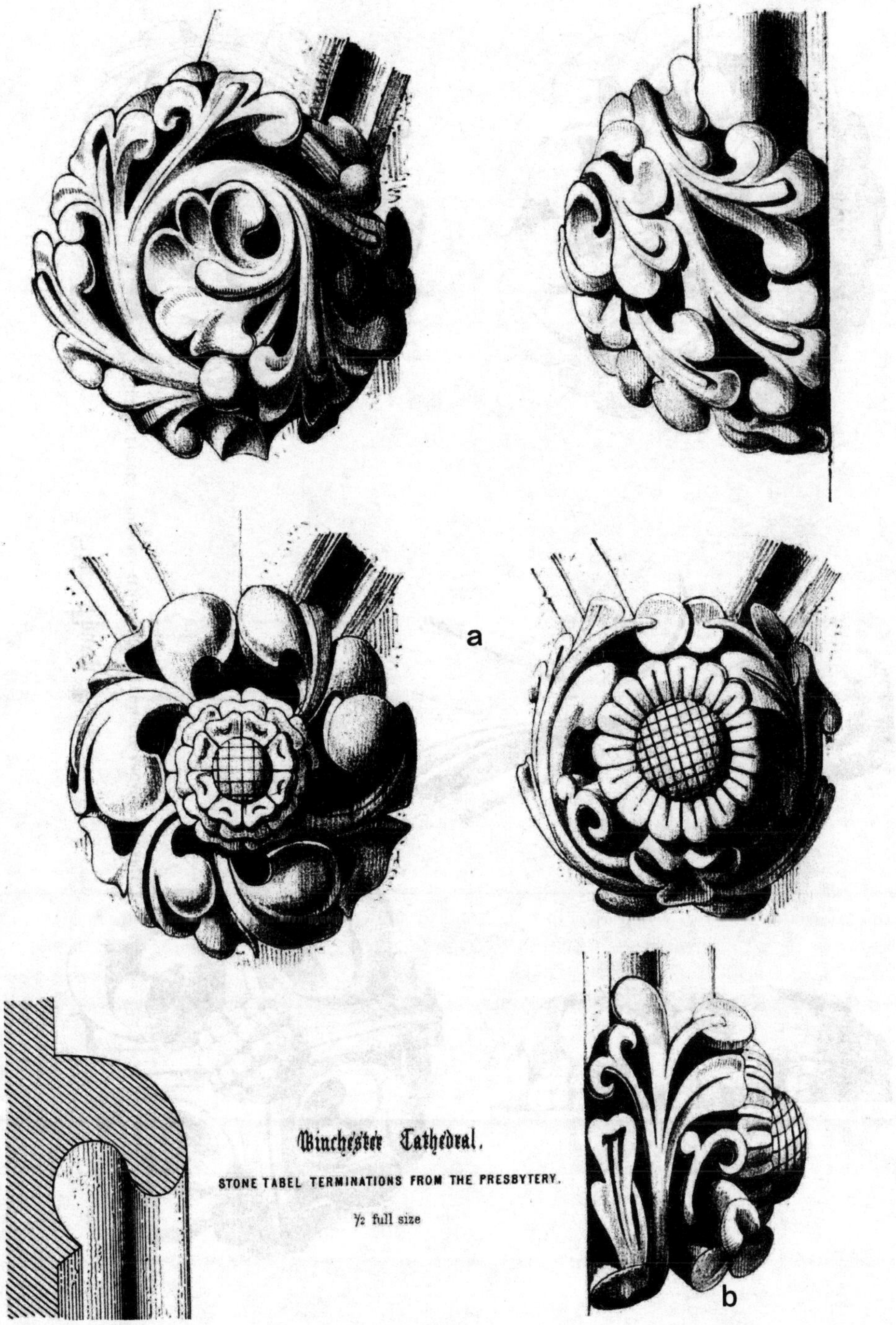

温彻斯特教堂：a. 长老会内披水石端部　b. 披水石侧面

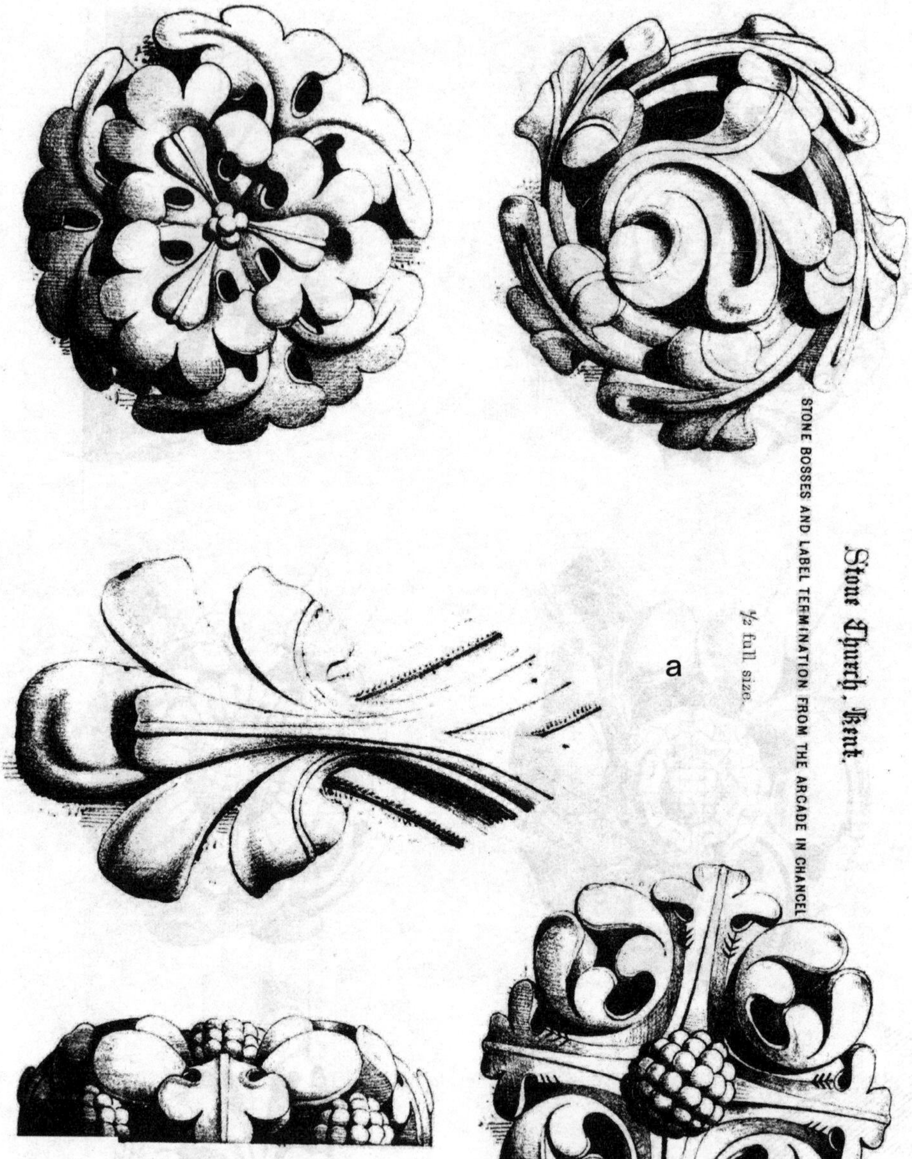

肯特郡石之教堂：a. 圣坛拱廊内石制凸饰及披水石端部　b. 侧面

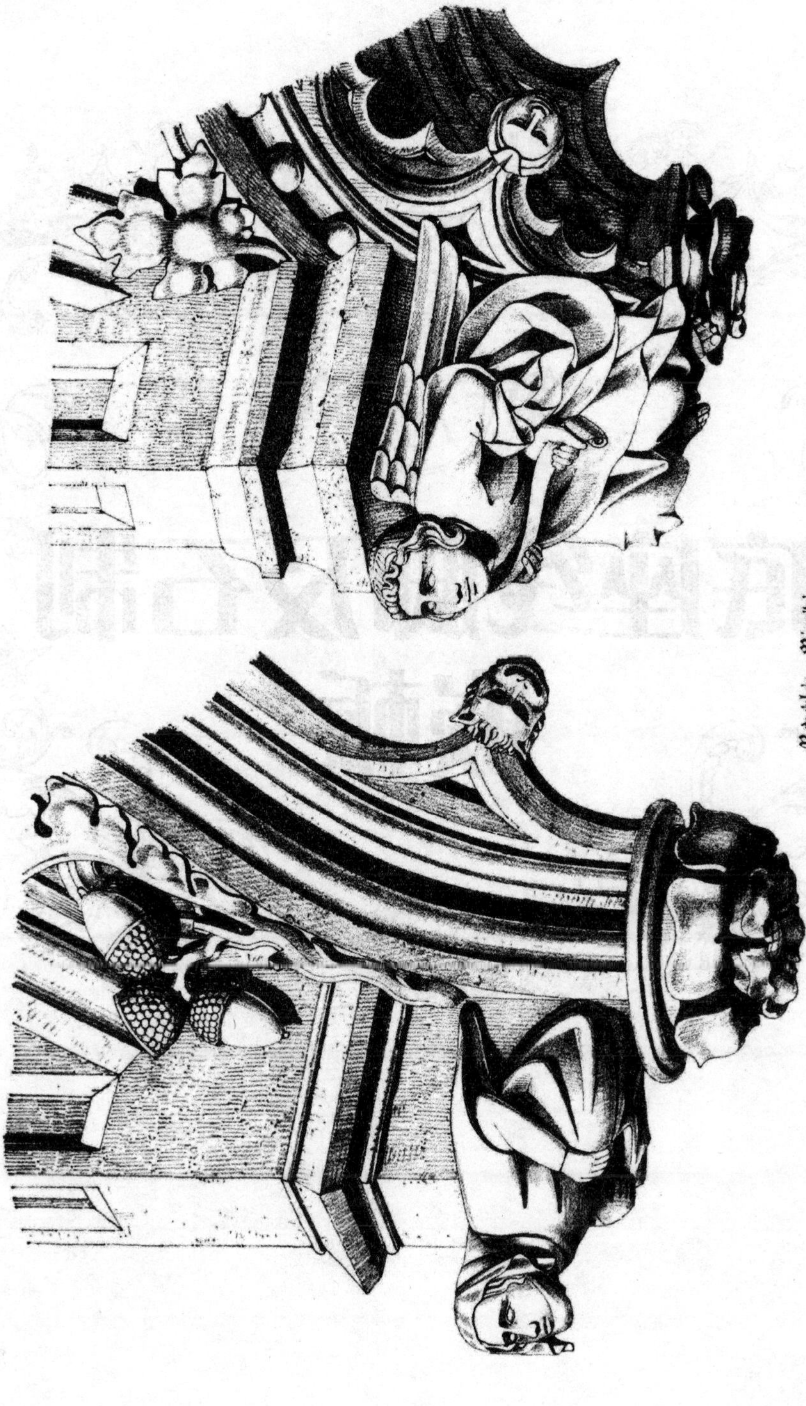

Beverley Minster.
STONE PENDANTS TO CANOPIES OF NICHES BACK OF ALTAR SCREEN
full size.

贝弗利修道院圣坛屏风背部壁龛石制小顶棚石制垂饰

第13章

底座线脚及石制嵌板

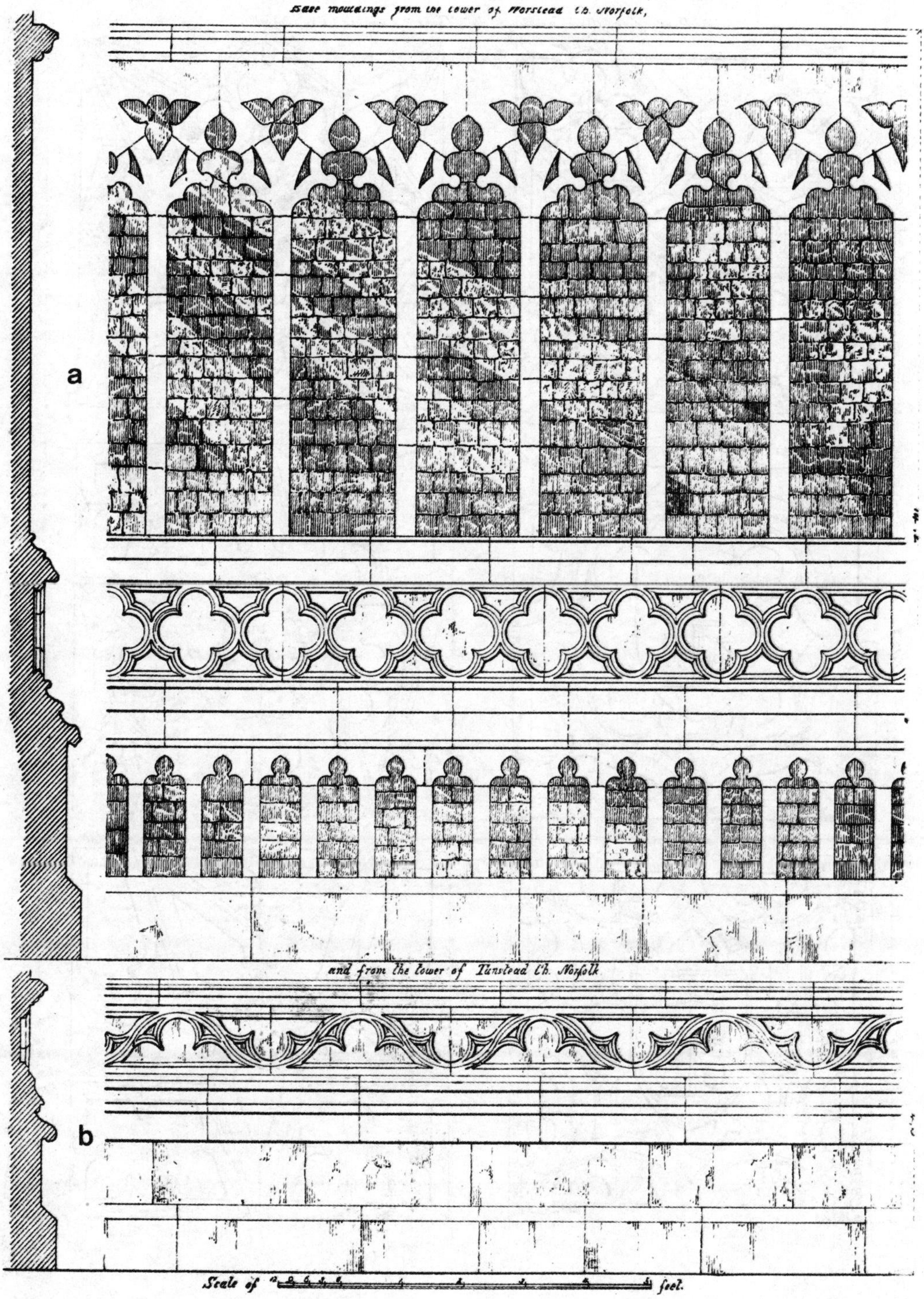

a. 诺福克郡沃斯戴德教堂塔楼底座线脚　b. 诺福克郡腾斯戴德教堂塔楼底座线脚

萨福克郡拉文纳姆教堂塔楼底座线脚上石制嵌板

石雕式样：a. 萨福克郡斯特拉特福德圣玛丽教堂　b. 萨福克郡艾教堂

第14章

尖拱及枪眼

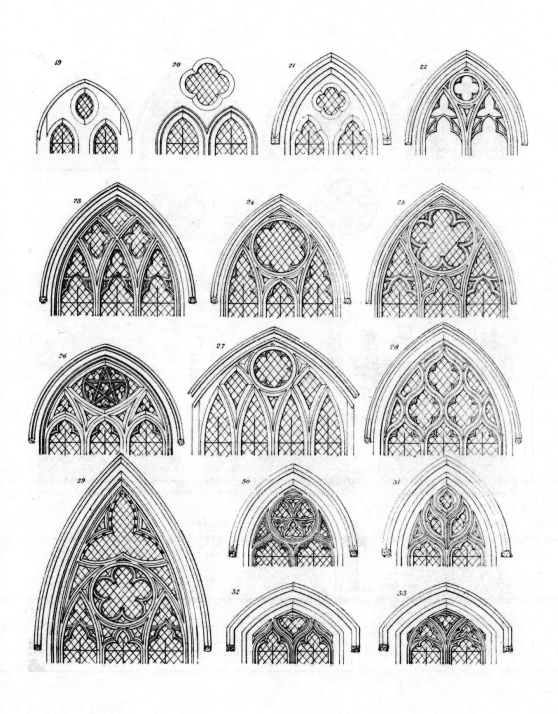

尖拱

第 14 章 尖拱及枪眼

尖拱

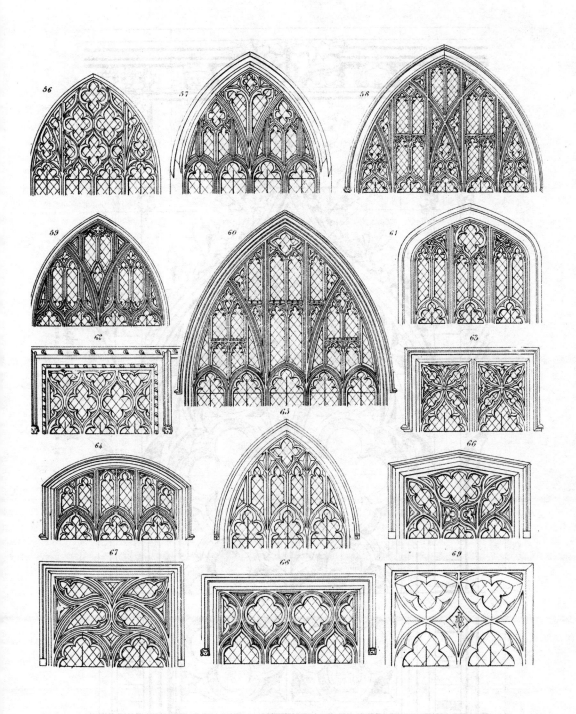

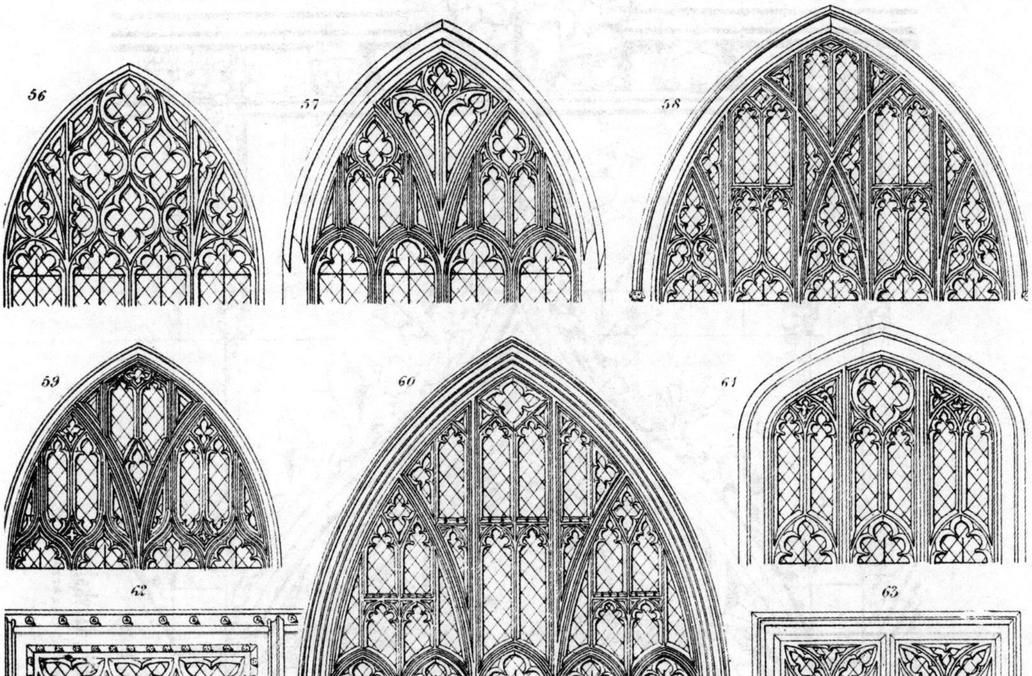

尖拱

约克郡豪登教堂牧师会礼堂里的石制拱形装饰

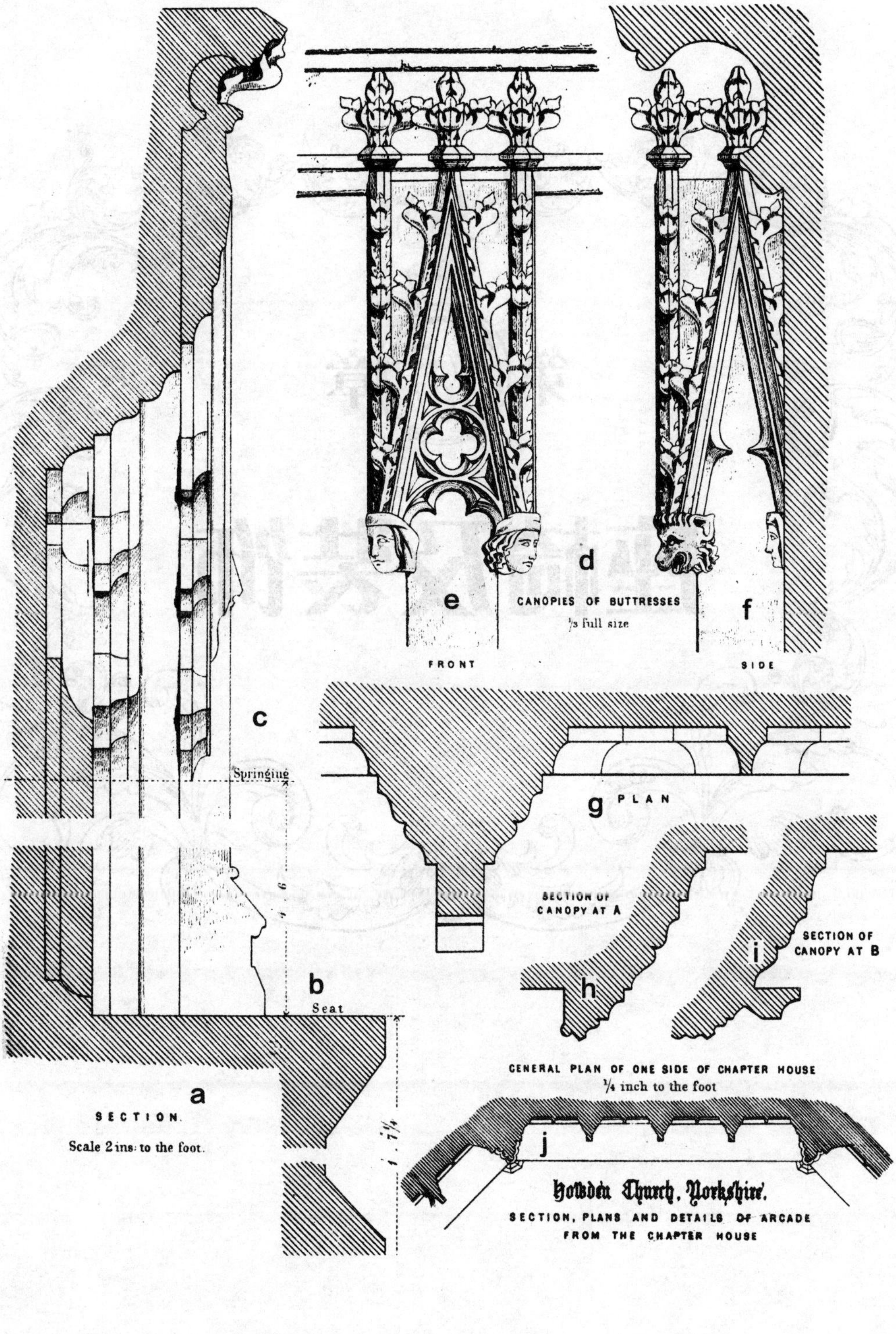

约克郡豪登教堂牧师会礼堂里石制拱形装饰剖面、平面及细部：a. 剖面　b. 座位　c. 起拱处　d. 扶壁顶盖　e. 正面　f. 侧面　g. 平面　h.A 处的顶盖剖面　i.B 处的顶盖剖面　j. 牧师会礼堂一侧总体平面

第15章

座椅及装饰

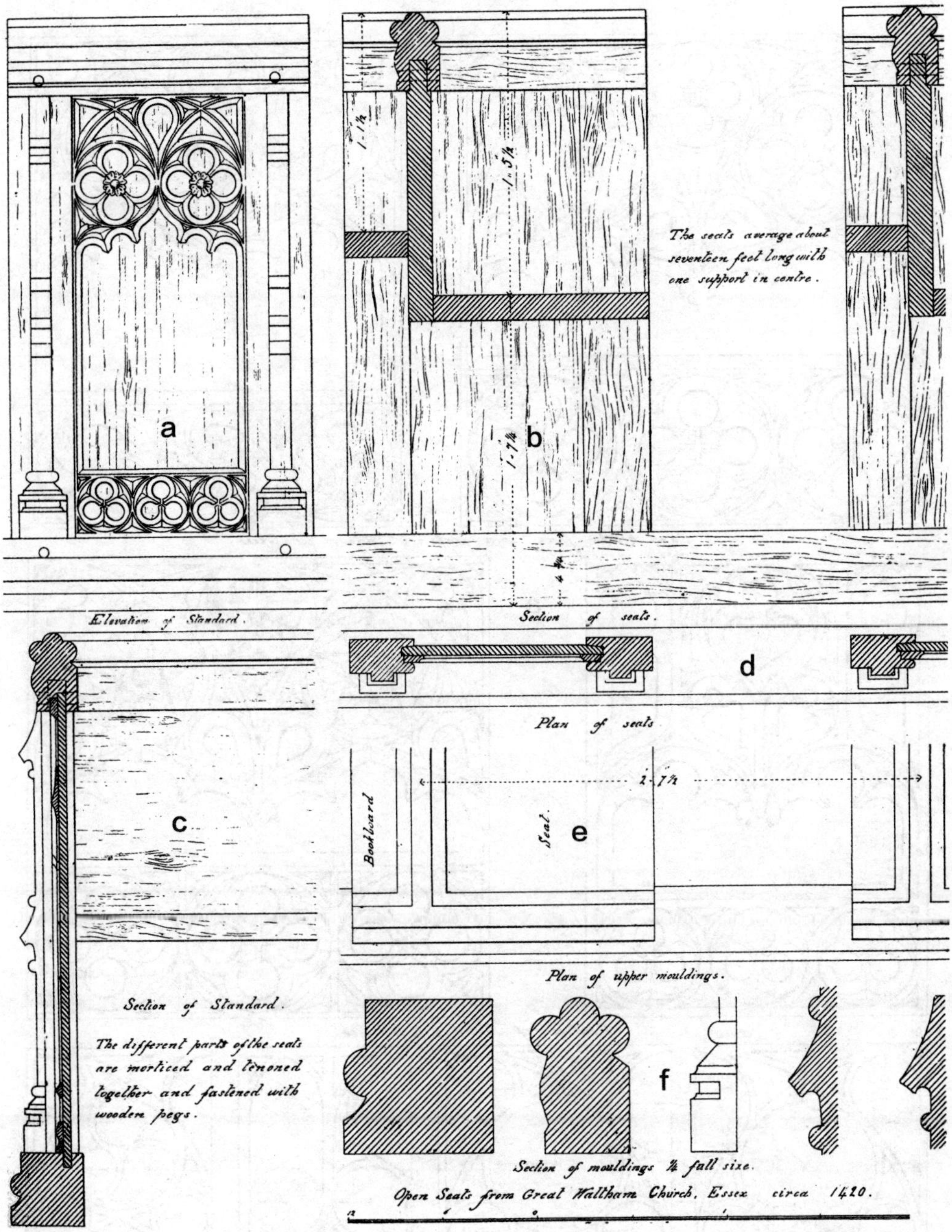

a. 埃塞克斯郡沃尔瑟姆教堂开敞座椅，座椅平均长十七尺，中心带一根支撑　b. 侧支板立面　c 侧支板剖面座椅不同部分之间采用榫接并且用木钉连接　d. 座椅剖面和平面　e. 座椅上部线脚平面　f. 线脚剖面

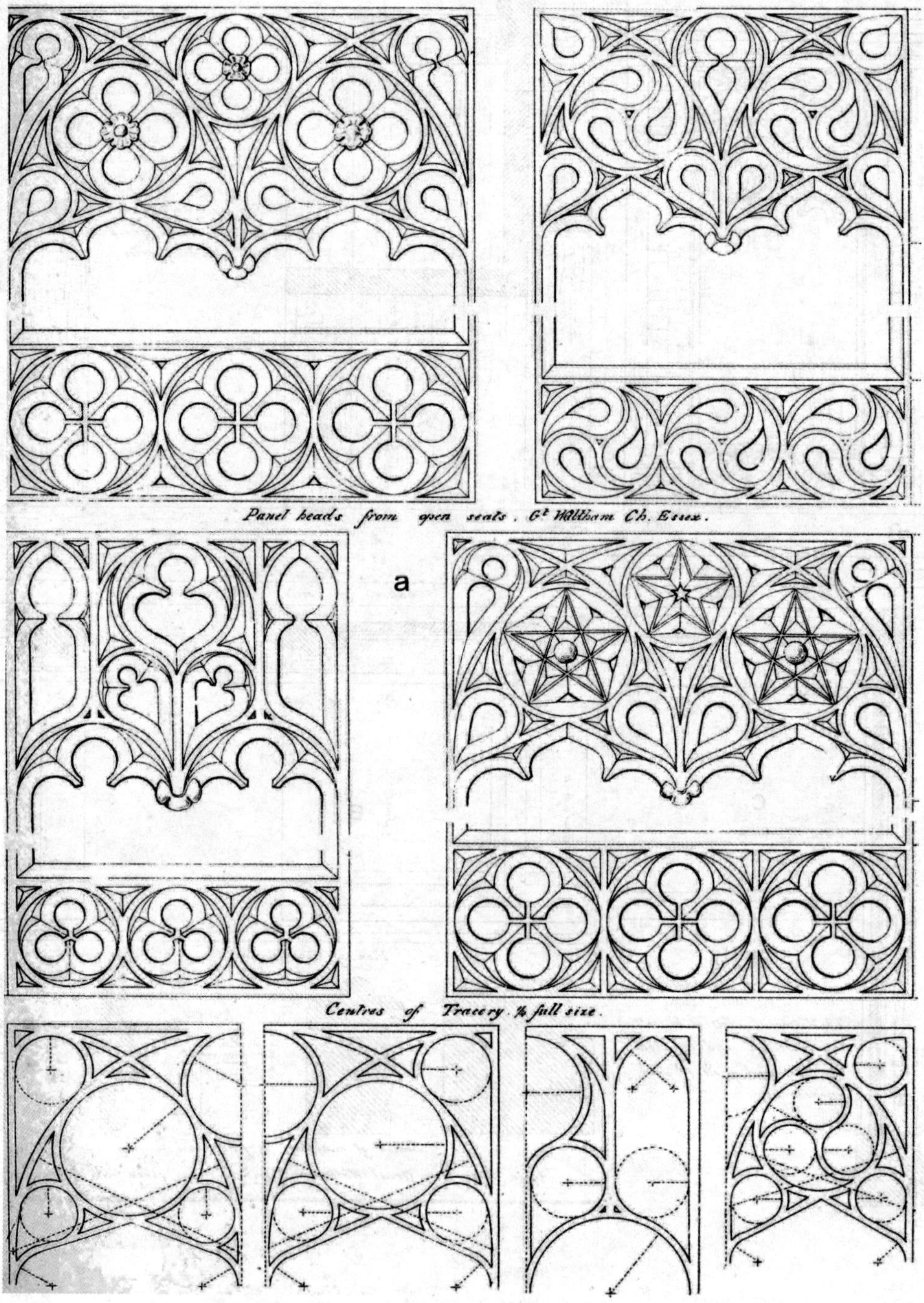

b

a. 埃塞克斯郡沃尔瑟姆教堂开敞座椅上嵌板　b. 花格饰中心点

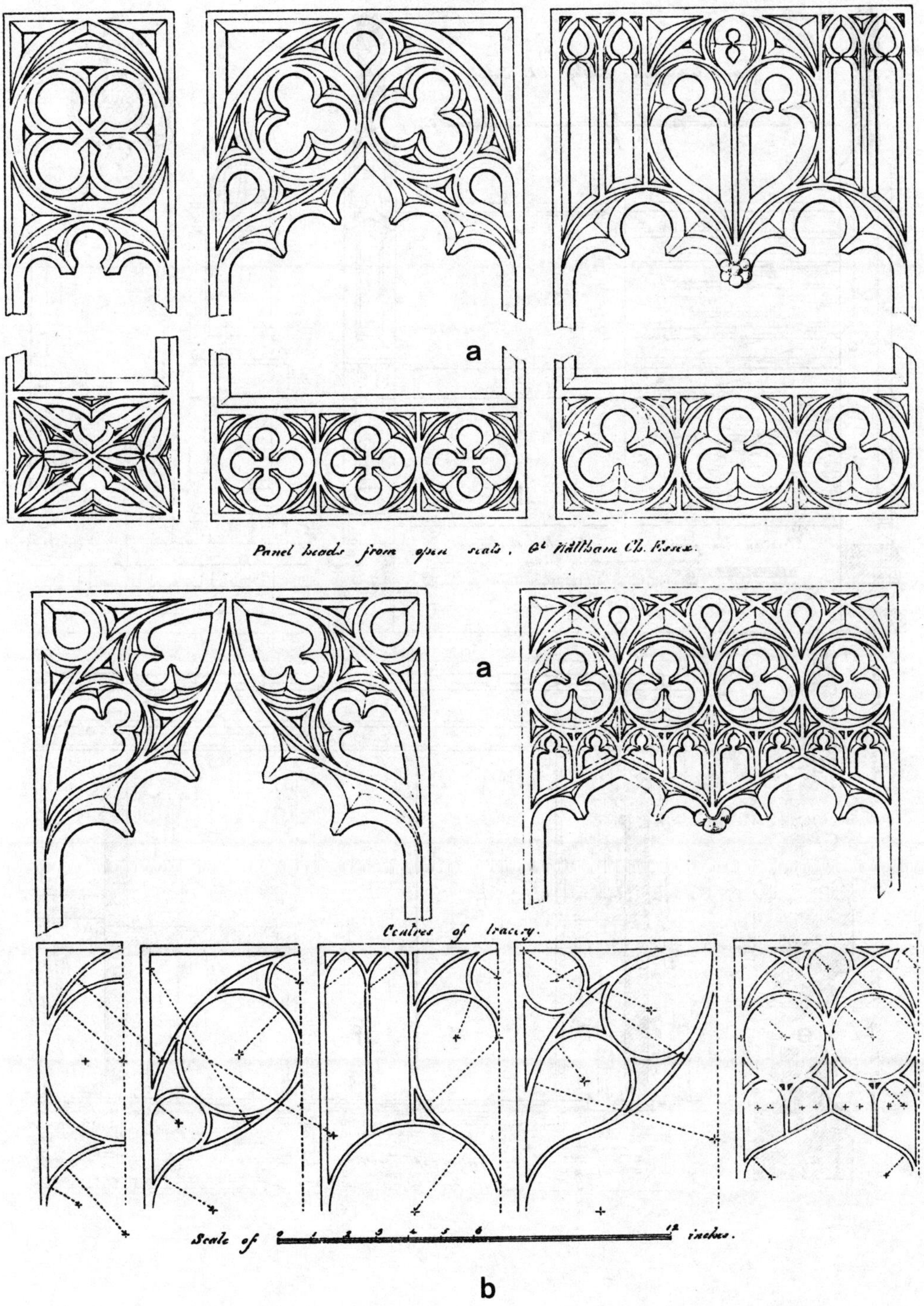

a. 埃塞克斯郡沃尔瑟姆教堂开敞座椅上嵌板　b. 花格饰中心点

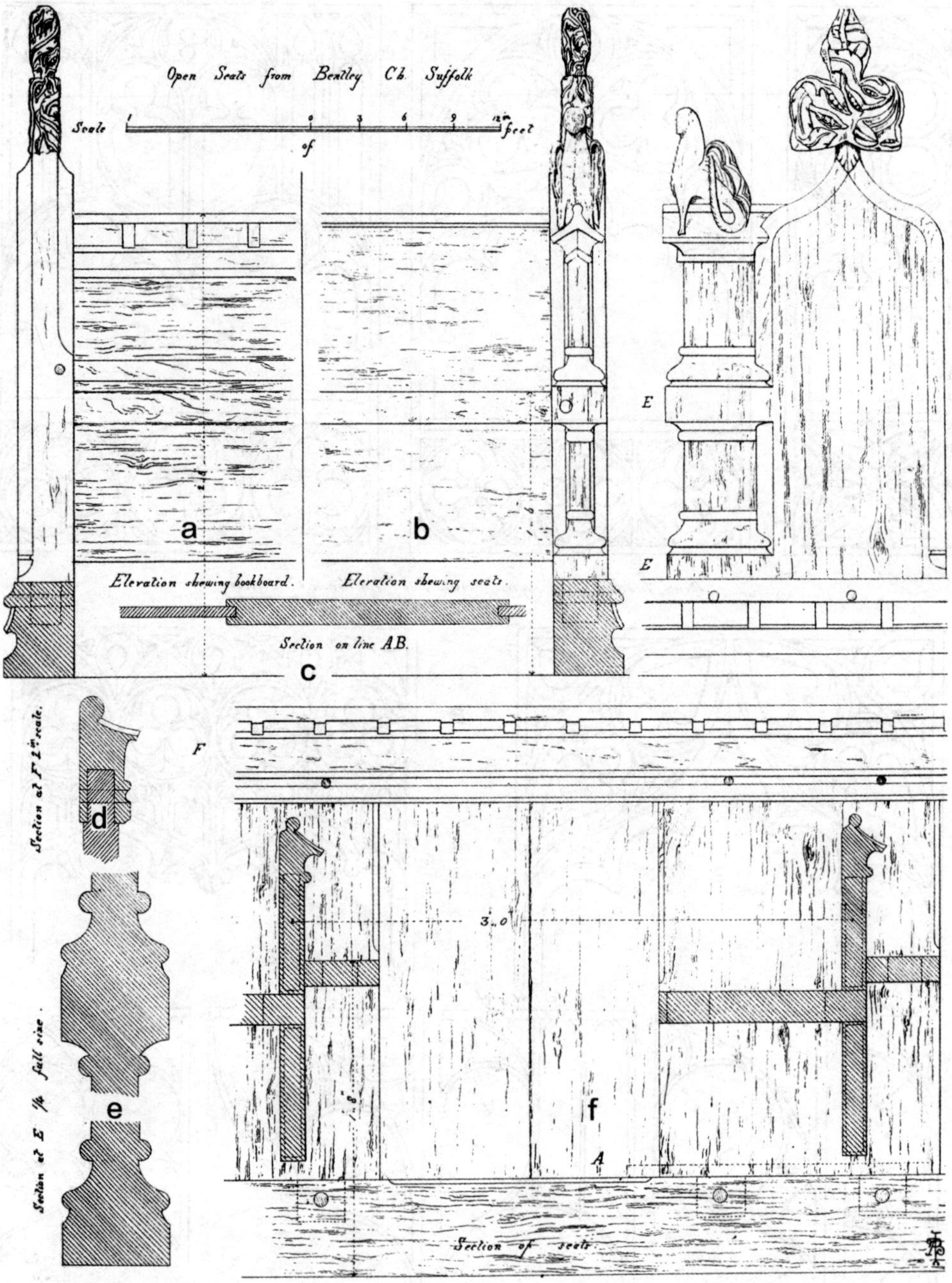

萨福克郡本特利教堂开敞座椅：a. 展示背板的立面　b. 展示座位的立面　c.A–B 处的剖面　d.F 处的剖面　e.E 处的剖面　f. 座椅剖面

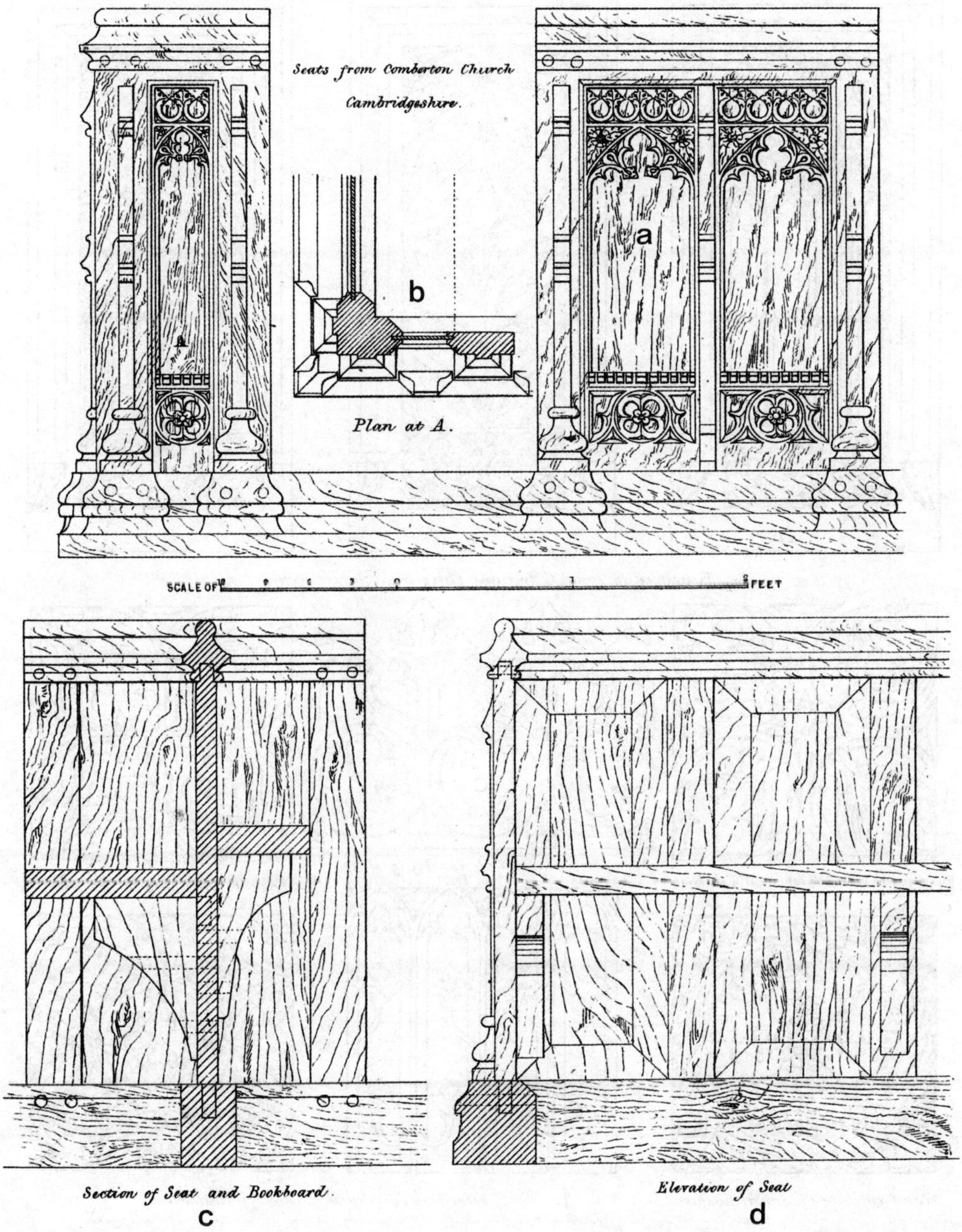

a. 剑桥郡康伯顿教堂座椅　b.A 处的平面　c. 座位及背板剖面　d. 座位立面

a. 索美塞特夏郡克劳库姆教堂座椅端部　b. 来自诺福克郡沃斯特德教堂　c. 来自剑桥郡小谢尔福德教堂　d. 来自诺福克郡沃斯特德教堂　e. 来自索美塞特夏郡克劳库姆教堂　f. 来自索美赛特夏郡李帝尔德主教堂

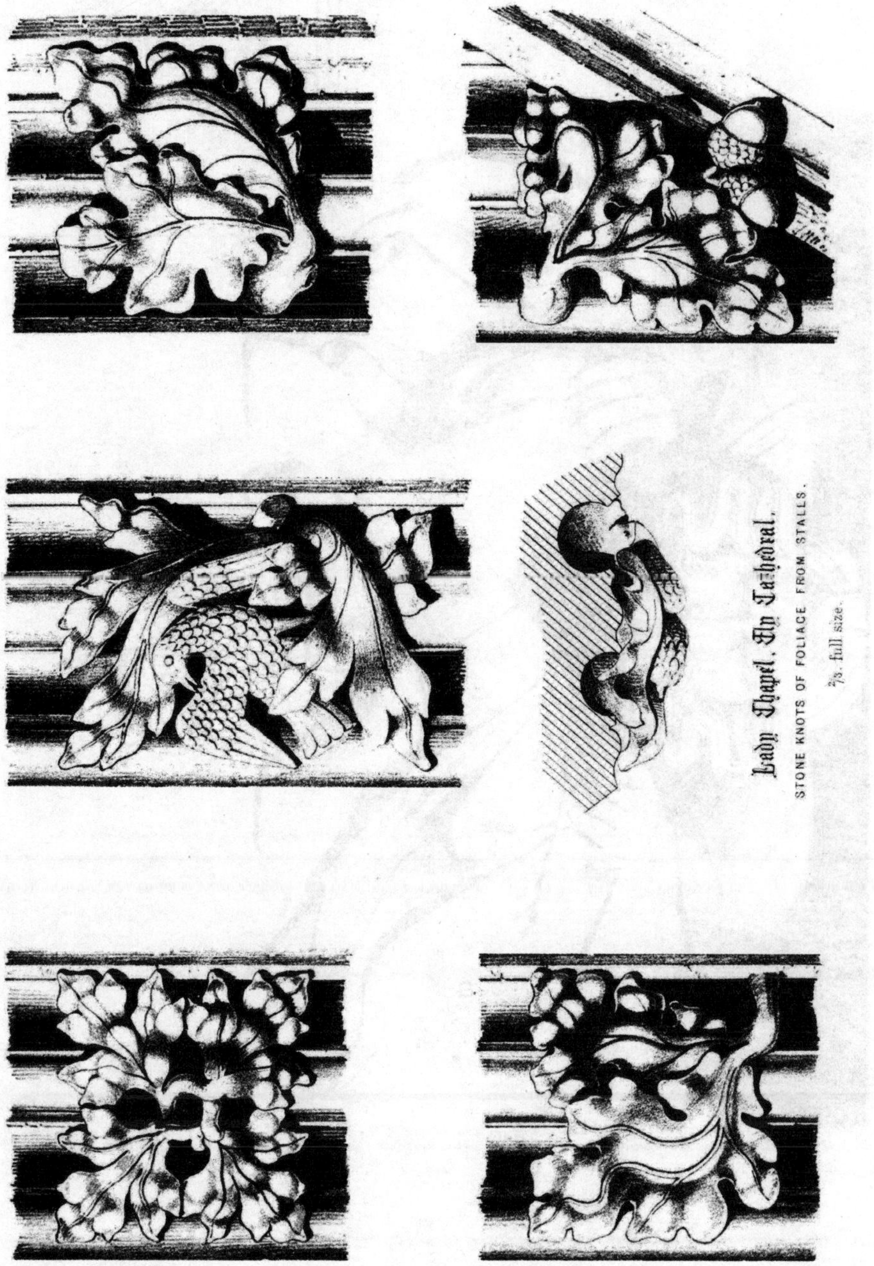

圣母堂座椅上石制叶状结饰

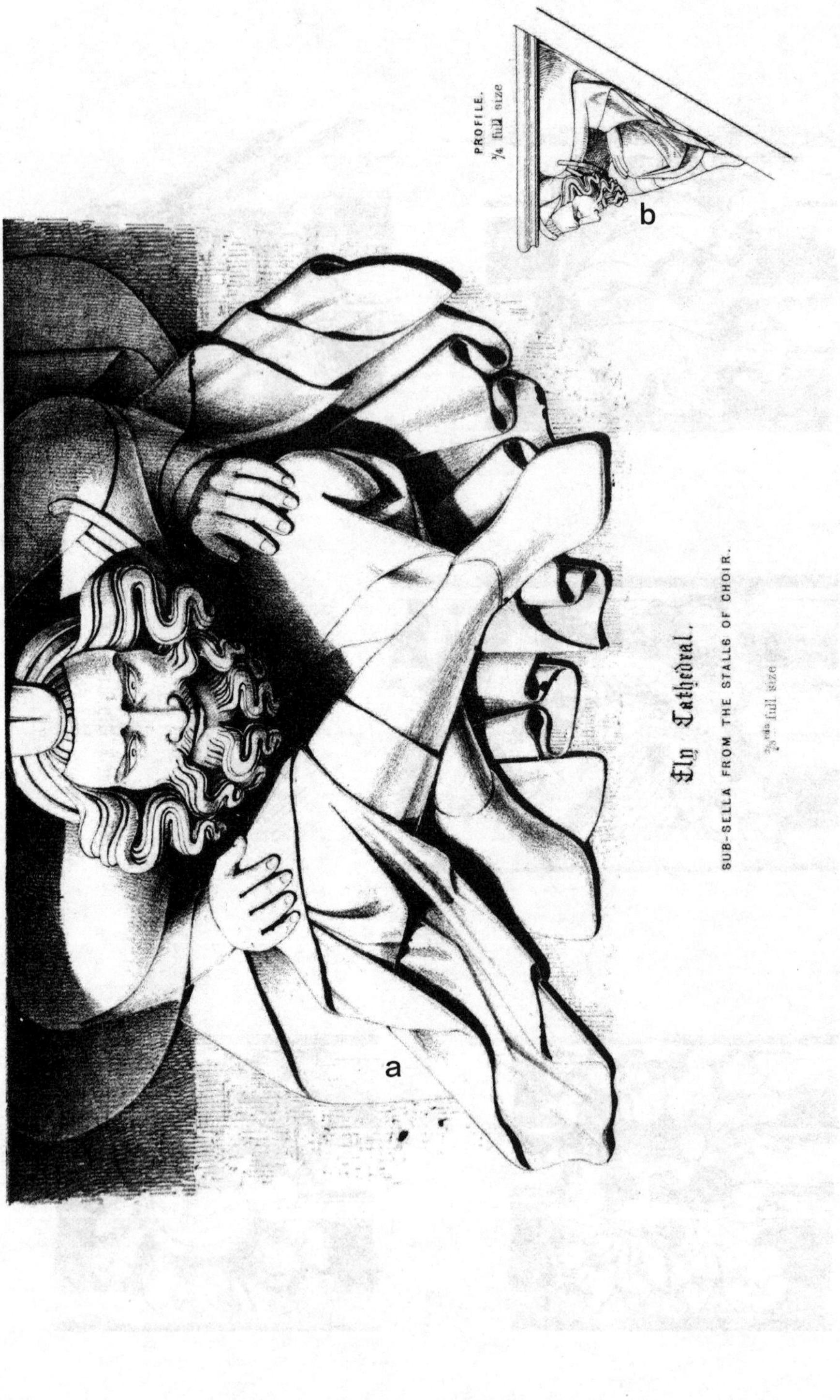

以利教堂：a. 唱诗室座椅上突出托板　b. 轮廓

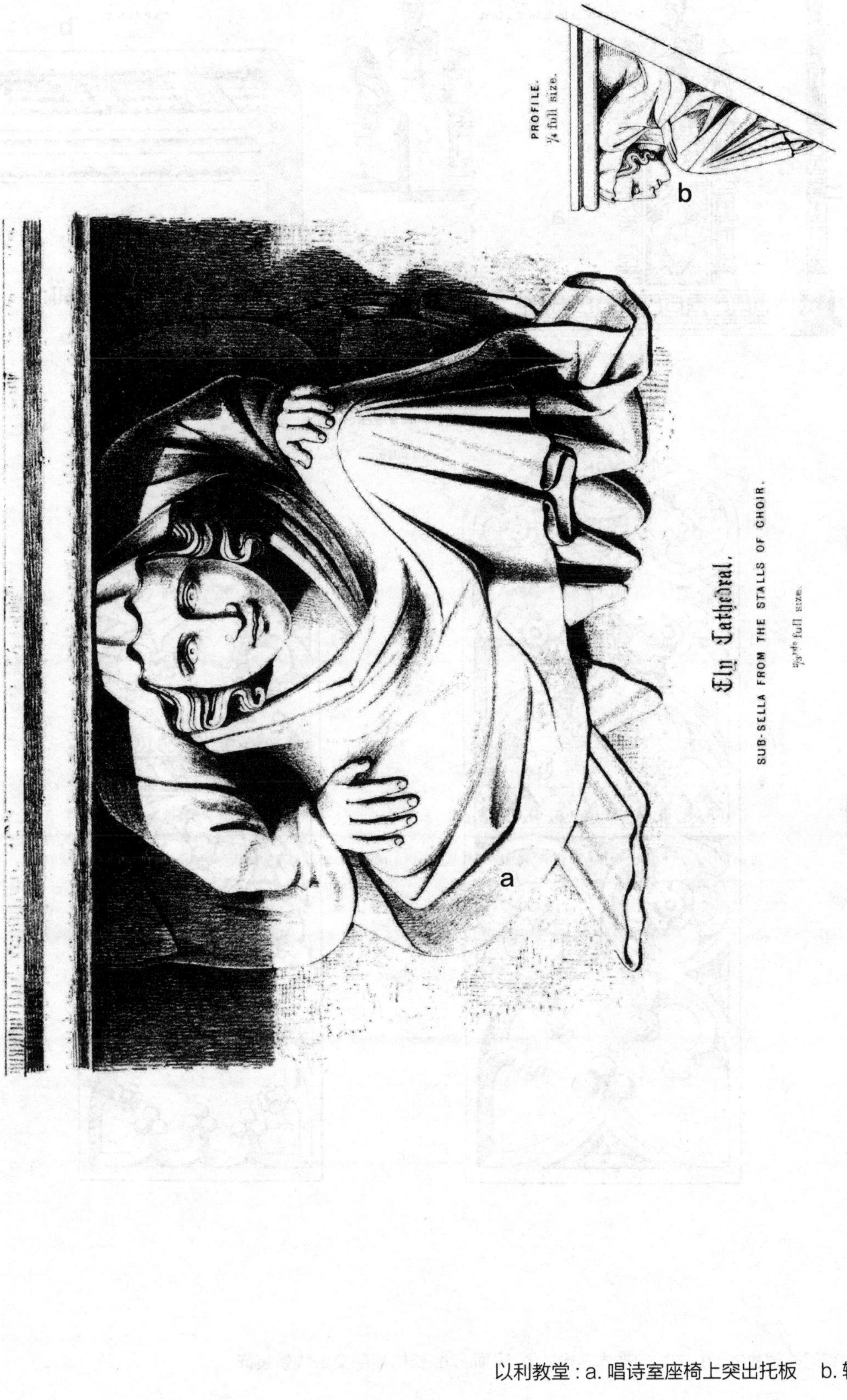

以利教堂：a. 唱诗室座椅上突出托板　b. 轮廓

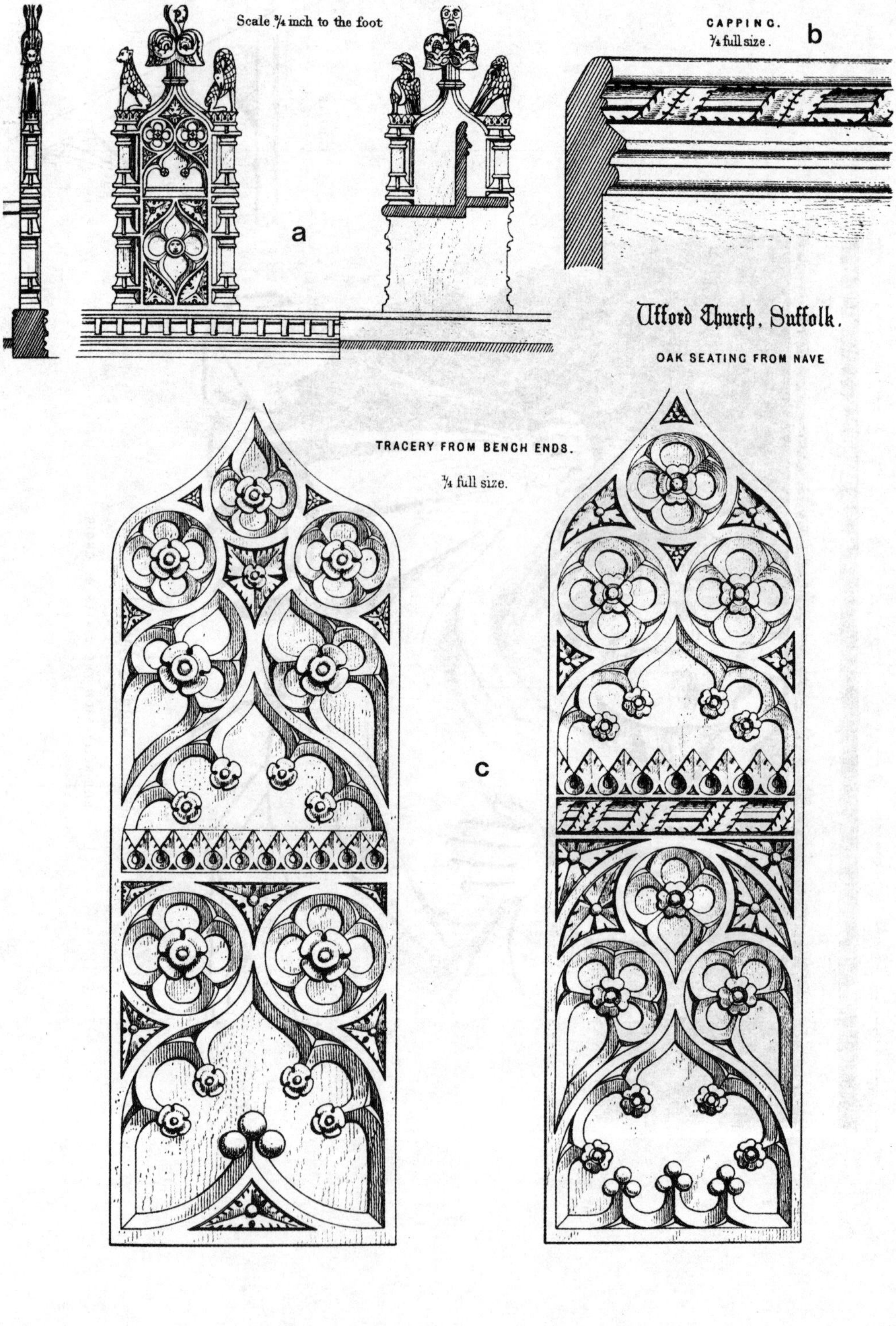

萨福克厄福德教堂：a. 中殿内橡木座椅　b. 压顶　c. 长椅端部交织线条装饰

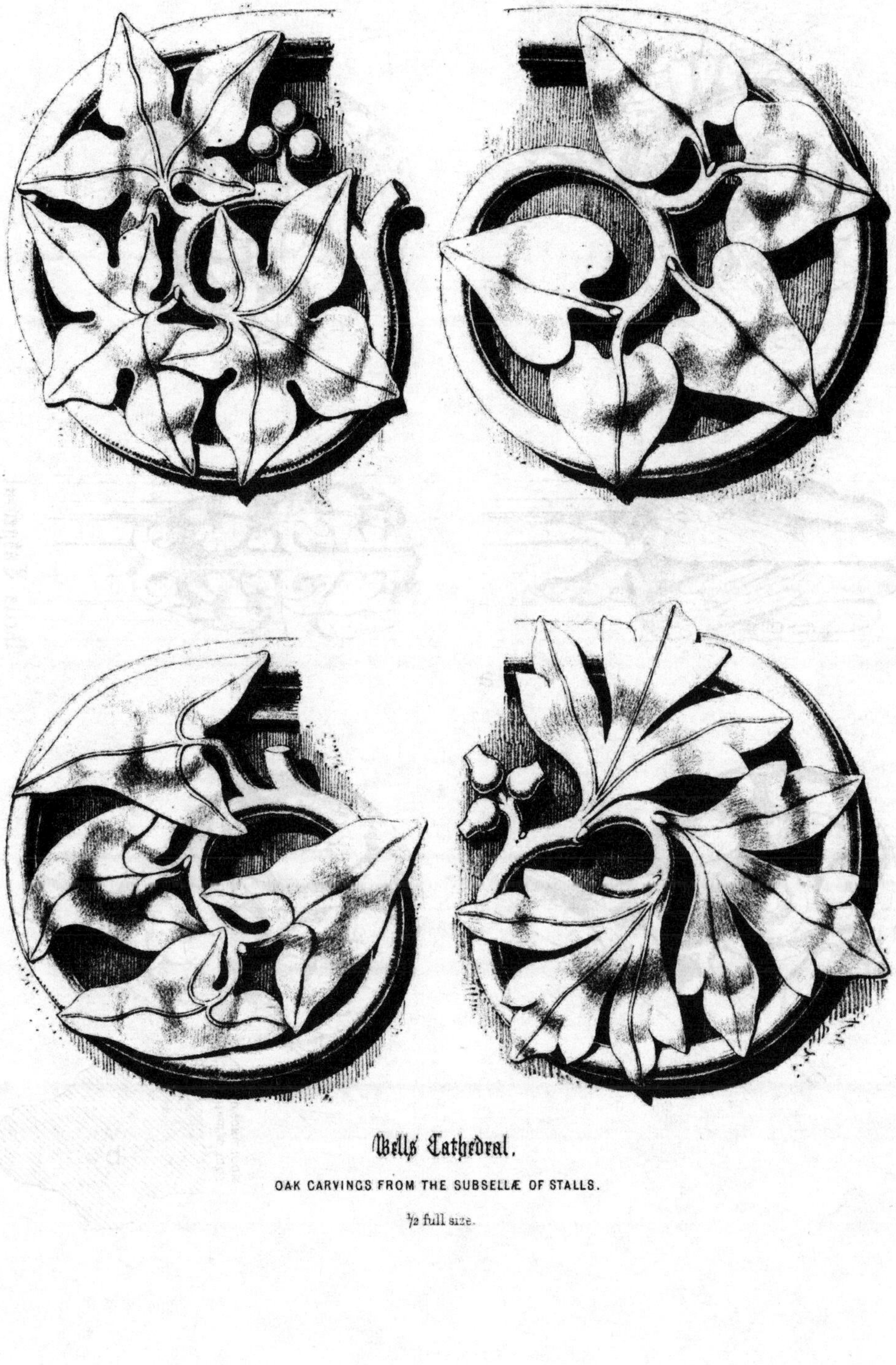

Wells' Cathedral.

OAK CARVINGS FROM THE SUBSELLÆ OF STALLS.

½ full size.

威尔斯教堂座椅凸出托板上橡木雕塑装饰

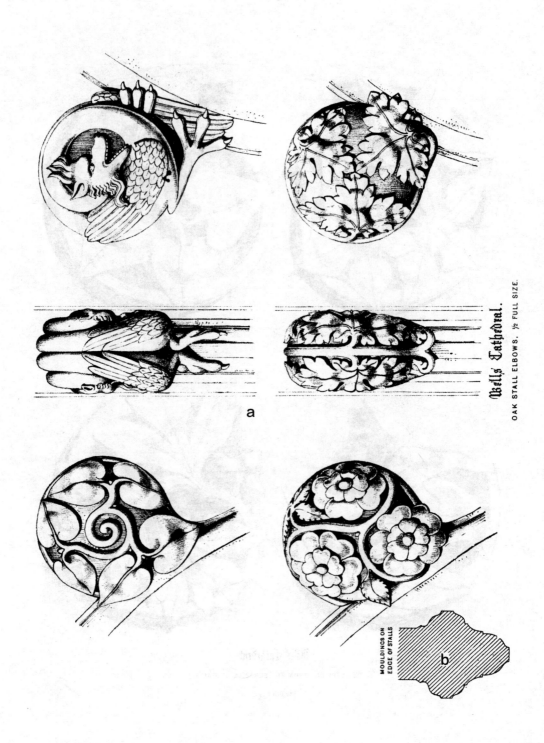

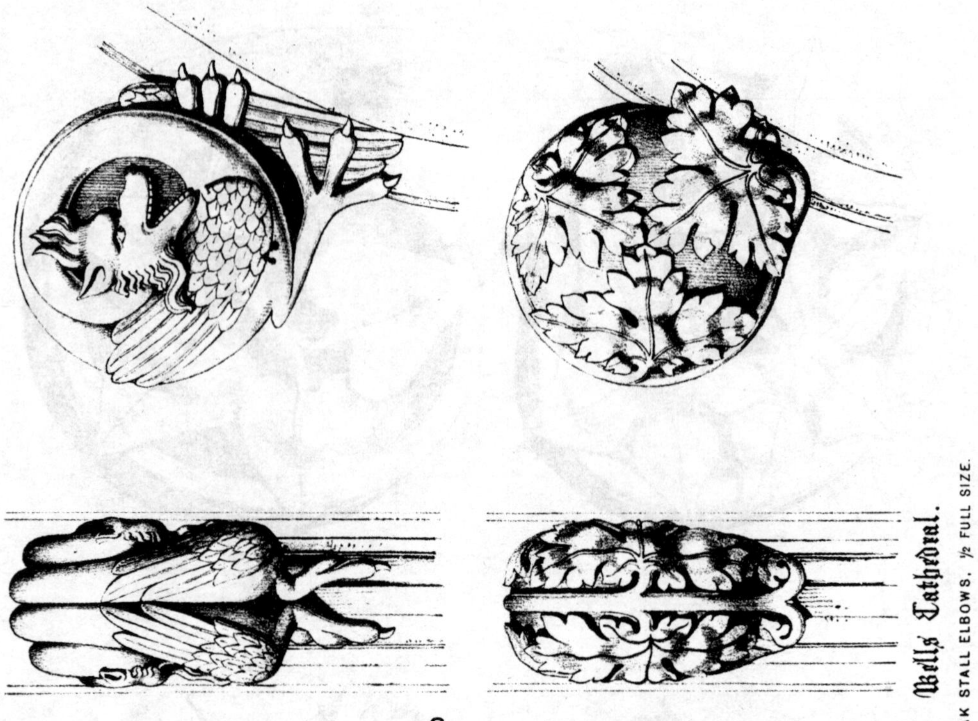

威尔斯教堂：a. 橡木座椅扶手　b. 座椅边缘线脚

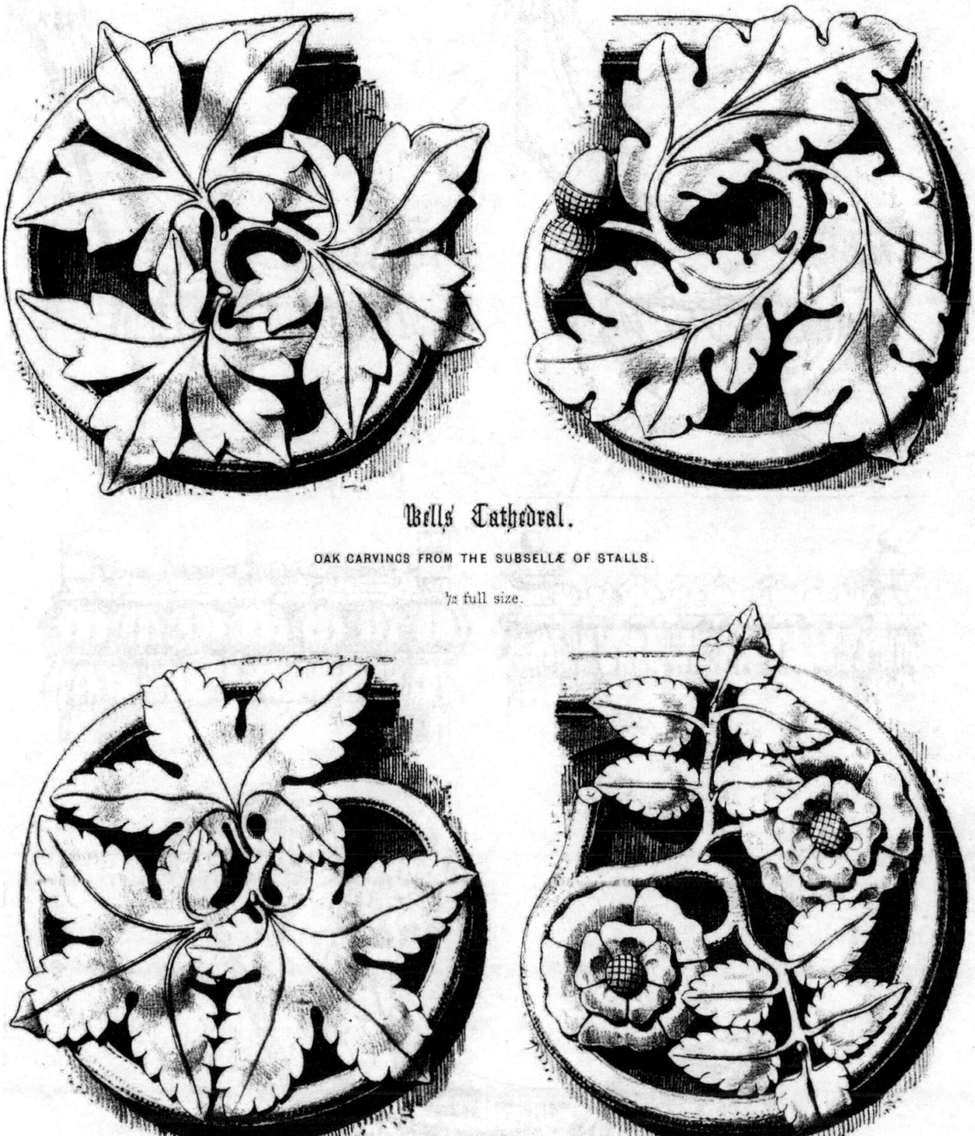

威尔斯教堂座椅凸出托板上的橡木雕装饰

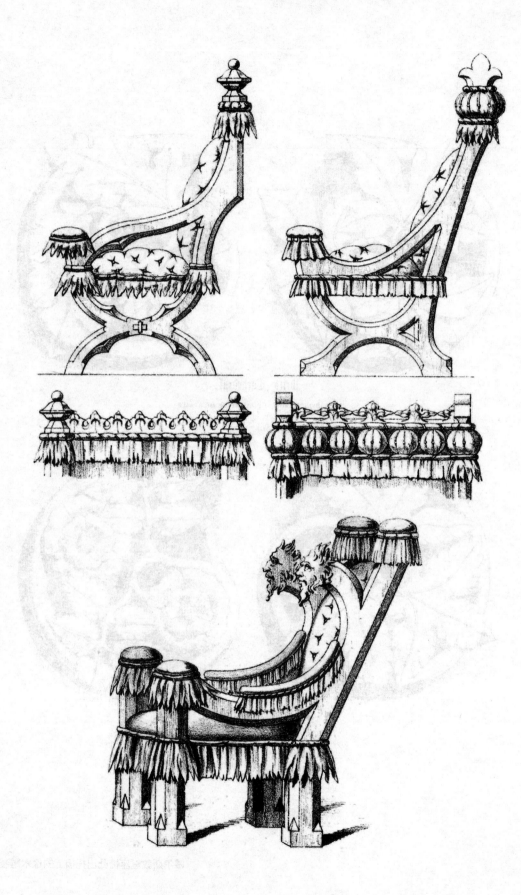

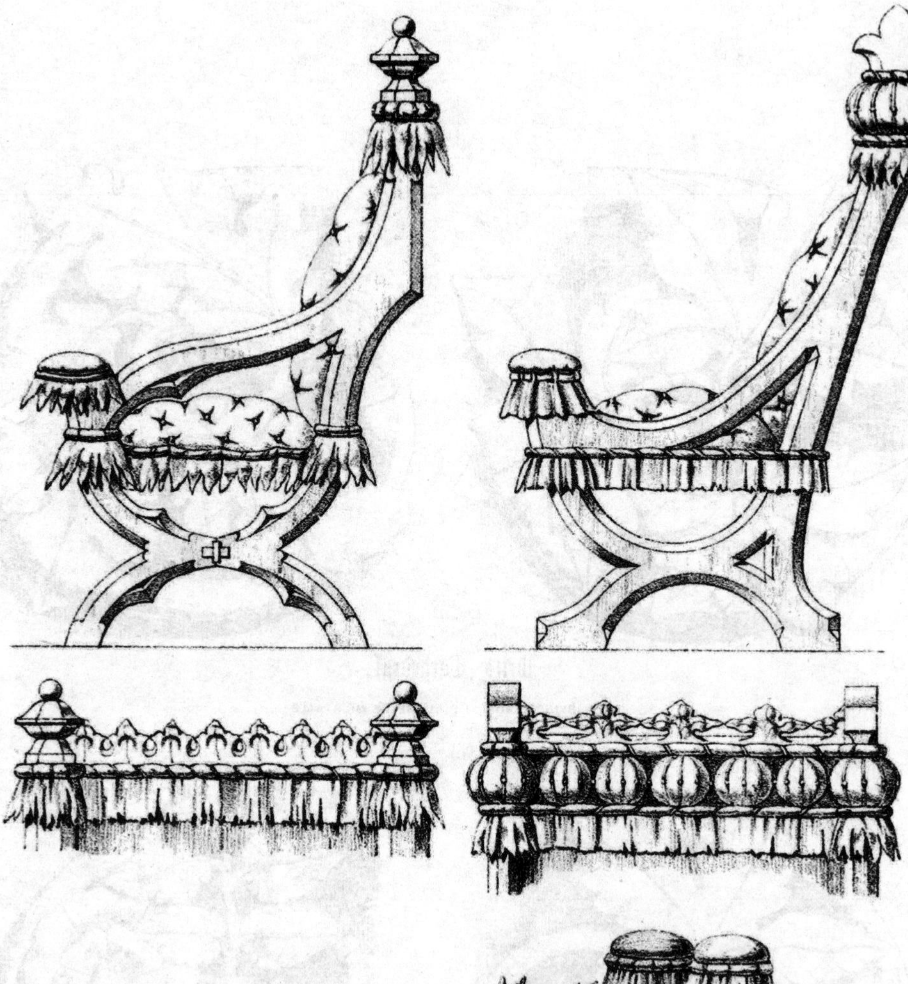

座椅

第16章

隔屏及嵌板

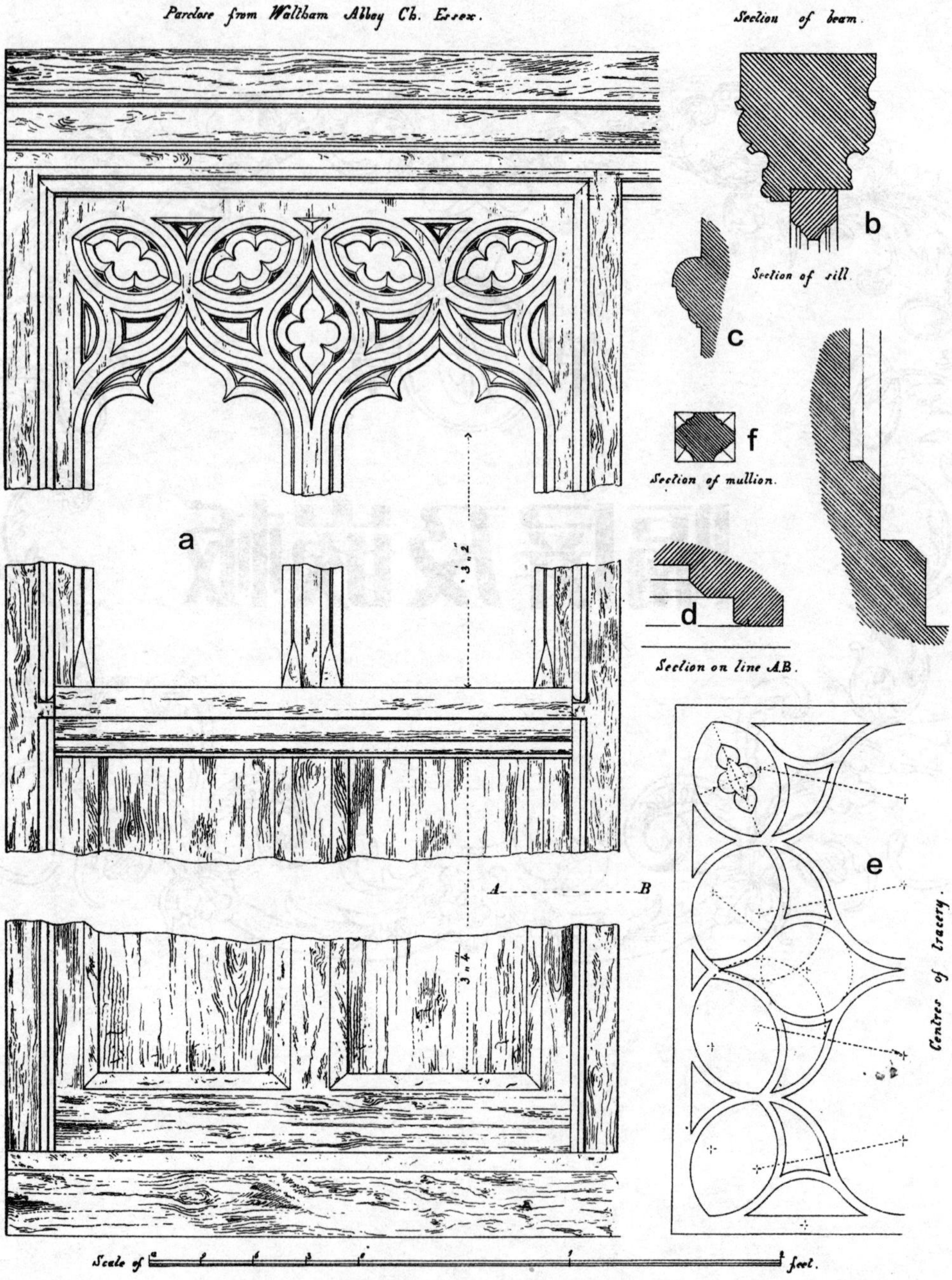

a. 埃塞克斯郡沃尔瑟姆修道院教堂隔屏　b. 横梁剖面　c. 底梁剖面　d. A–B 处的剖面　e. 花格饰中心点　b. 横梁剖面

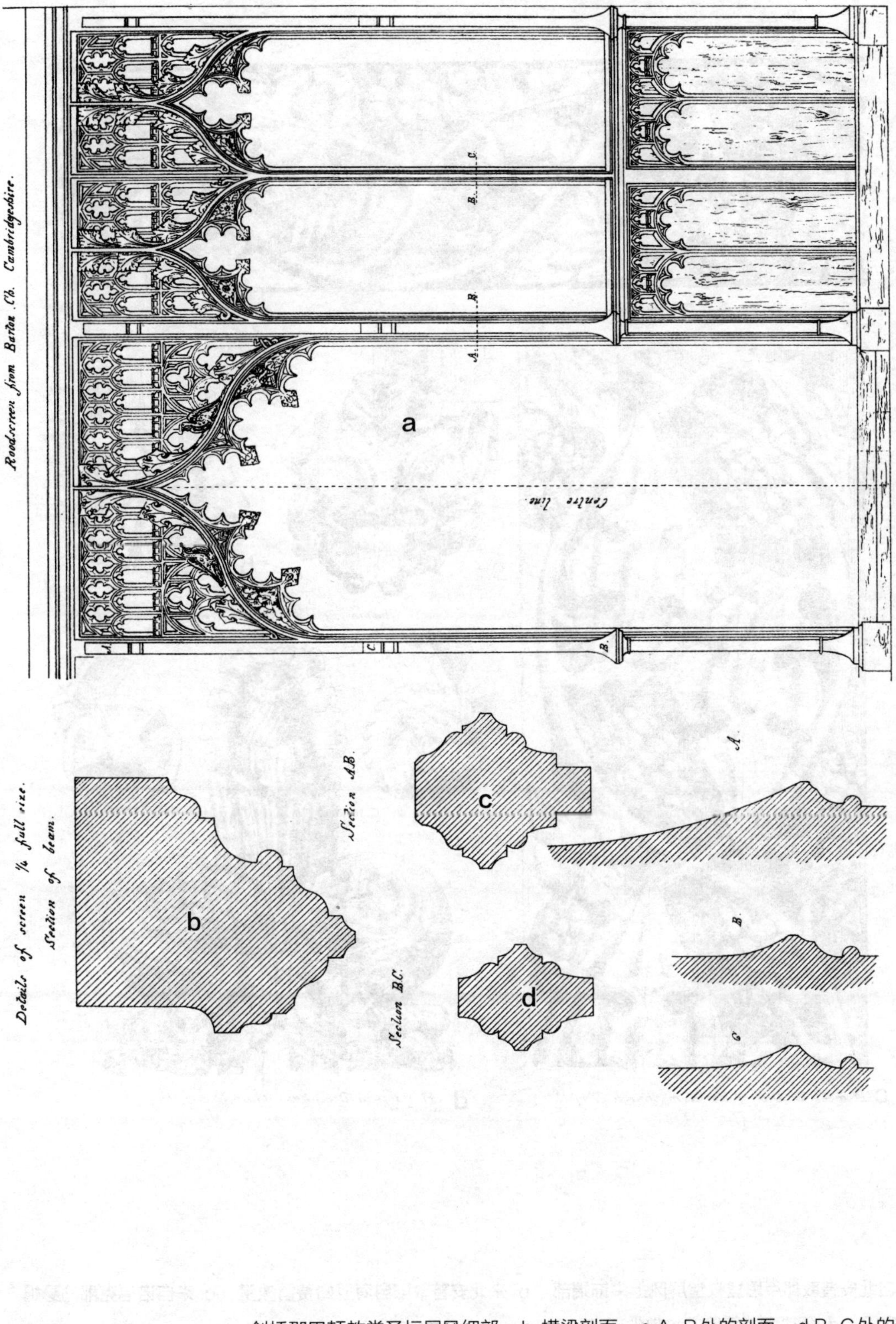

a. 剑桥郡巴顿教堂圣坛屏风细部　b. 横梁剖面　c. A–B 处的剖面　d. B–C 处的剖面

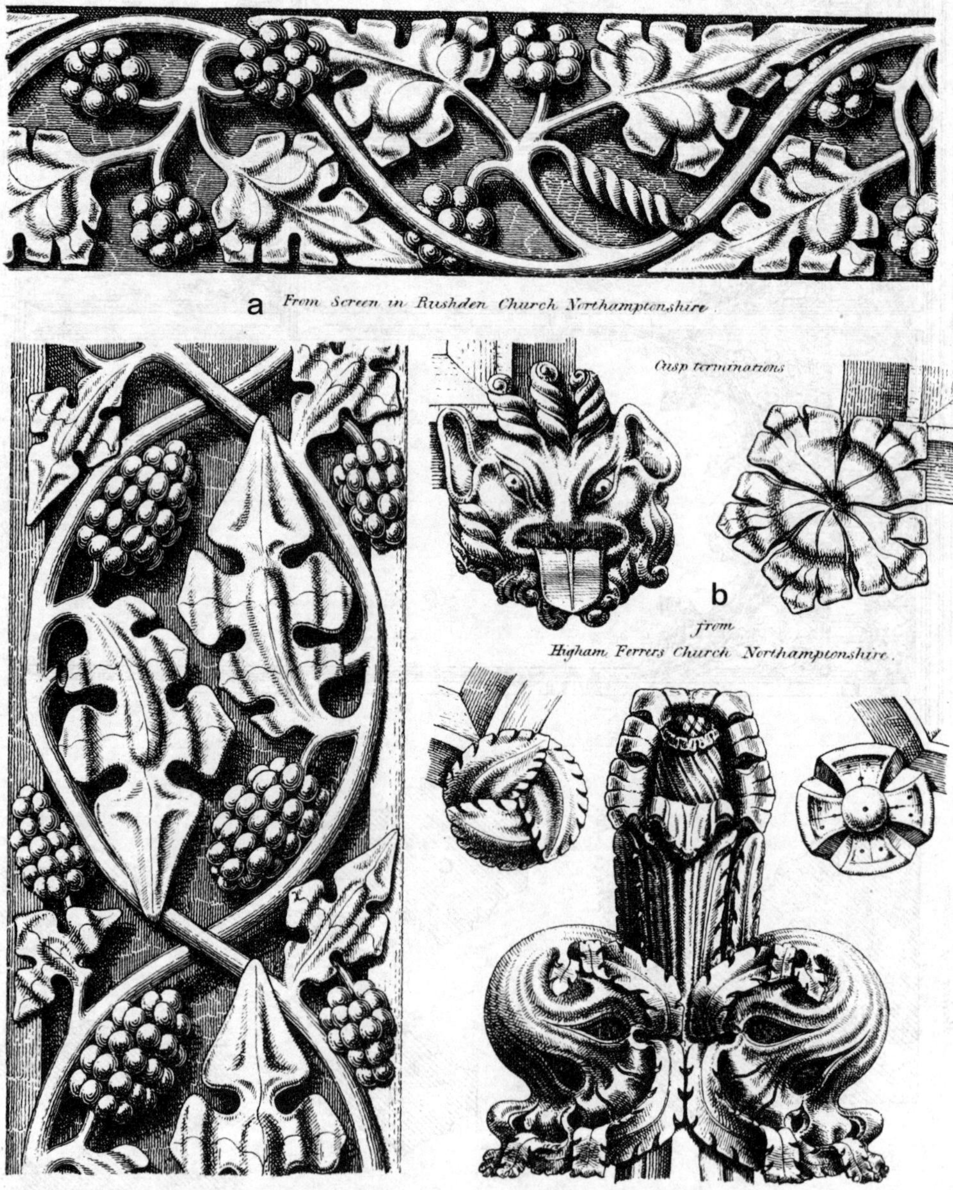

a. 来自北安普敦郡卢思登教堂屏风上尖饰端部　b. 来北安普敦郡自海厄姆费雷教堂　c. 来自诺福克郡马瑟姆教堂南门　d. 萨福克郡德贝汉教堂尖顶饰

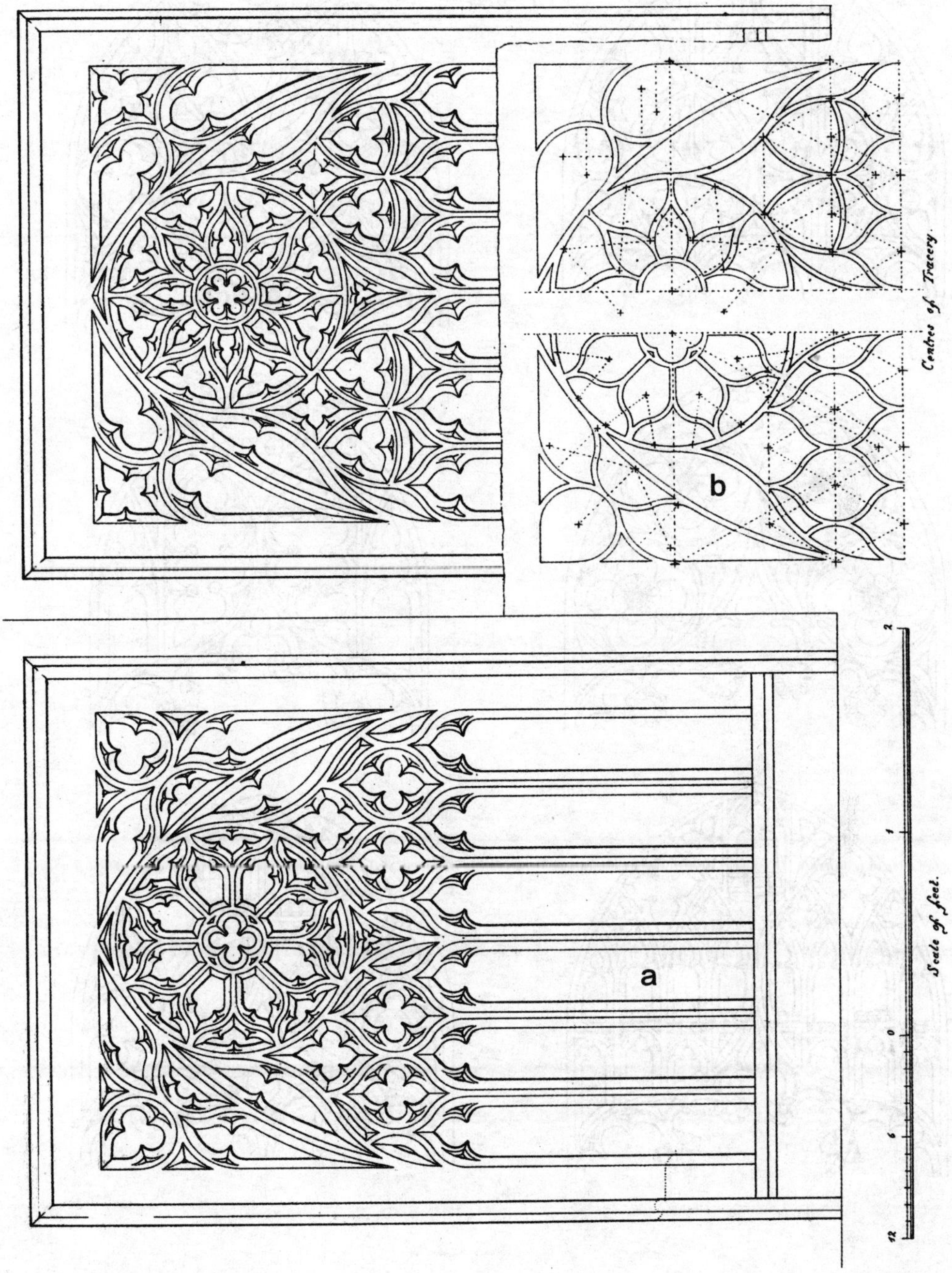

a. 威尔斯教堂门下部嵌板　b. 花格饰中心点

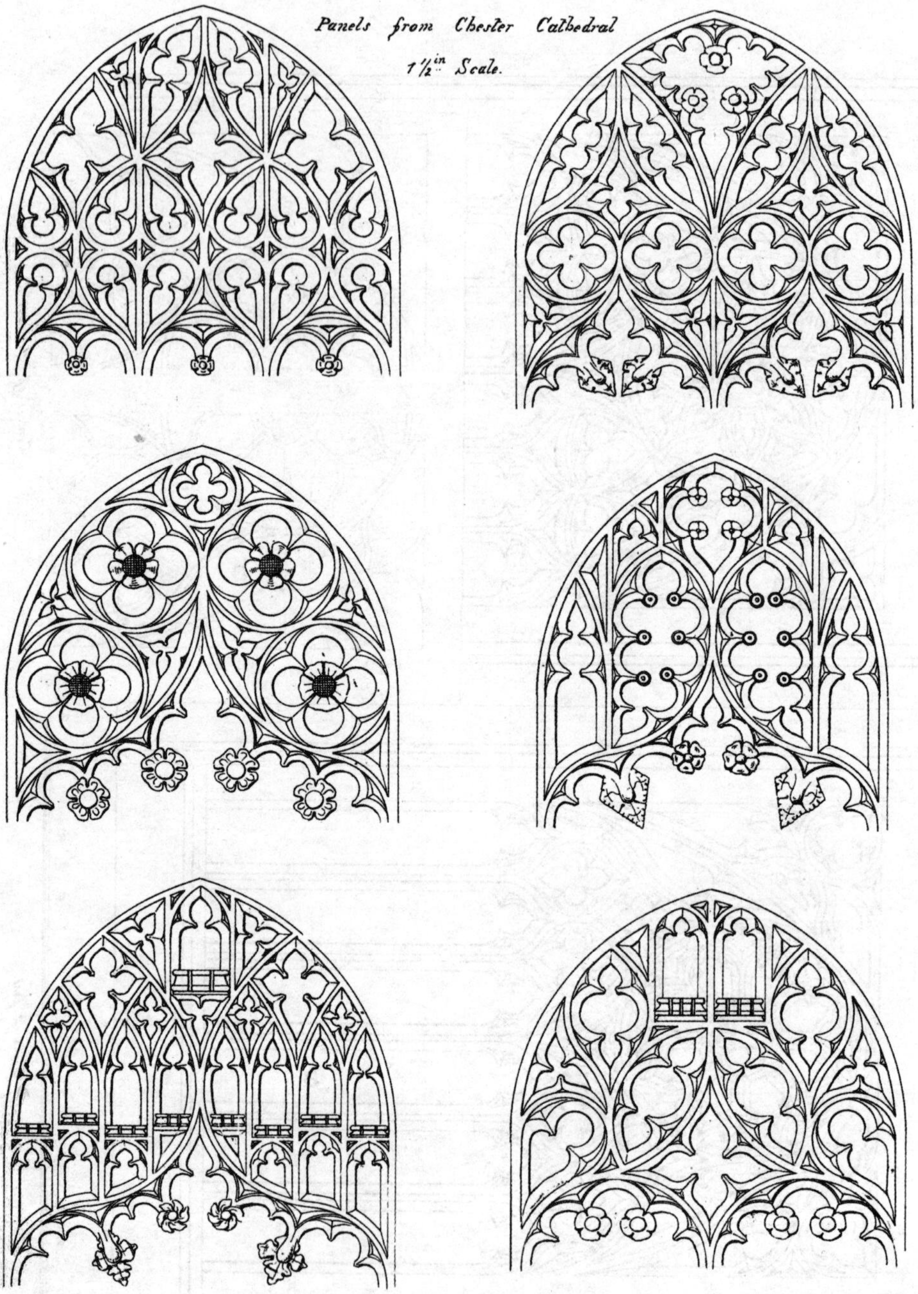
Panels from Chester Cathedral
1½in Scale.

切斯特教堂嵌板

a

Wells Cathedral.

STONE PANELLING FROM BISHOP BECKINGTONS SHRINE.

1/3 full size.

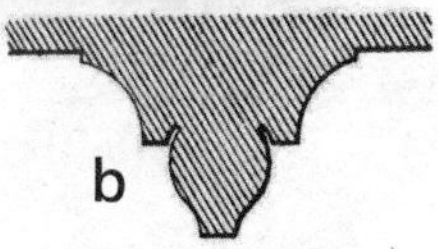

b

SECTION OF MULLION.

威尔斯教堂：a. 贝金顿主教圣堂内石制嵌板　b. 竖框剖面

第17章

拱心石及肋上的凸饰

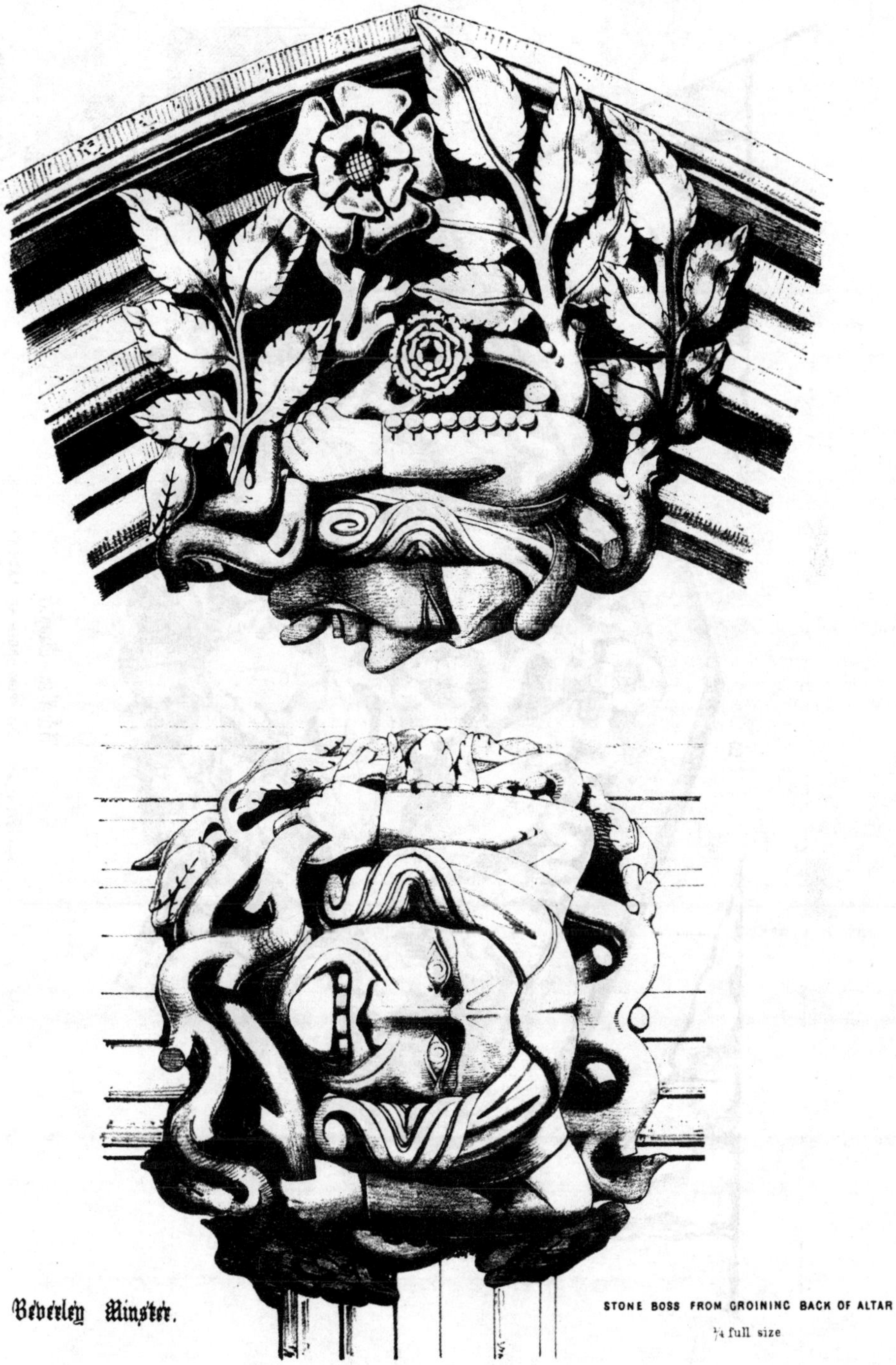

贝弗利修道院圣坛背部拱肋上石制凸饰

贝弗利修道院：a. 圣坛背部拱肋上拱心石及凸饰　b. 拱线脚剖面

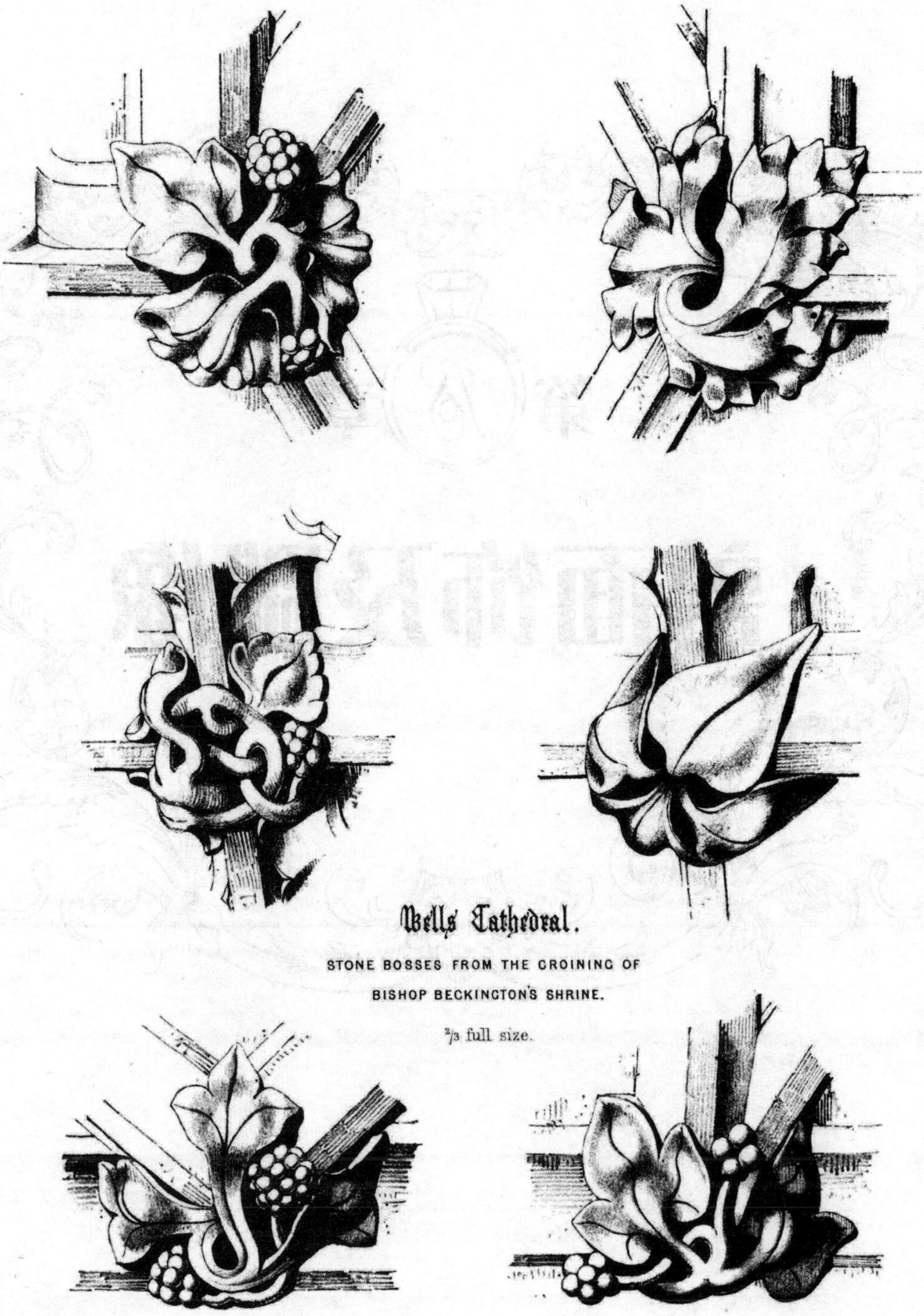

威尔斯教堂贝金顿主教圣堂石制穹顶上凸饰

第18章

兽面饰及雕像

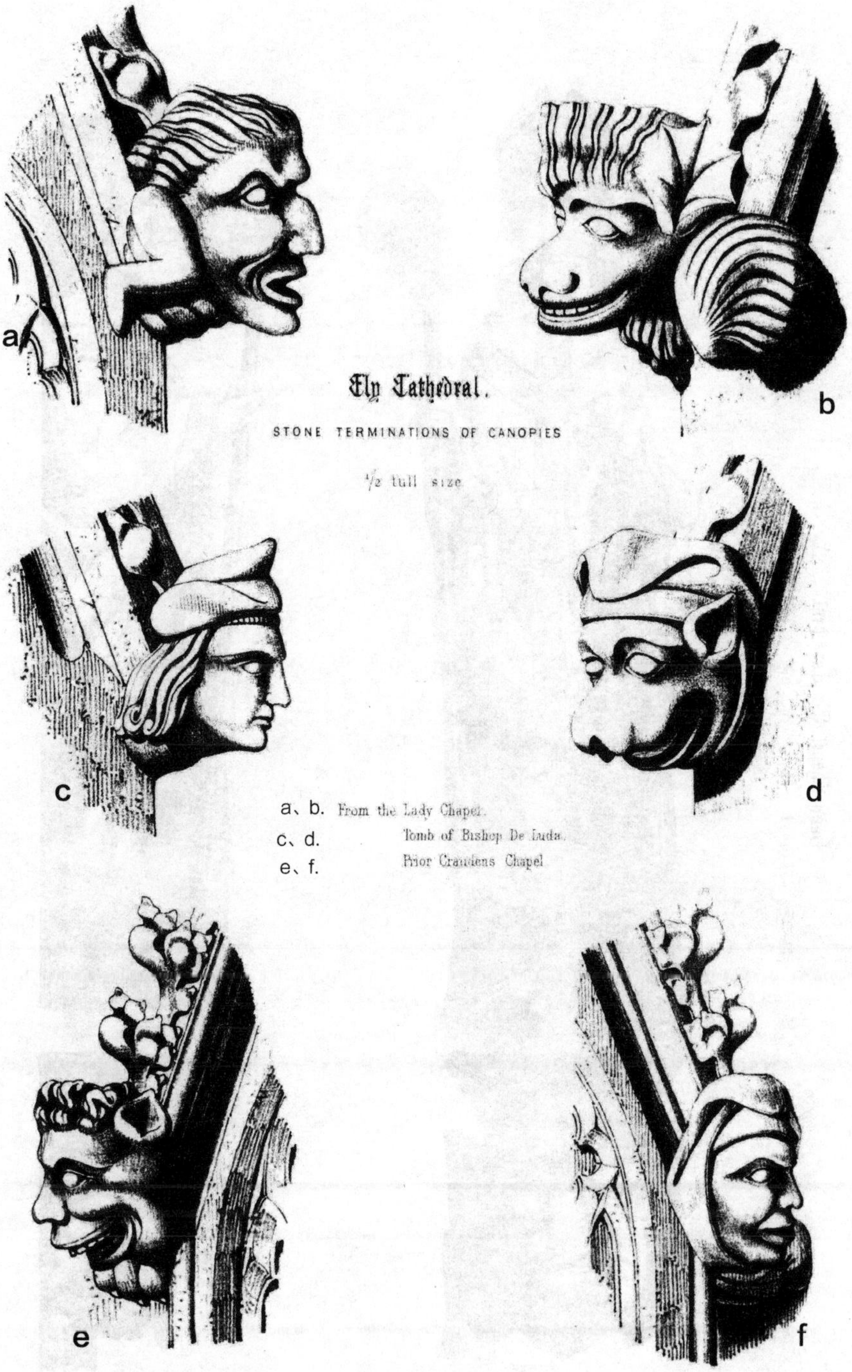

以利教堂顶棚石制端部： a、b. 来自圣母堂 c、d. 来自德鲁达主教墓地 e、f. 来自伊利普赖尔克兰登教堂

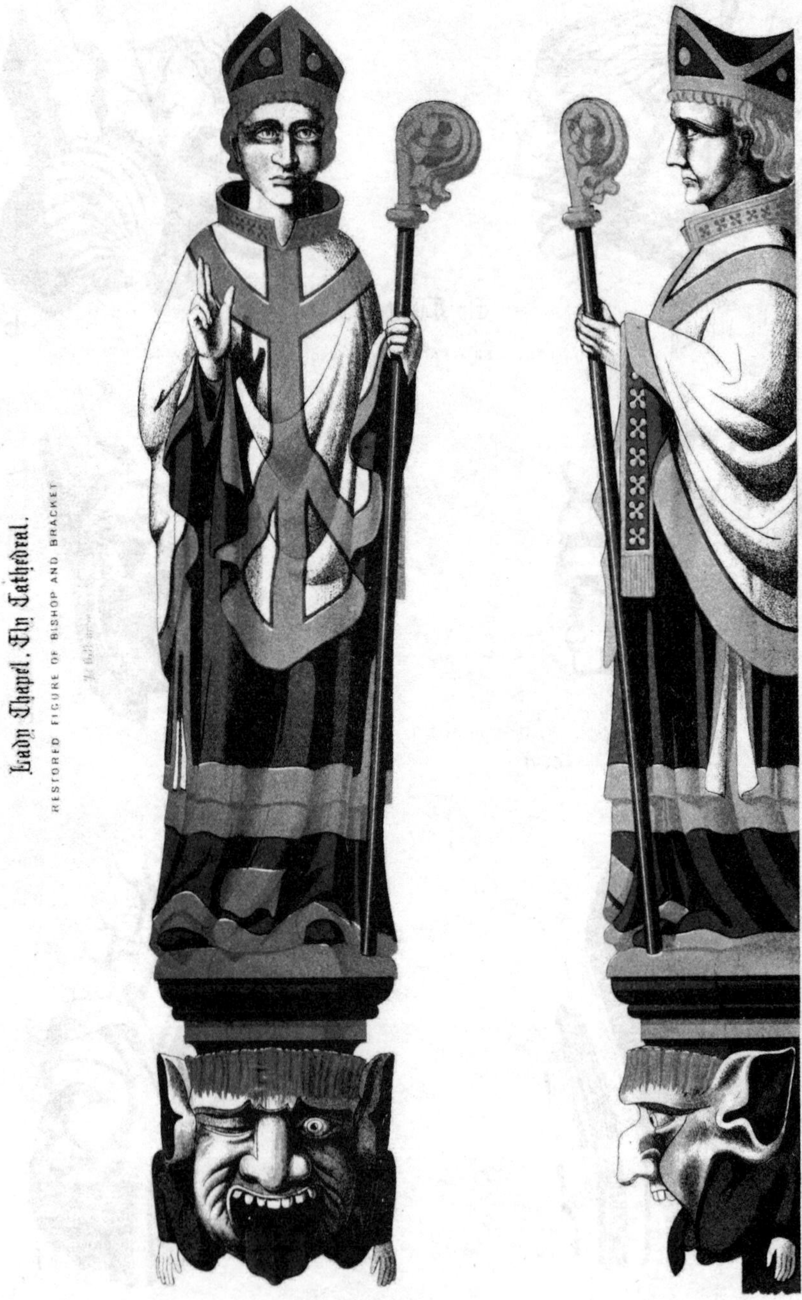

以利教堂圣母堂修复的主教像及托架

贝弗利修道院：a. 珀西神殿内石制雕像　b. 雕像背面

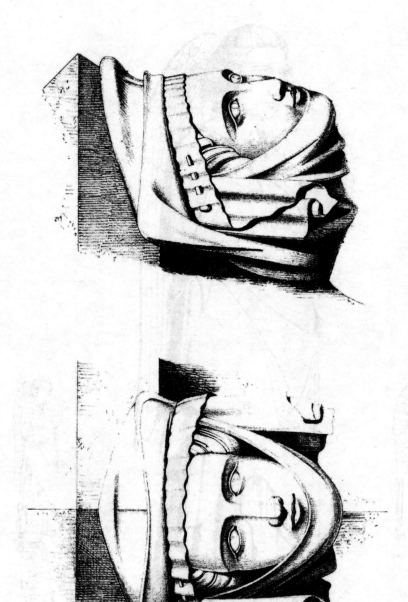

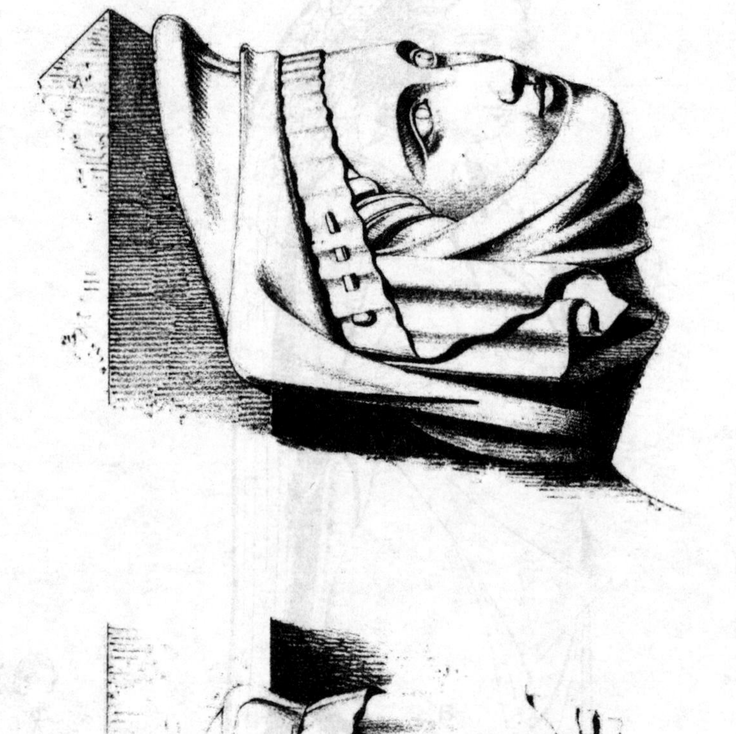

Beverley Minster, Yorkshire.

STONE HEAD FROM THE ANGLE OF BUTTRESS FROM THE PERCY SHRINE.

Full size.

约克郡贝弗利修道院珀西神殿扶壁转角处石制头像

第19章

圣堂穹顶及洗礼盆装饰

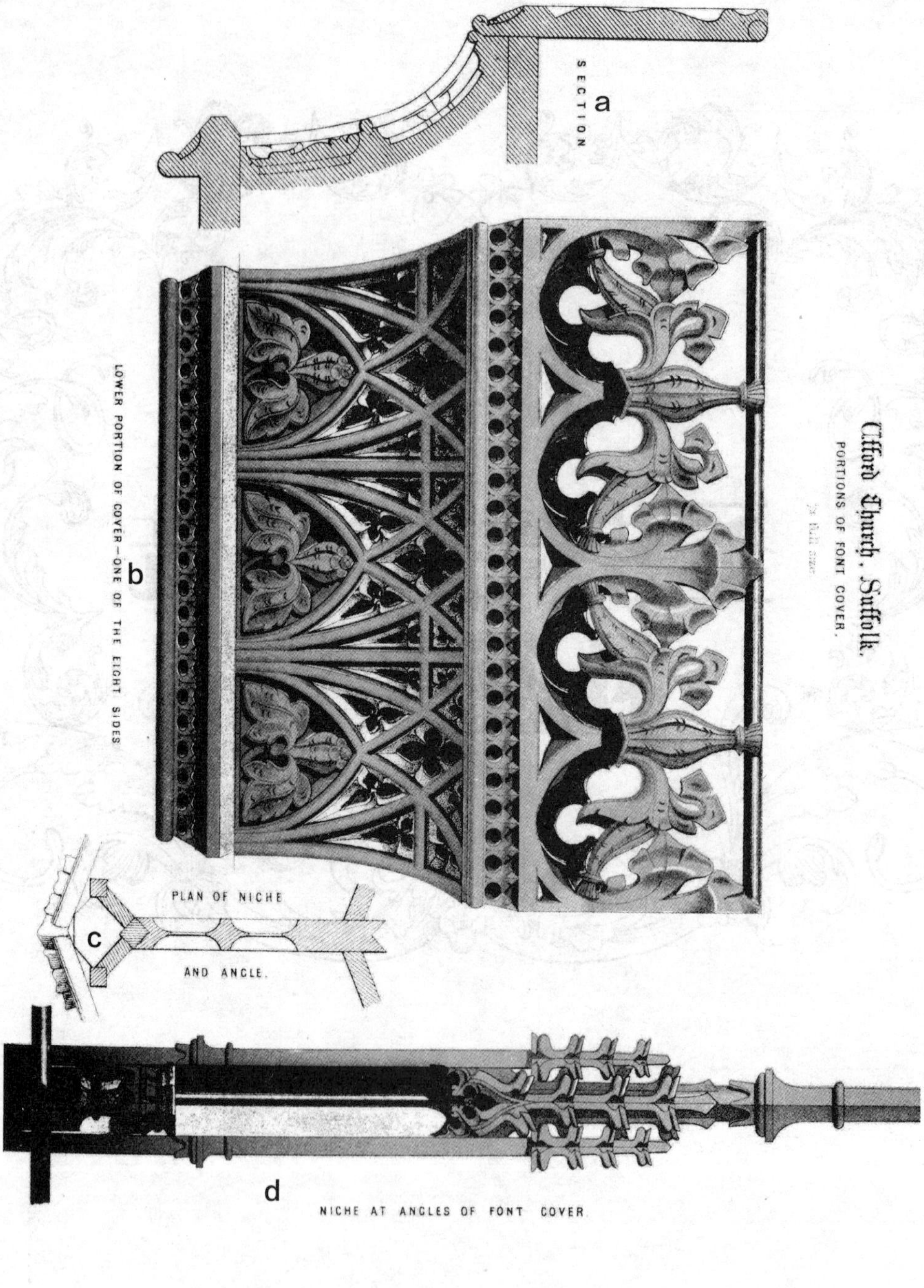

萨福克厄福德教堂洗礼盆顶盖局部：a. 剖面　b. 顶盖低处局部——八个面其中之一　c. 壁龛及转角平面　d. 洗礼盆顶盖转角处壁龛

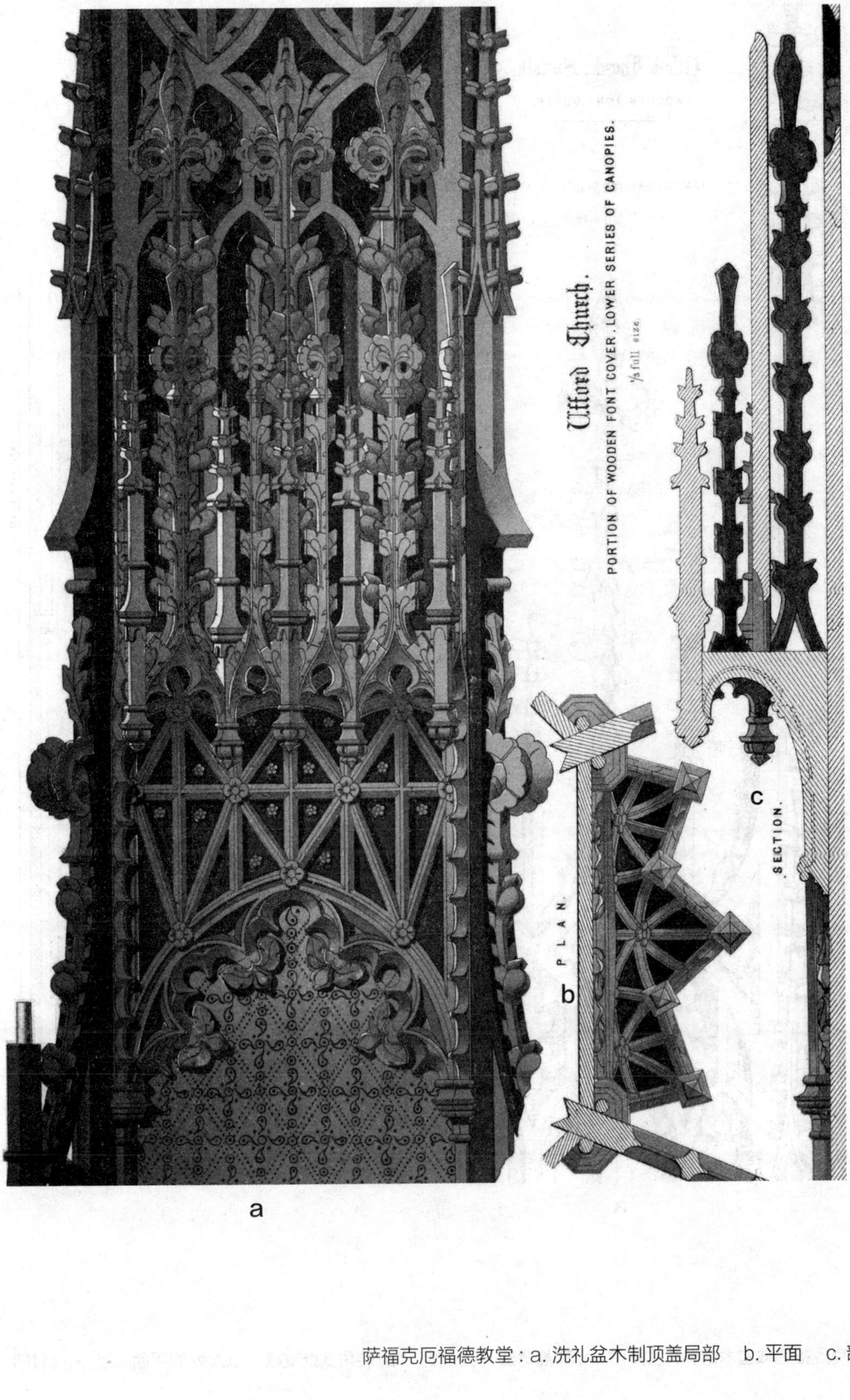

a

萨福克厄福德教堂：a. 洗礼盆木制顶盖局部　b. 平面　c. 剖面

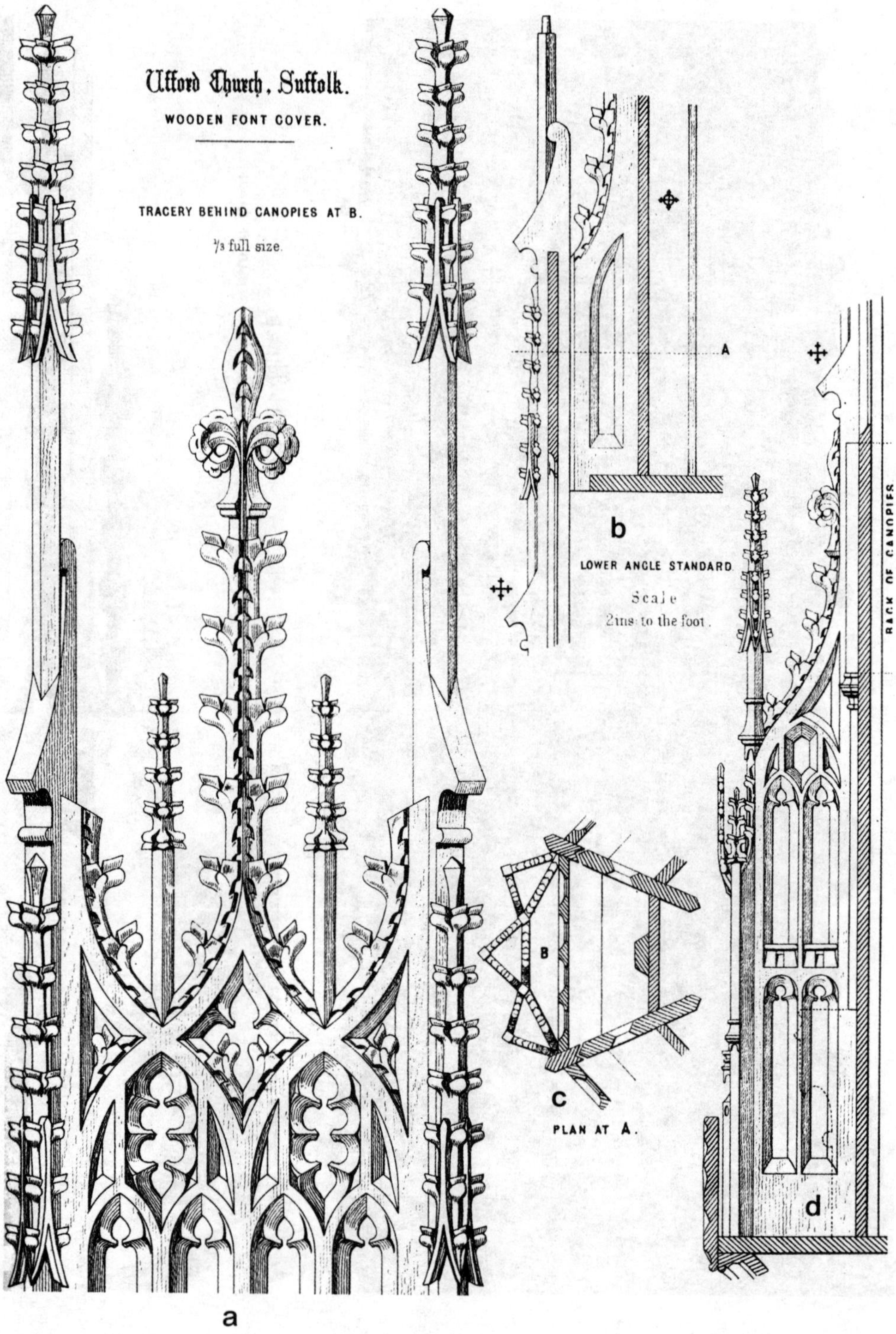

萨福克厄福德教堂木制洗礼盆顶盖：a. 顶盖背部 B 处窗饰　b. 转角低处支柱　c.A 处的平面　d. 顶盖背部

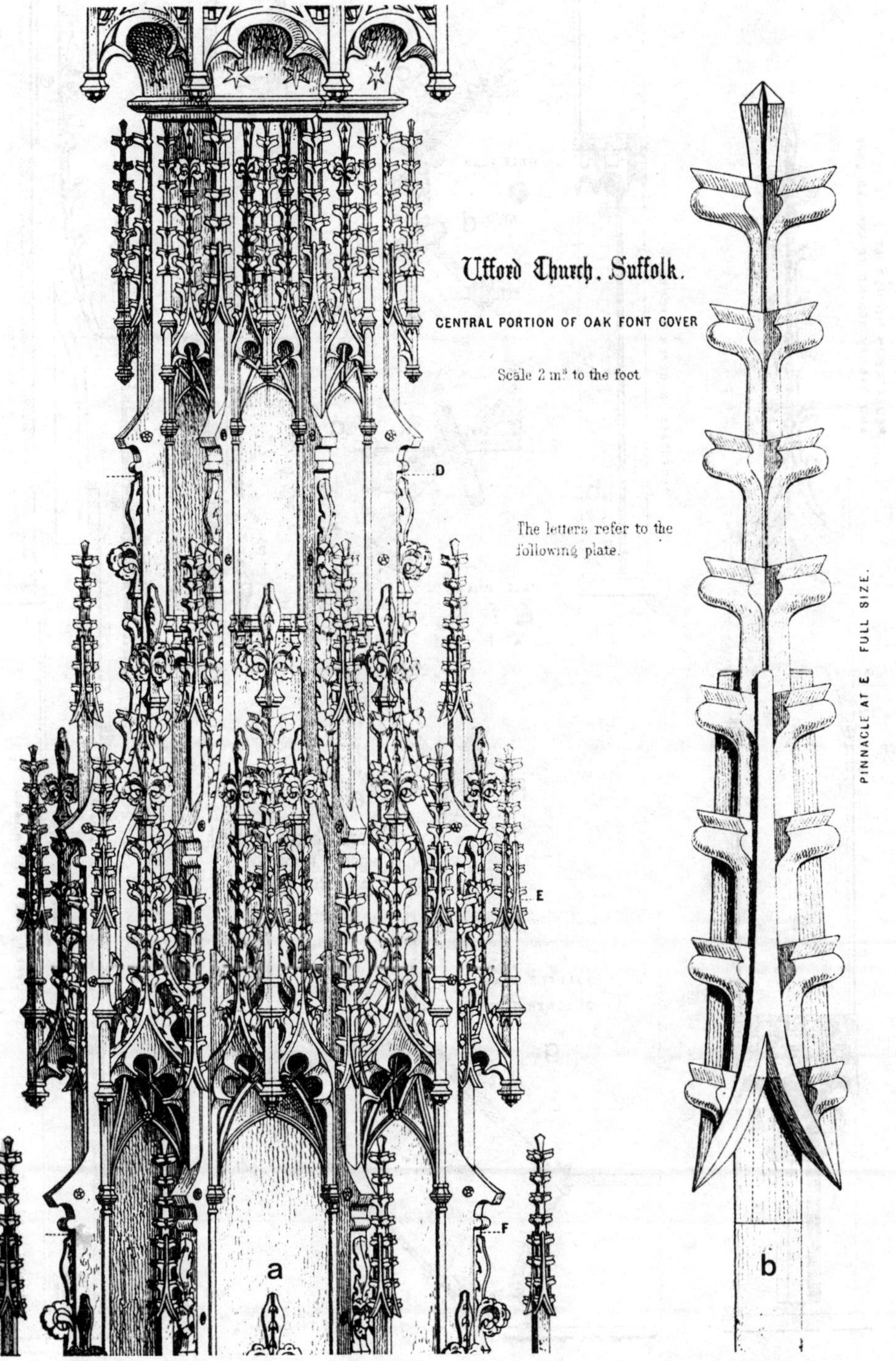

萨福克厄福德教堂：a. 洗礼盆橡木顶盖中间局部　b.E 处的小尖塔

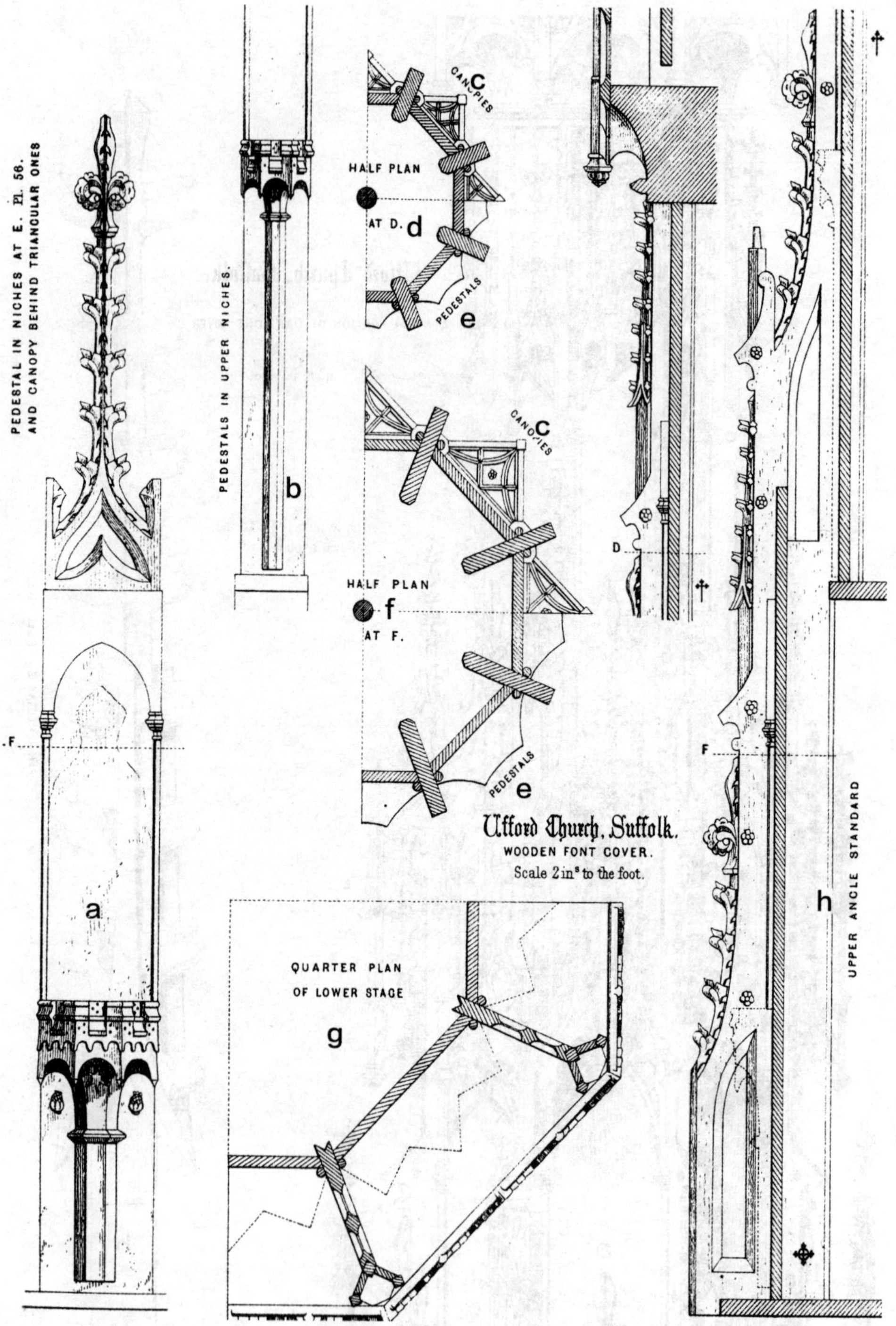

萨福克厄福德教堂木制洗礼盆顶盖：a. 壁龛内基座及在三角形顶盖背后的顶盖　b. 高出壁龛内基座　c. 顶盖　d.D 处的半平面　e. 基座　f.F 处的半平面　g. 低层的四分之一平面　h. 上部转角处支柱

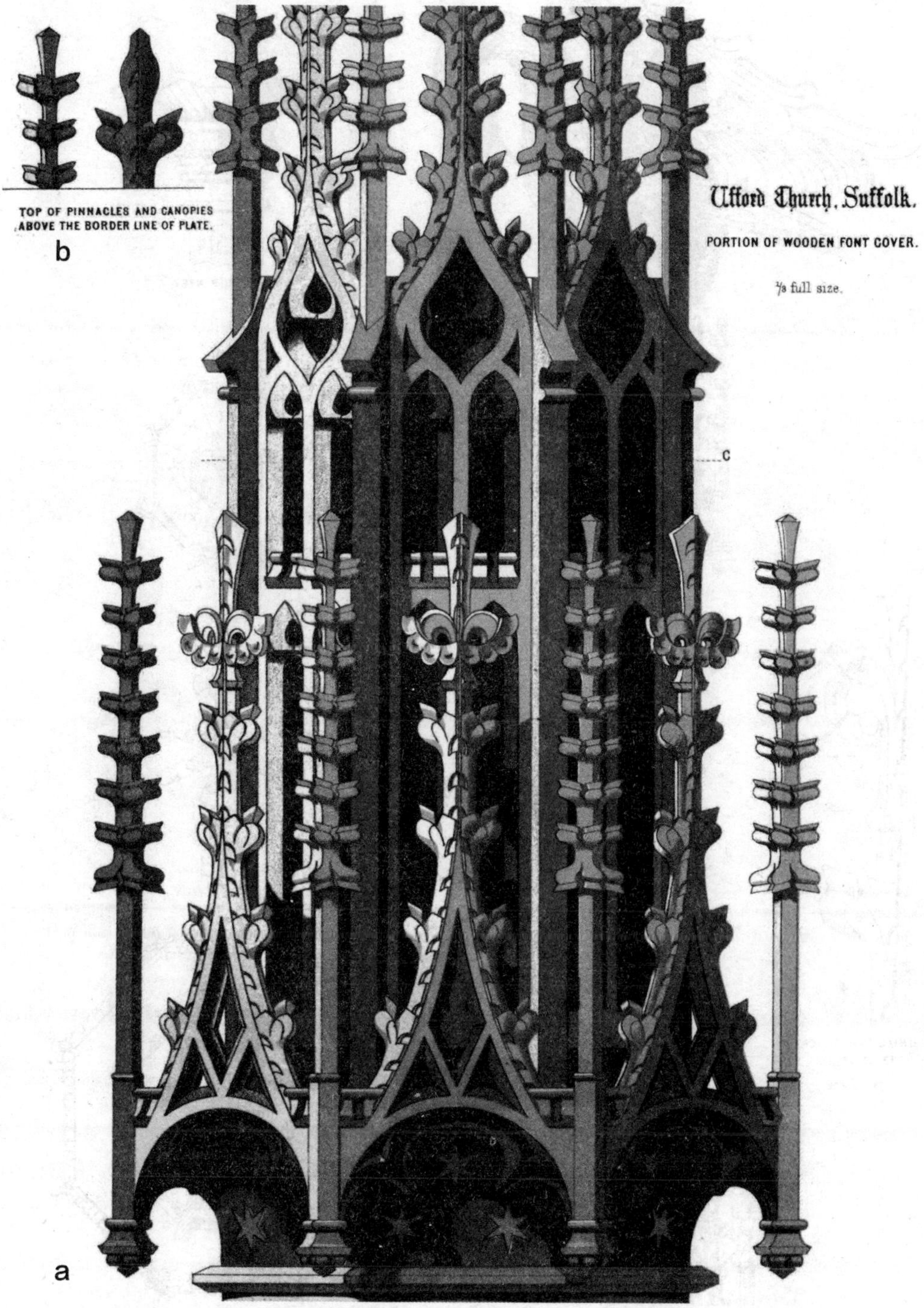

萨福克厄福德教堂：a. 木制洗礼盆顶盖局部　b. 小尖塔顶部及横木边缘线以上的顶盖

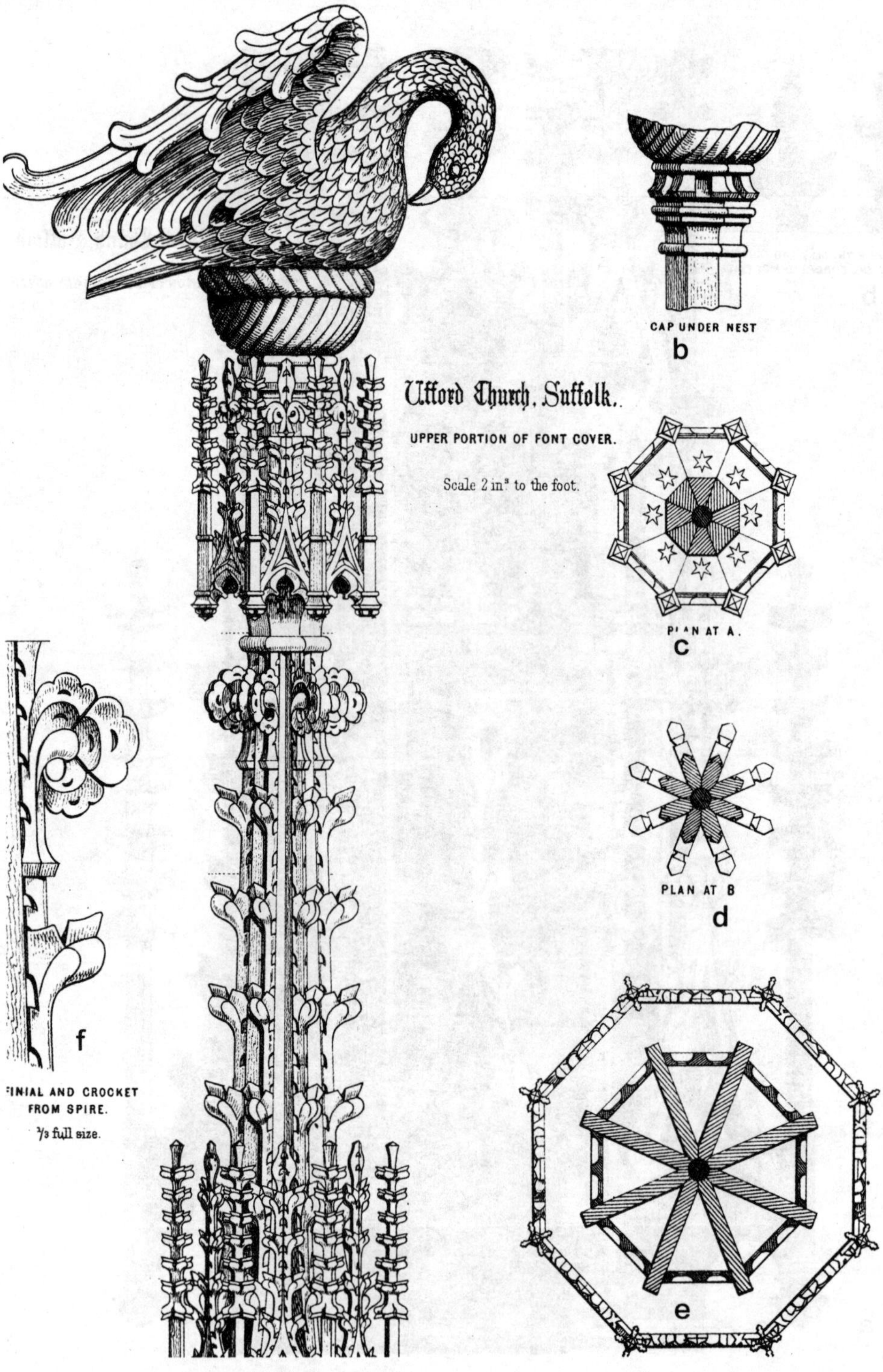

萨福克厄福德教堂：a. 洗礼盆顶盖高处局部　b. 巢下面柱头　c.A处的平面　d.B处的平面　e.C处的平面　f. 尖塔上尖顶饰及卷叶形花饰

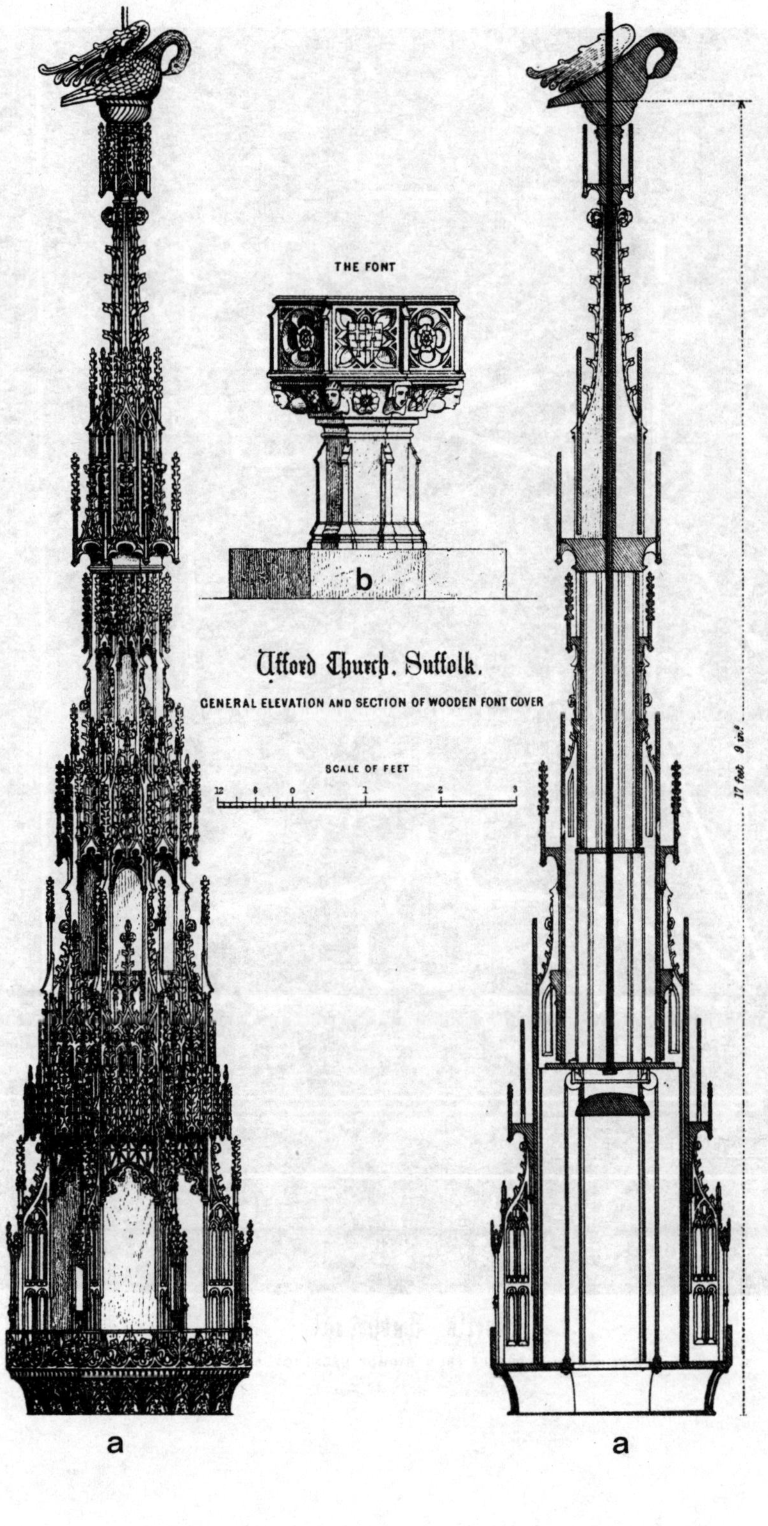

萨福克厄福德教堂：a. 木制洗礼盆顶盖总体立面及剖面　b. 洗礼盆

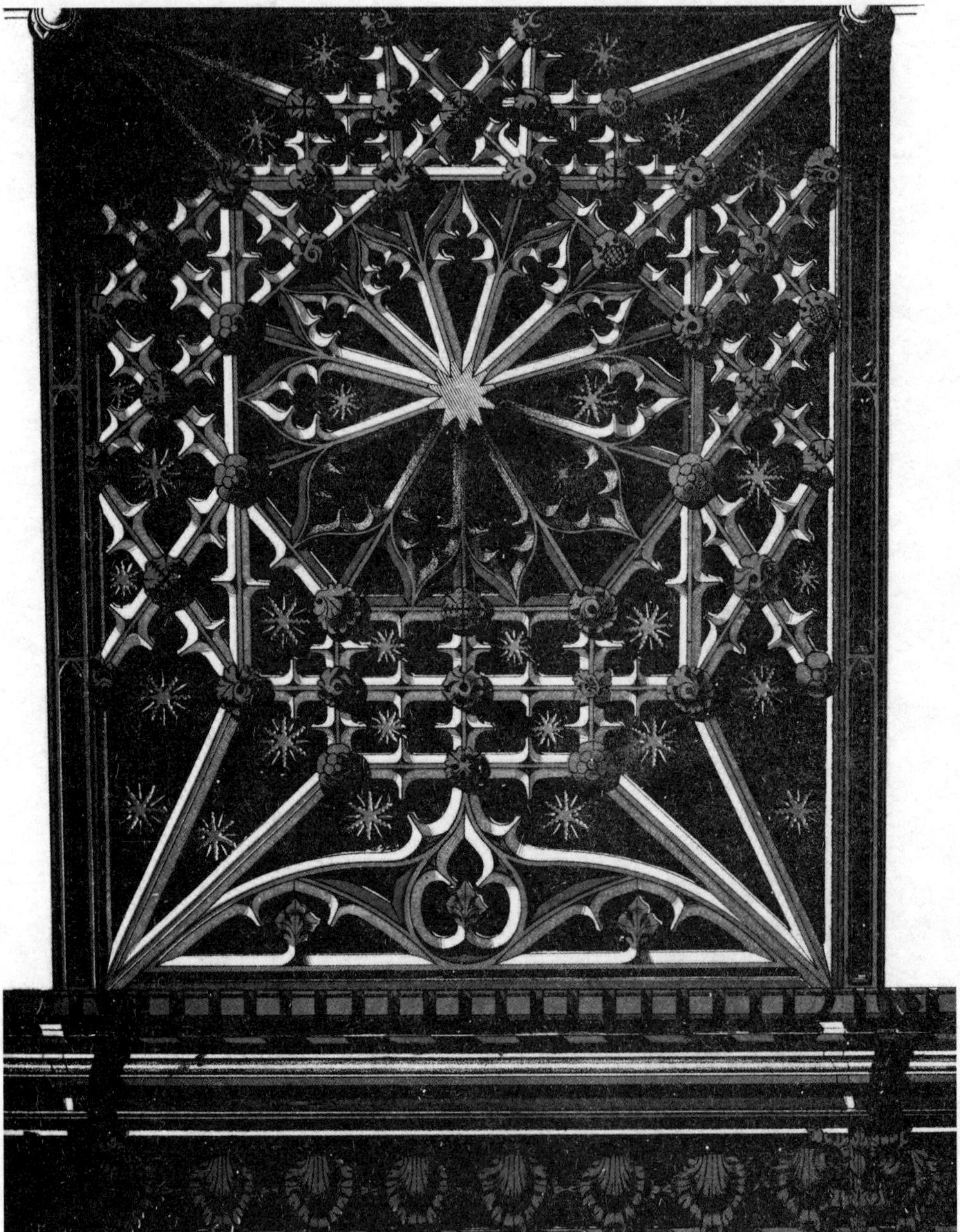

Wells Cathedral.

CROINING AND CORNICE FROM BISHOP BECKINGTON'S SHRINE.

Scale 2 in.s to the foot.

威尔斯教堂贝金顿主教圣堂内穹顶及檐口

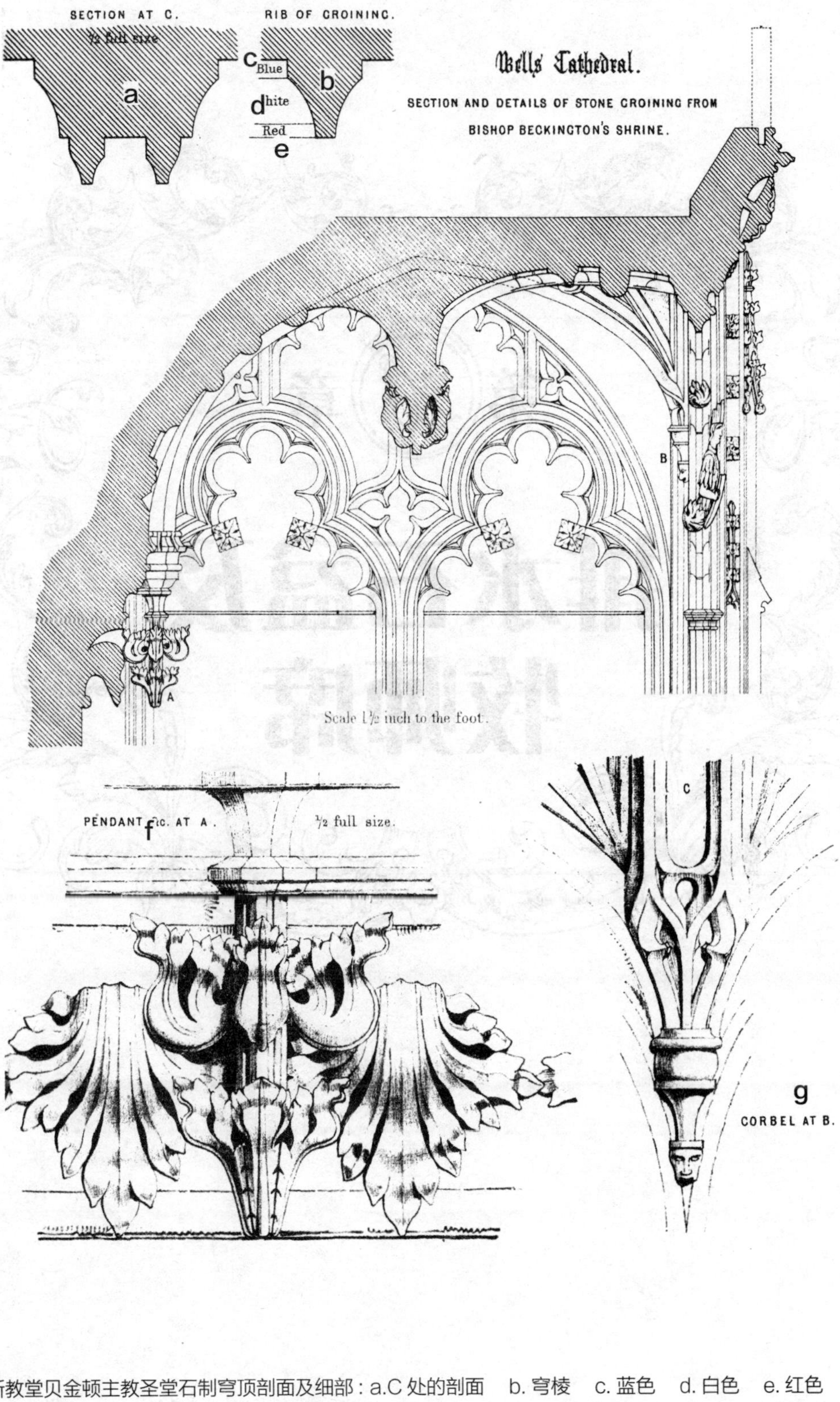

威尔斯教堂贝金顿主教圣堂石制穹顶剖面及细部：a.C 处的剖面　b. 穹棱　c. 蓝色　d. 白色　e. 红色　f.A 处的垂饰　g.B 处的托架

第20章

排水石盆及牧师席

第 20 章 排水石盆及牧师席

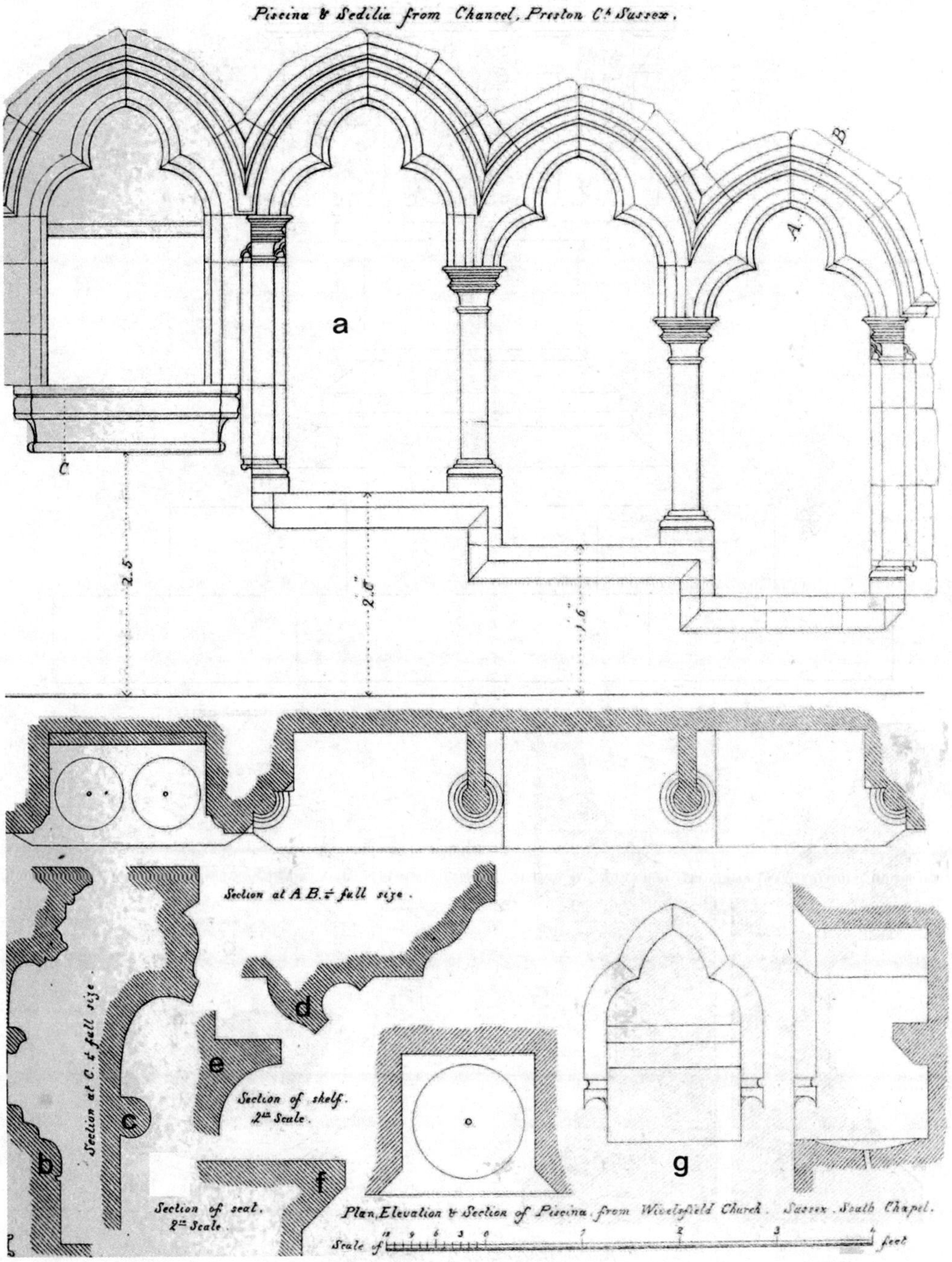

a. 萨塞克斯郡普勒斯顿教堂圣坛内排水石盆及牧师席　b. 牧师席上部柱头及柱础线脚　c.C 处的剖面　d.A–B 处的剖面　e. 搁板剖面　f. 座位剖面　g. 萨塞克斯郡威伍斯菲尔德教堂南侧排水石盆平面、立面及剖面

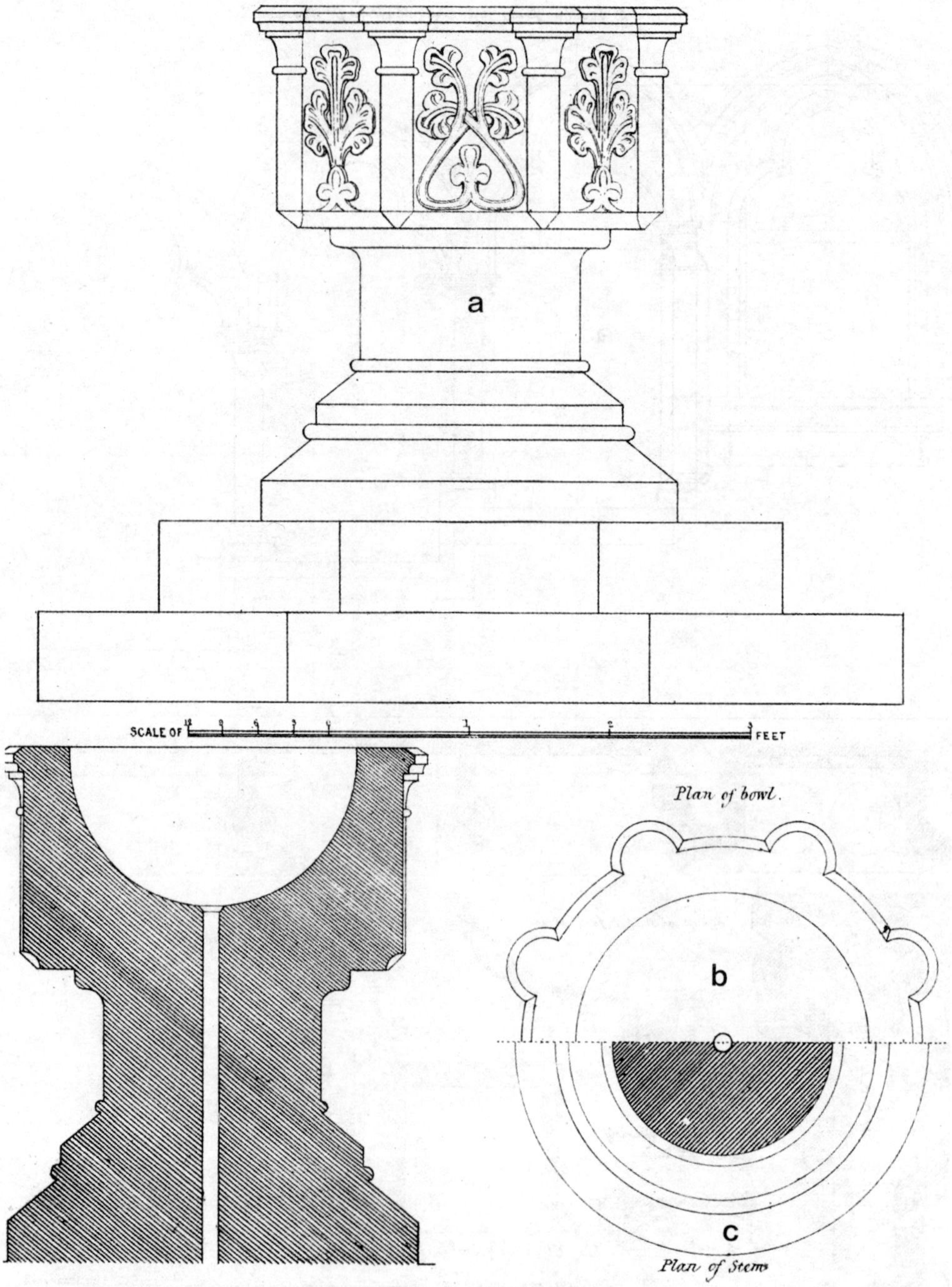

a. 林肯郡韦斯顿教堂洗礼盆　b. 盆体平面　c. 支柱剖面

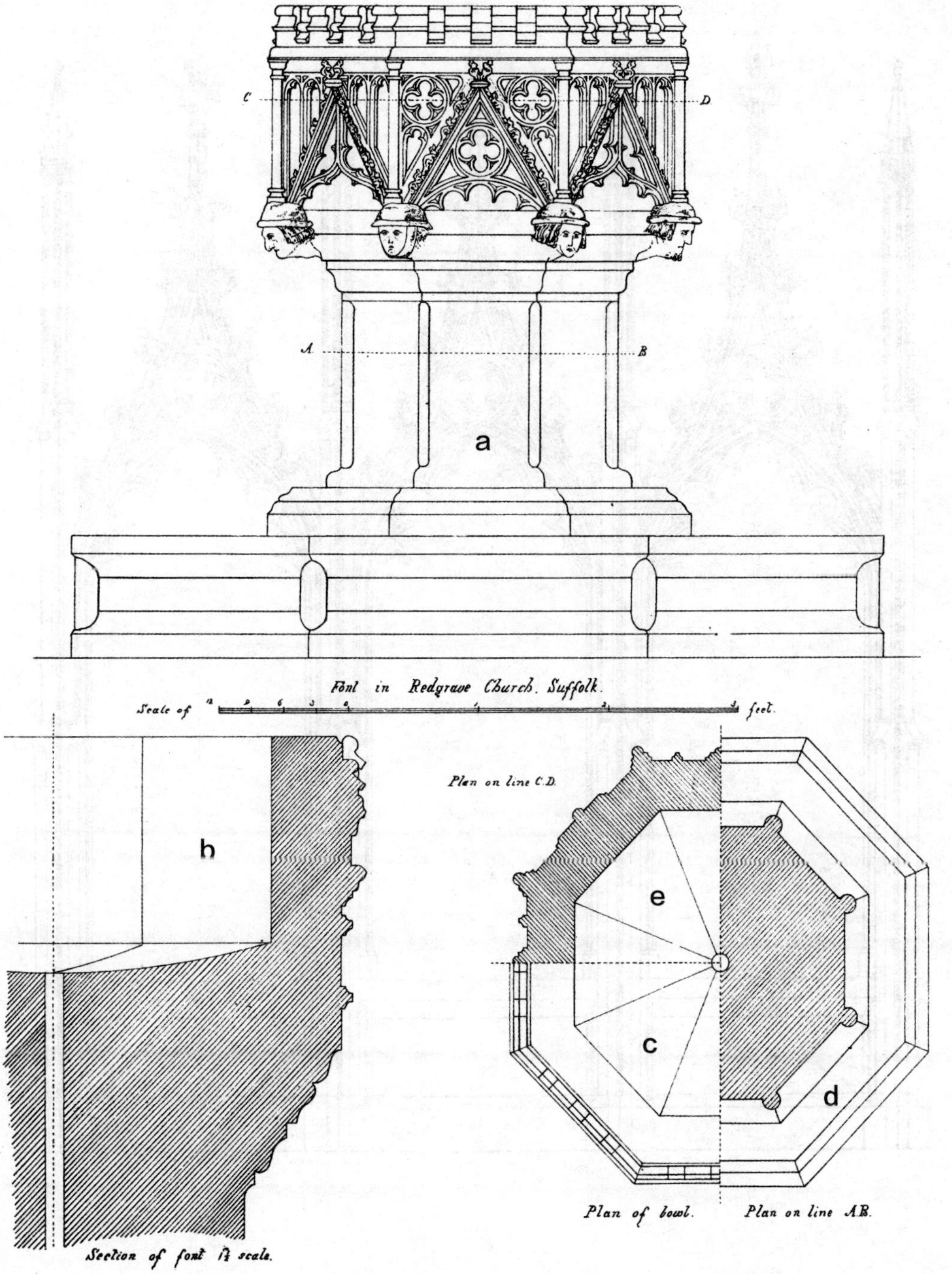

a. 萨福克郡雷德格雷夫教堂洗礼盆　b. 洗礼盆剖面　c. 盆体平面　d. A–B 处的平面　e. C–D 处的平面

肯特郡科巴姆教堂圣坛内牧师席

第 20 章 排水石盆及牧师席

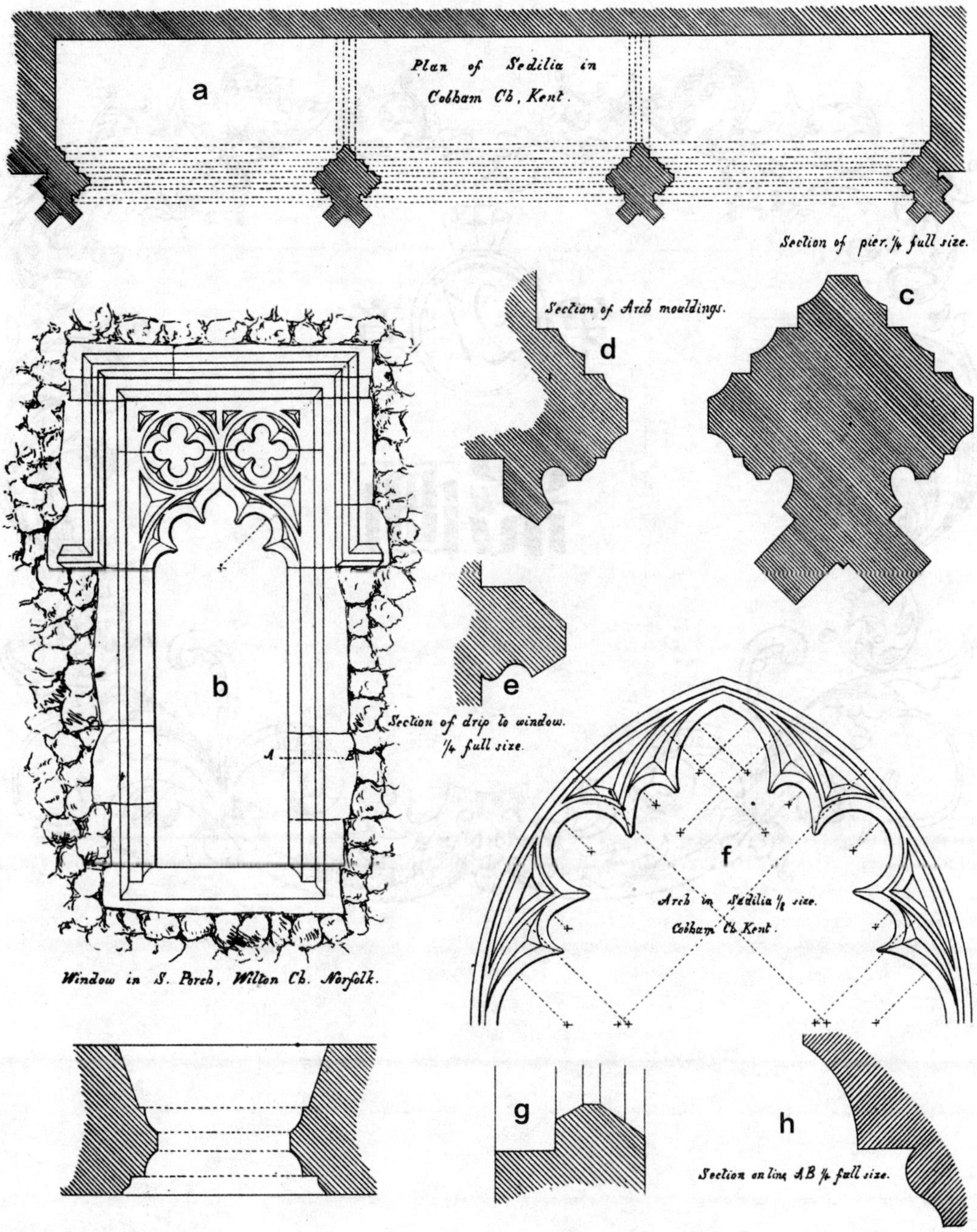

a. 肯特郡科巴姆教堂牧师席平面　b. 诺福克郡威尔顿教堂南门廊窗户　c. 柱子剖面　d. 拱线脚剖面　e. 窗户上披水石剖面　f. 肯特郡科巴姆教堂牧师席上拱　g. 窗台剖面　h.A-B 处的剖面

第21章

彩画

St Mary's Church. Guildford. Surrey.

VIEW OF APSE, EAST END OF NORTH AISLE, SHEWING THE PAINTING ON GROIN &c.

萨里郡吉尔福德圣玛丽教堂北侧廊东尽端半圆形后殿内拱顶上的彩画

萨里郡吉尔福德圣玛丽教堂北侧廊东尽端半圆形后殿主拱上彩画：a. 拱腹　b. 正面　c. 拱剖面

萨里郡吉尔福德圣玛丽教堂：a. 北侧廊东尽端半圆形后殿穹顶上彩画　b. 穹棱剖面　c. 画的粗糙的图案

萨里郡吉尔福德圣玛丽教堂：a. 北侧廊东尽端半圆形后殿穹顶上彩画　b. 画的粗糙的图案

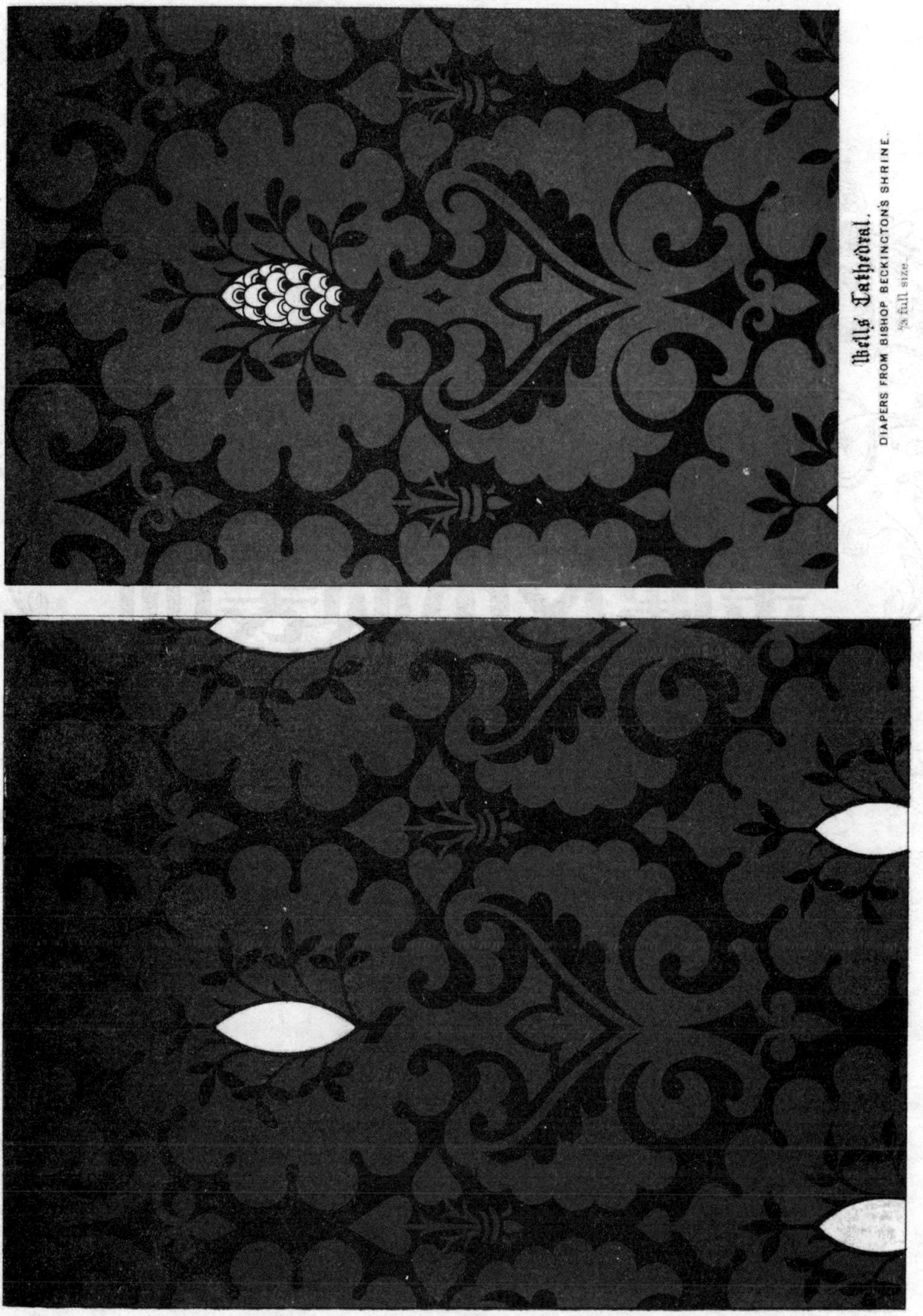

威尔斯教堂贝金顿主教圣堂内花纹装饰

第22章

金属及木质装饰

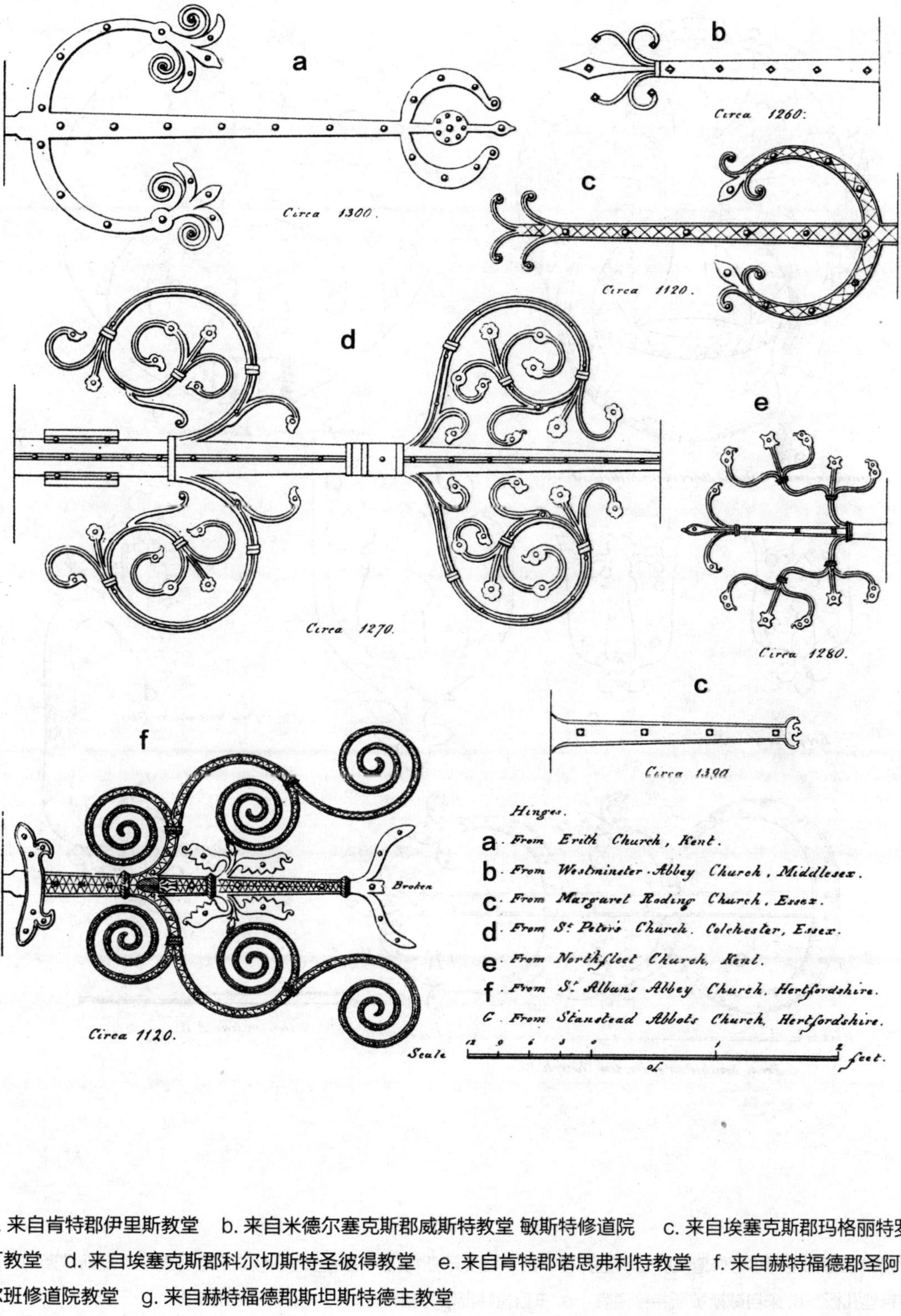

a. 来自肯特郡伊里斯教堂　b. 来自米德尔塞克斯郡威斯特教堂 敏斯特修道院　c. 来自埃塞克斯郡玛格丽特罗丁教堂　d. 来自埃塞克斯郡科尔切斯特圣彼得教堂　e. 来自肯特郡诺思弗利特教堂　f. 来自赫特福德郡圣阿尔班修道院教堂　g. 来自赫特福德郡斯坦斯特德主教堂

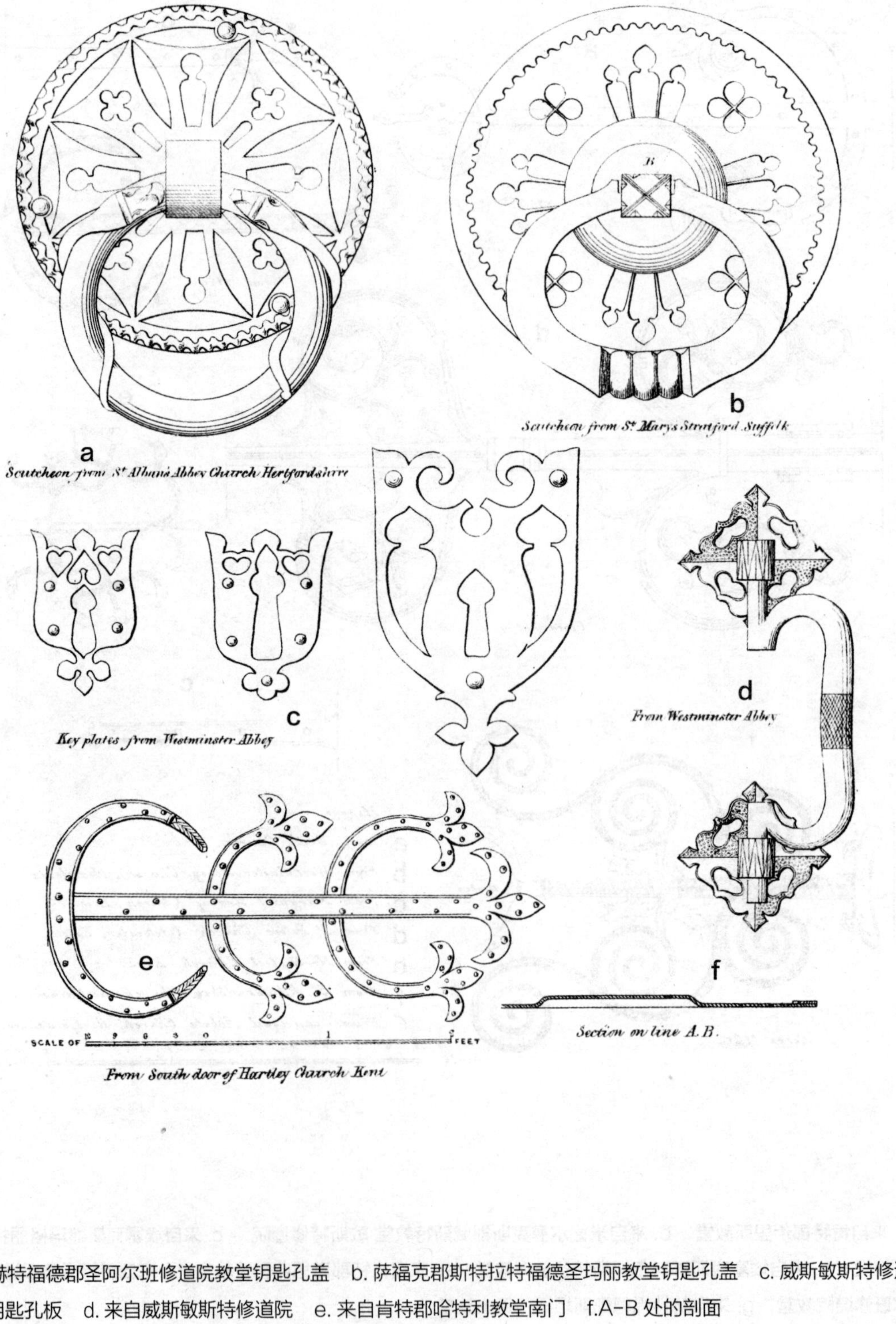

a. 赫特福德郡圣阿尔班修道院教堂钥匙孔盖　b. 萨福克郡斯特拉特福德圣玛丽教堂钥匙孔盖　c. 威斯敏斯特修道院钥匙孔板　d. 来自威斯敏斯特修道院　e. 来自肯特郡哈特利教堂南门　f.A-B 处的剖面

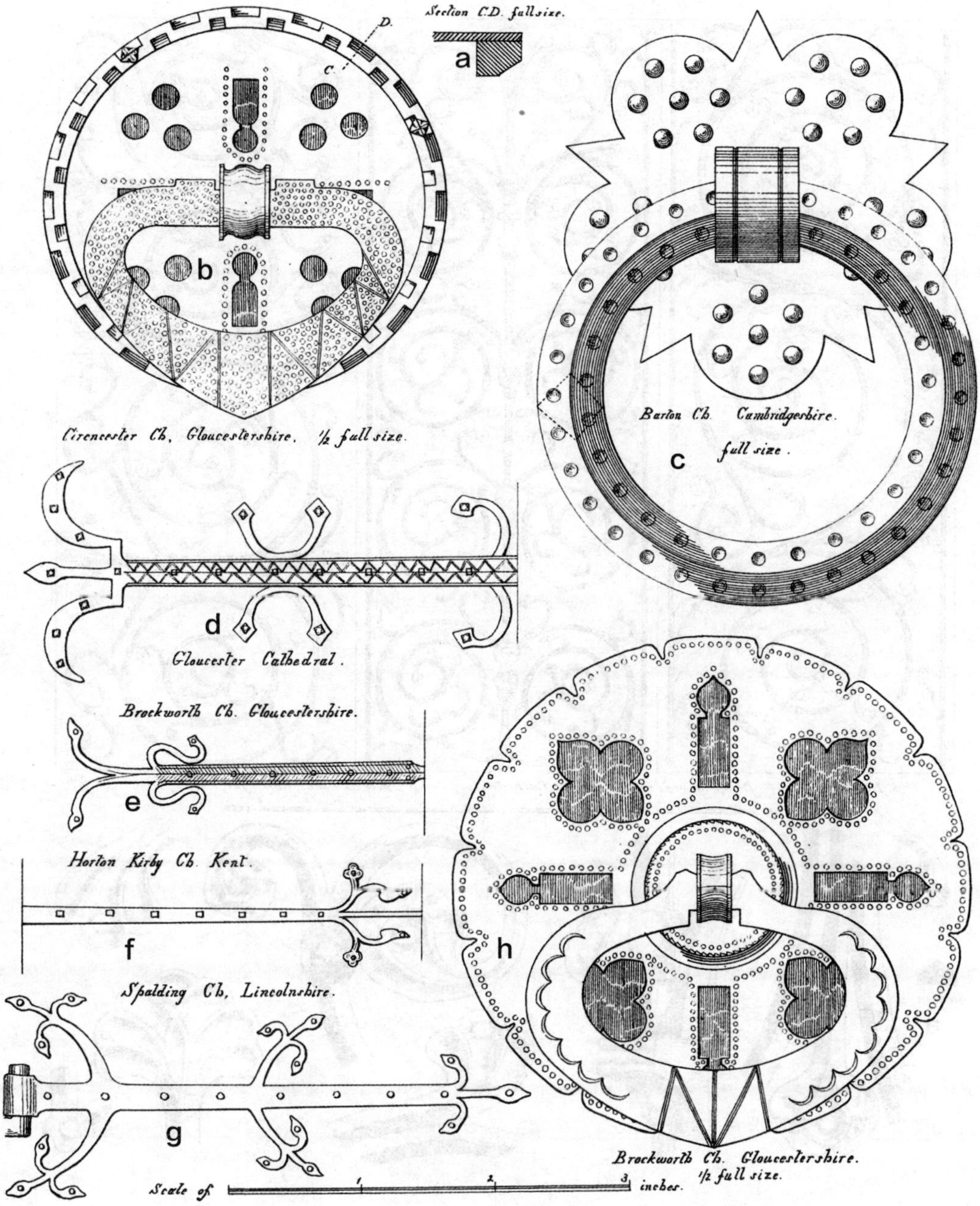

a.C–D 处的剖面　b. 格洛斯特郡赛伦塞斯特教堂　c. 剑桥郡巴顿教堂　d. 格洛斯特教堂　e. 格洛斯特郡布罗克沃斯教堂　f. 肯特郡霍顿卡比教堂　g. 林肯郡斯伯丁教堂　h. 格洛斯特郡布罗克沃斯教堂

切斯特教堂门上铁制卷饰及细部：a. 剖面图

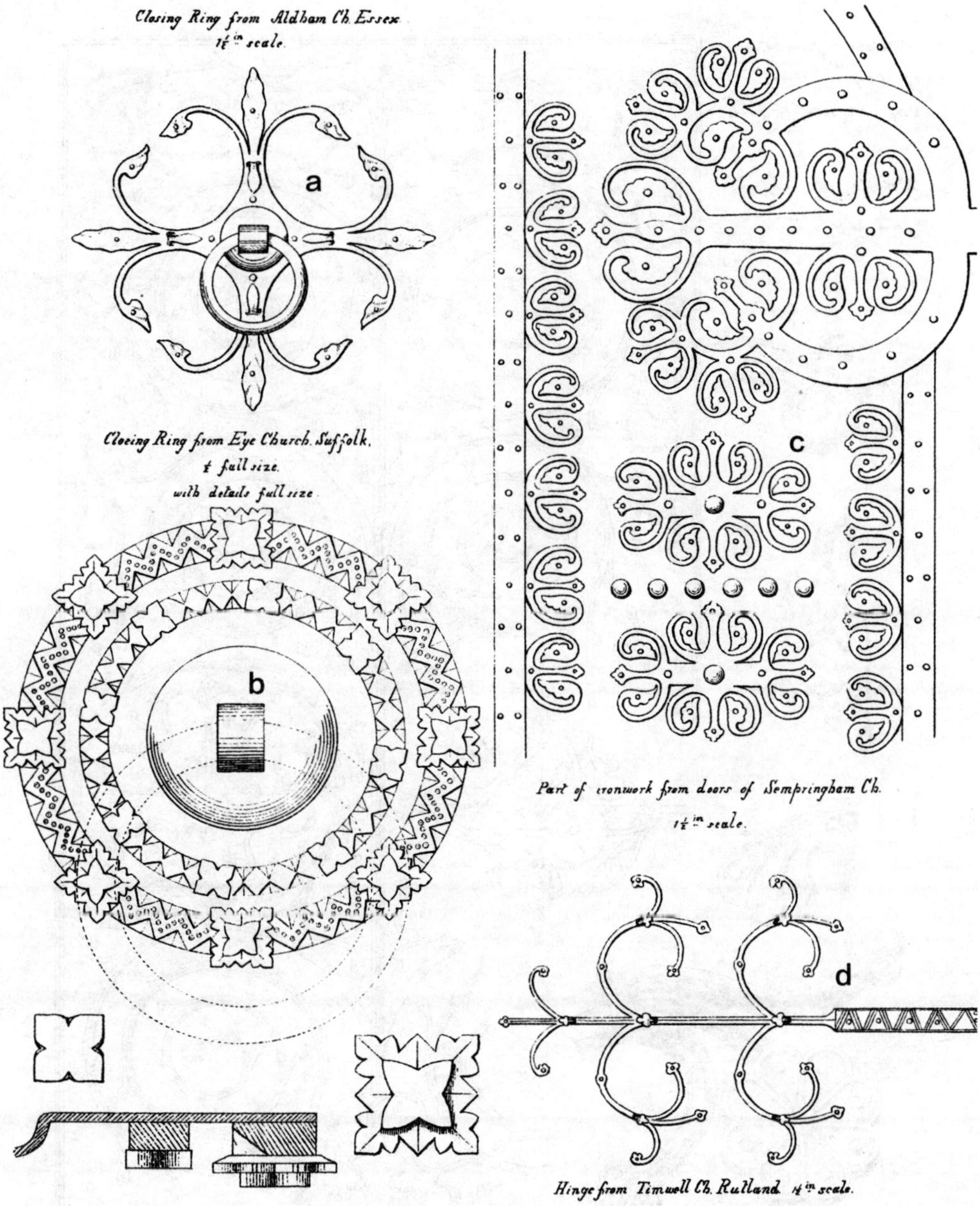

a. 埃塞克斯郡奥尔德姆教堂门环　b. 萨福克郡艾教堂门环　c. 瑟布淩汉教堂门上铁饰局部　d. 拉特兰郡蒂姆维尔教堂铰链

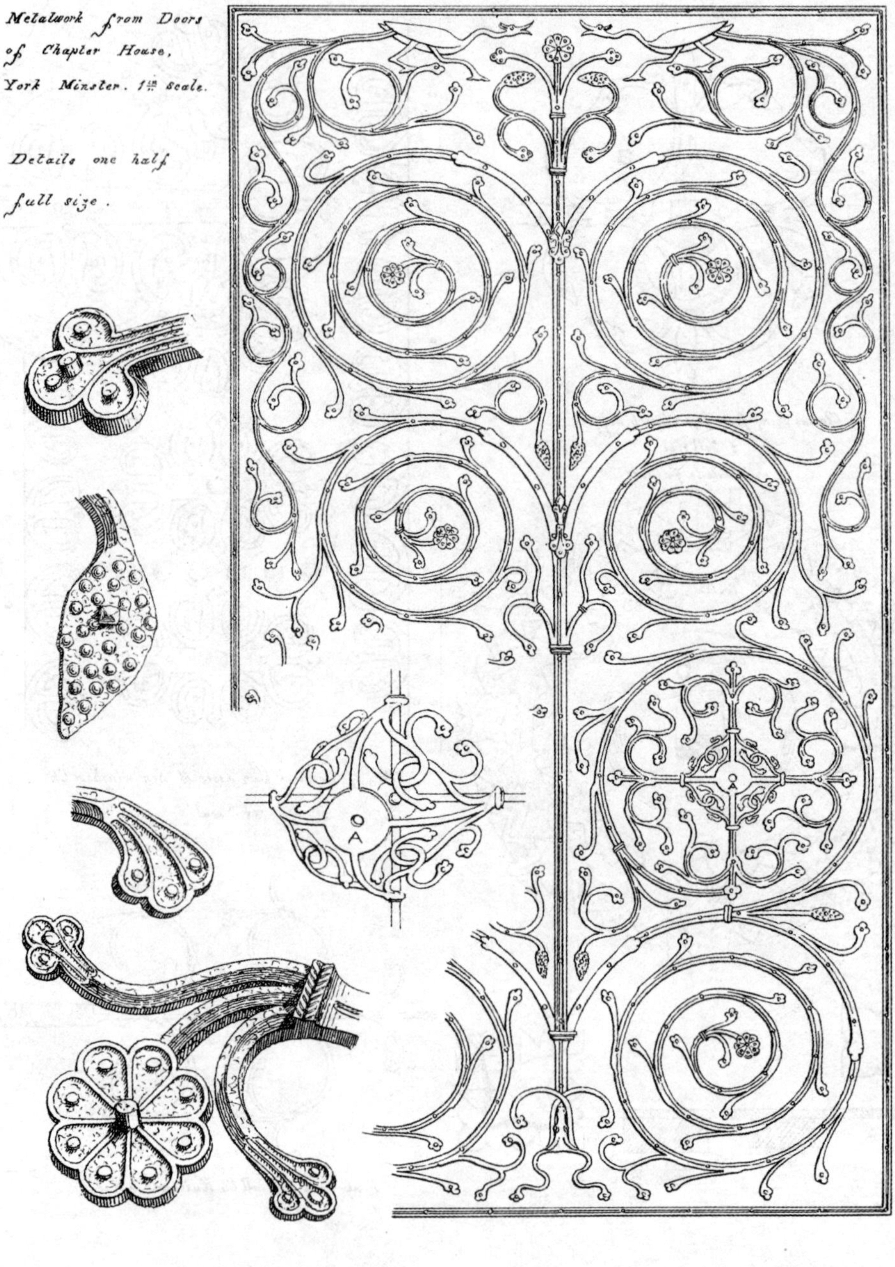

约克大教堂牧师会礼堂大门上金属饰

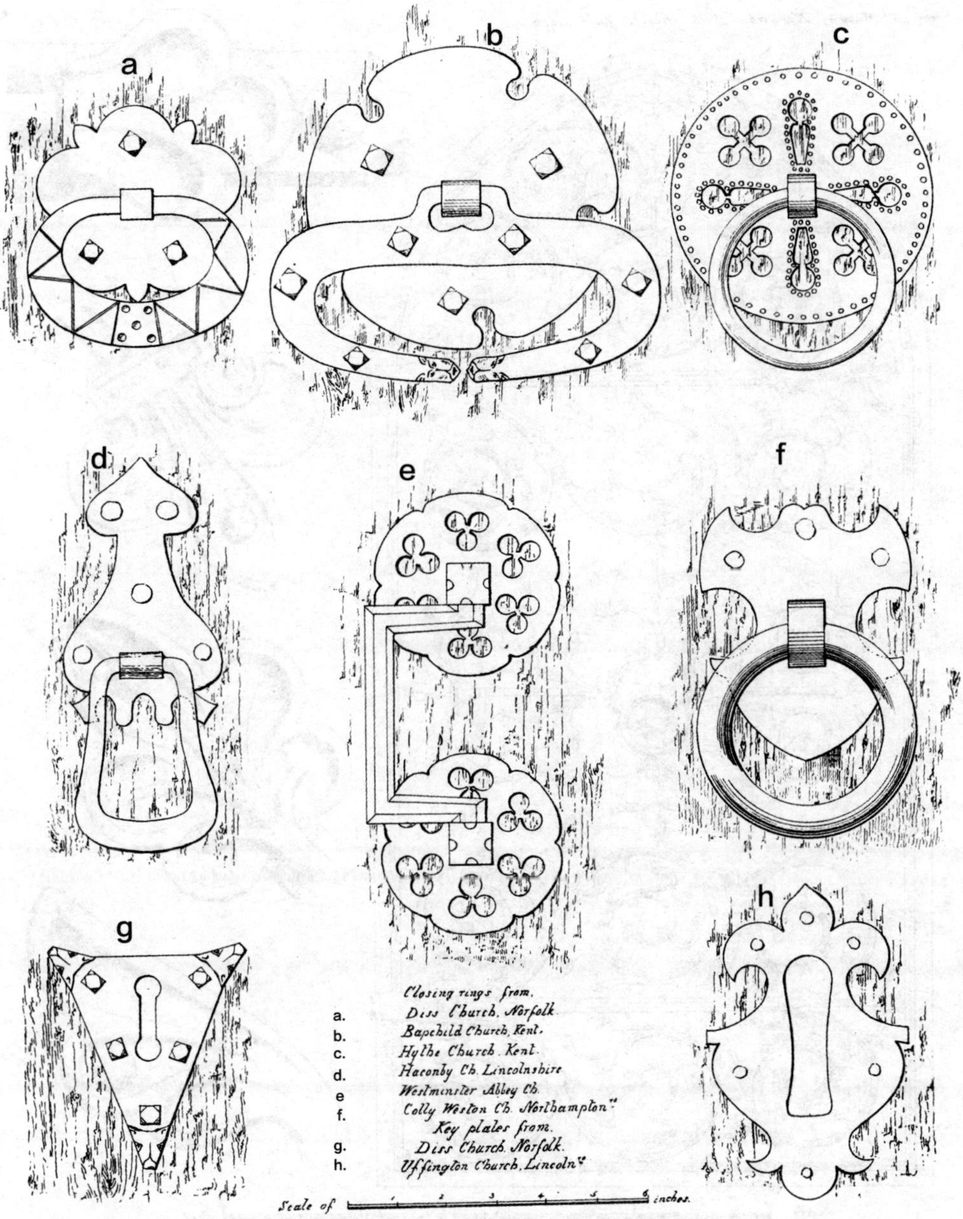

门环来自：a. 诺福克郡迪丝教堂　b. 肯特郡巴普蔡尔德教堂　c. 肯特郡海斯教堂　d. 林肯郡哈肯比教堂　e. 威斯敏斯特修道院　f. 钥匙孔板北安普敦郡科利韦斯顿教堂　g. 诺福克郡迪丝教堂　h. 林肯郡阿芬顿教堂

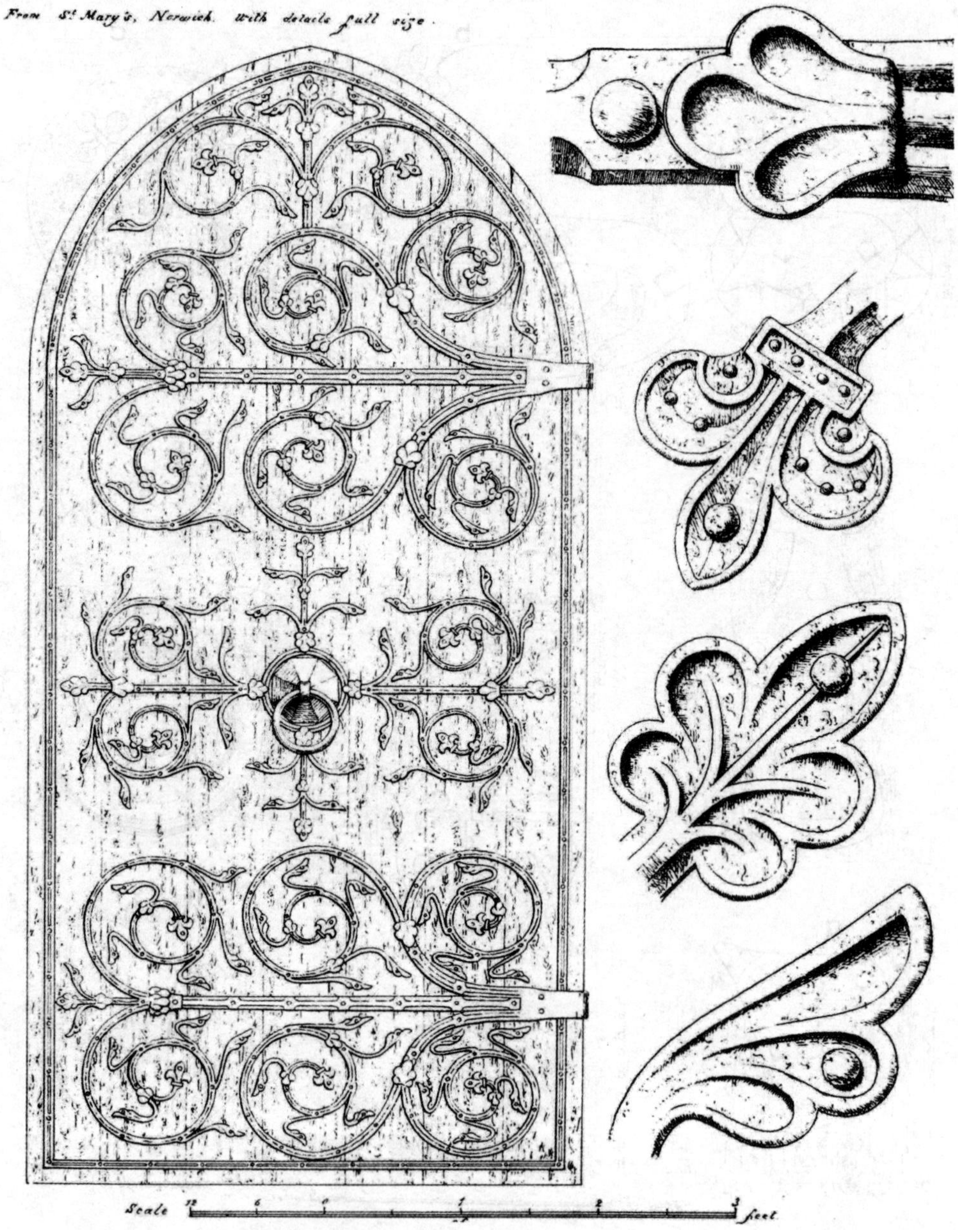
From S.t Mary's, Norwich. with details full size.
Scale
12
6
0
1
2
3
feet

来自诺威奇圣玛丽教堂

第 22 章 金属及木质装饰

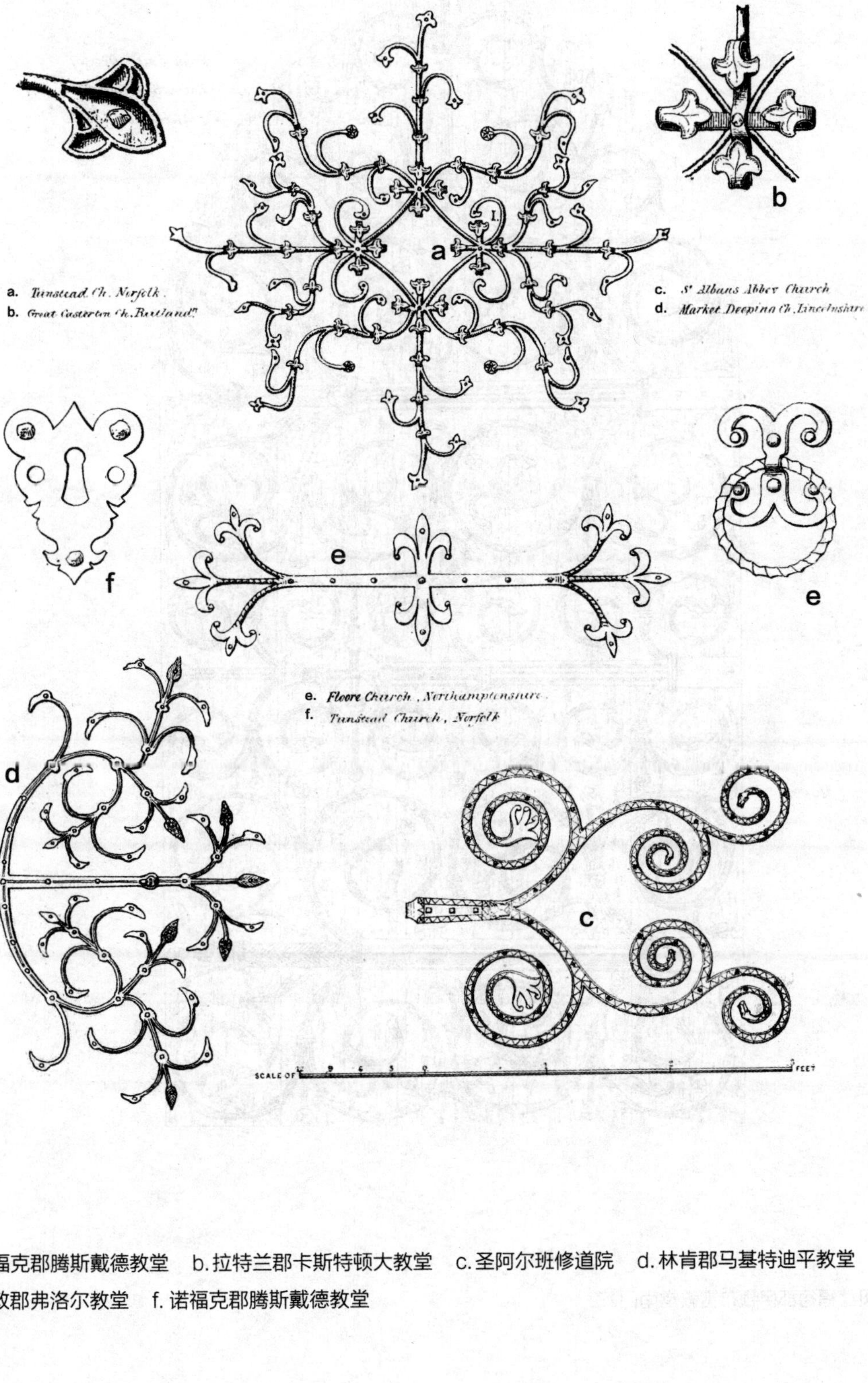

a. 诺福克郡腾斯戴德教堂　b. 拉特兰郡卡斯特顿大教堂　c. 圣阿尔班修道院　d. 林肯郡马基特迪平教堂　e. 北安普敦郡弗洛尔教堂　f. 诺福克郡腾斯戴德教堂

From South Door of
Eaton Bray Church,
Bedfordshire.

来自贝德福德郡伊顿布雷教堂南门

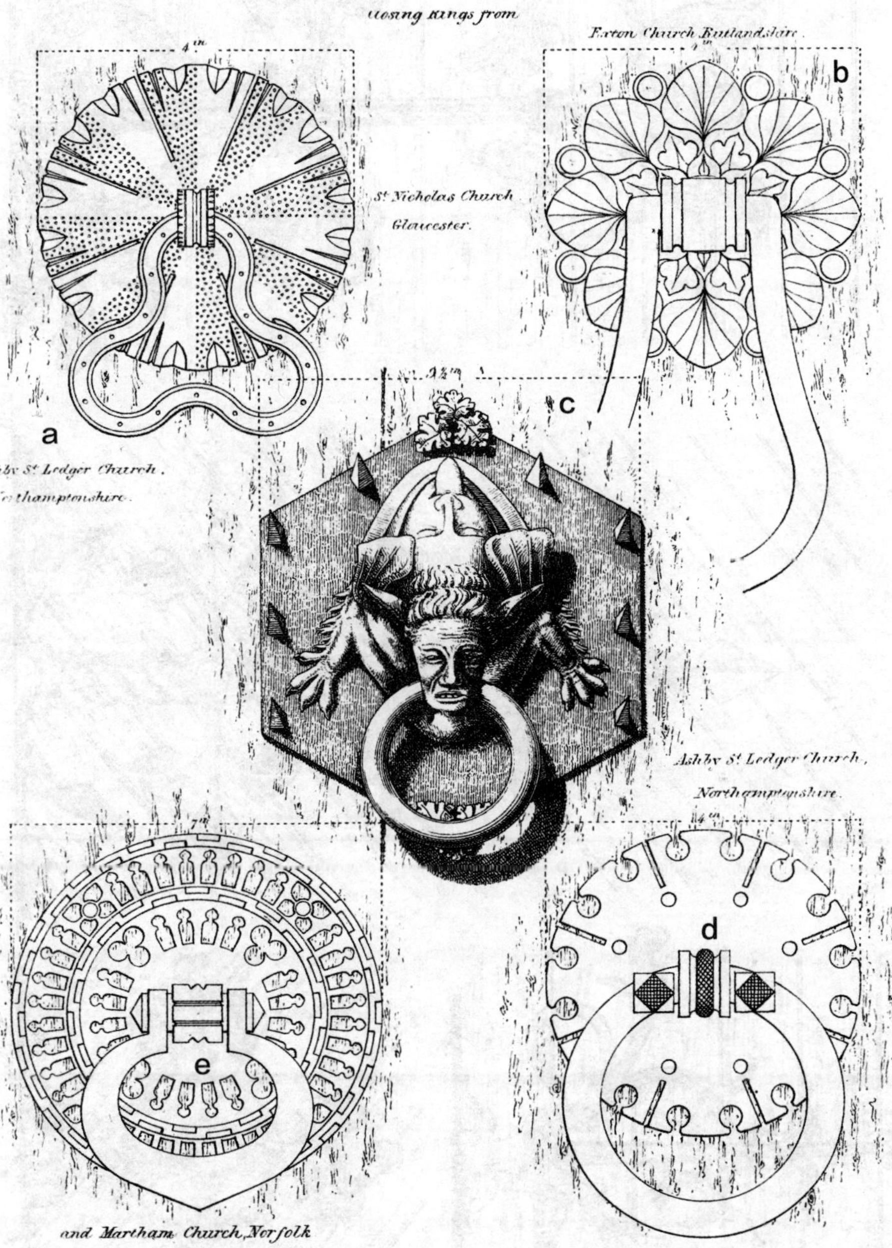

门环来自：a. 北安普敦郡阿什比圣莱杰教堂 b. 拉特兰郡菲尔顿教堂 c. 格洛斯特郡圣尼古拉斯教堂 d. 拉特兰郡埃克斯顿教堂 e. 诺福克郡马瑟姆教堂

金属装饰

金属装饰

金属装饰

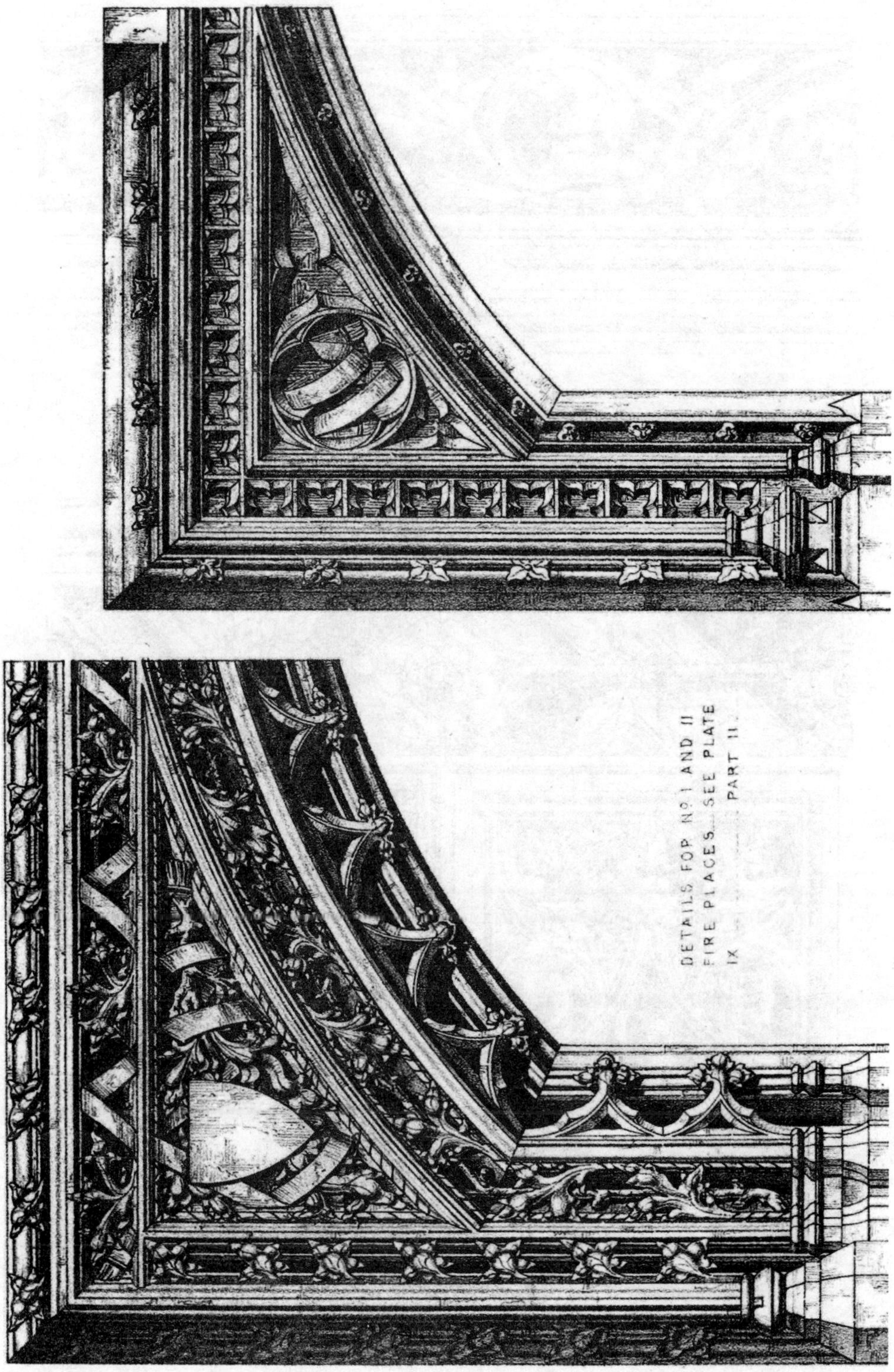
DETAILS FOP Nº I AND II
FIRE PLACES... SEE PLATE
IX ... PART II.

金属装饰

金属及木质装饰

金属及木质装饰

金属及木质装饰

金属及木质装饰

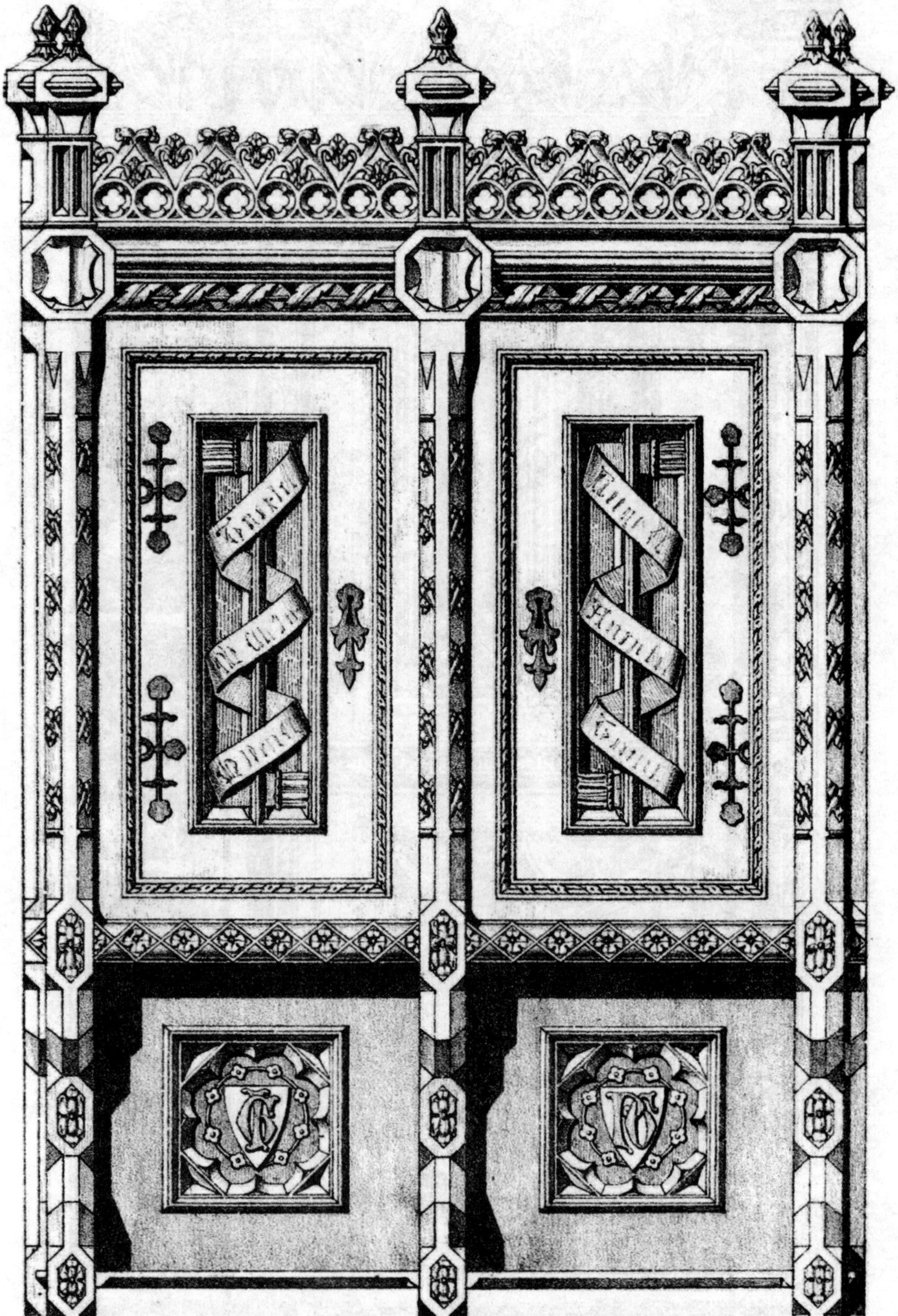

金属及木质装饰

金属及木质装饰

金属及木质装饰

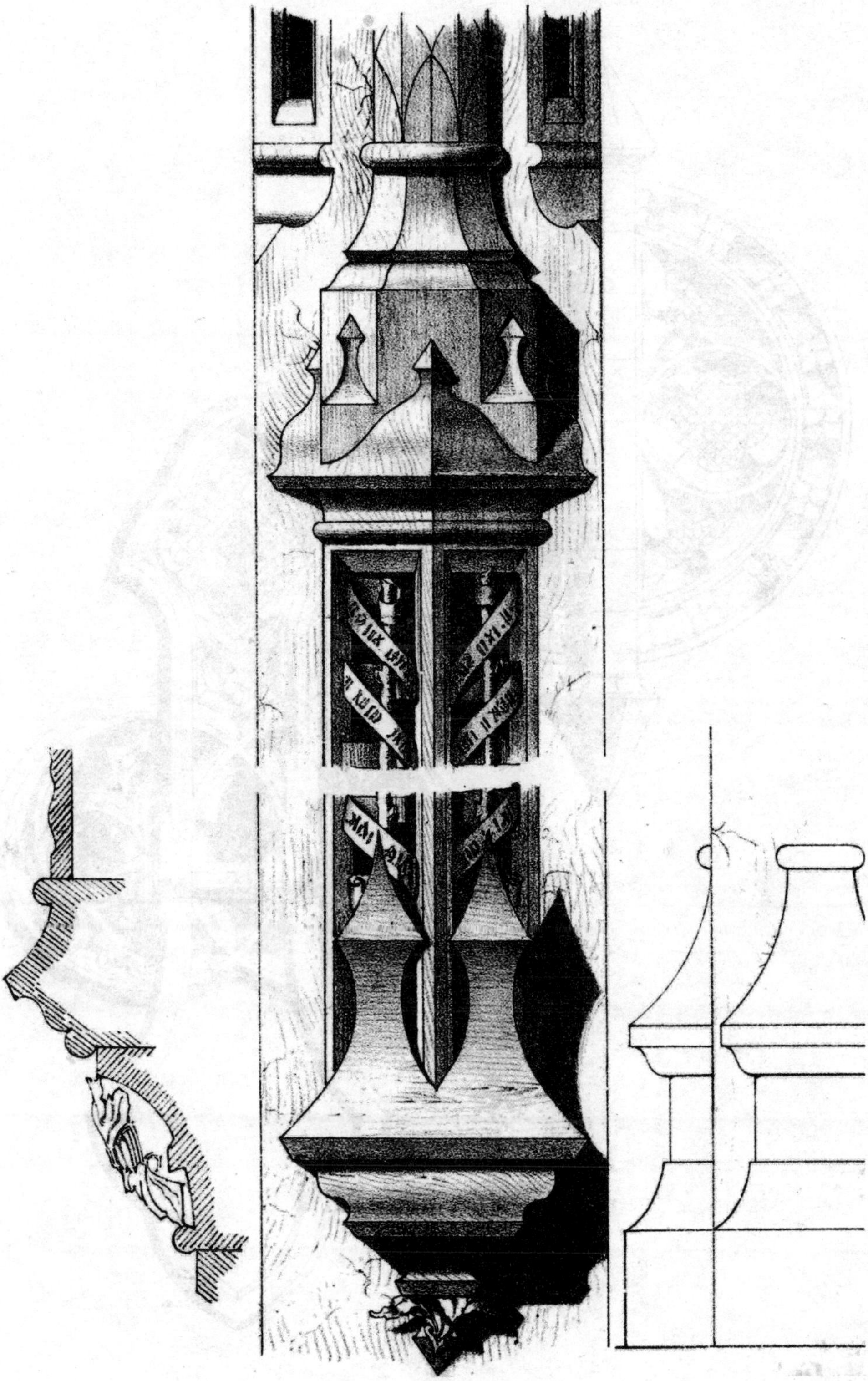

金属及木质装饰

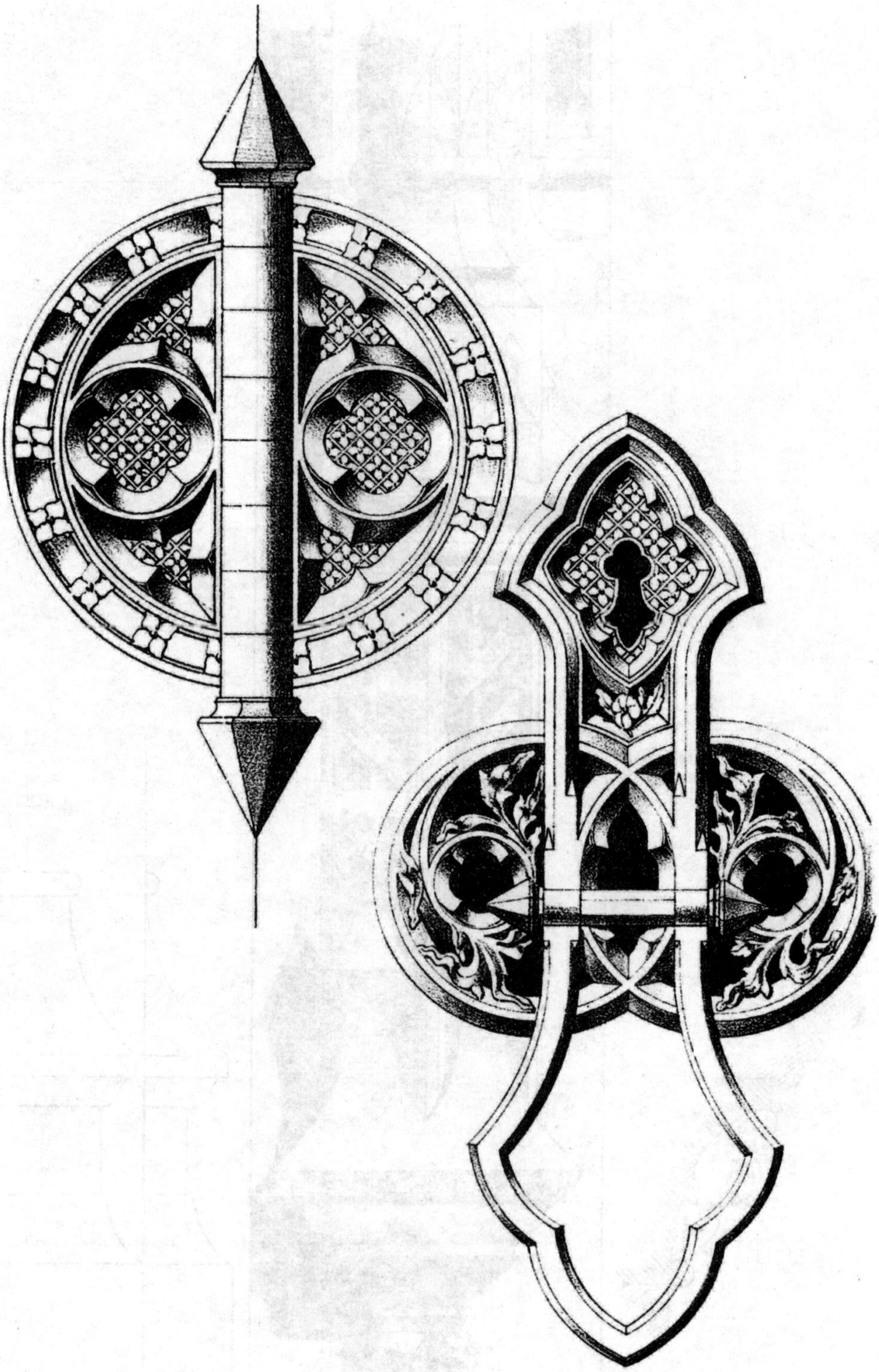

金属装饰

PEACE
be within thy
walls and
PLENTEOUSNESS
within thy
PALACES

金属装饰

金属装饰

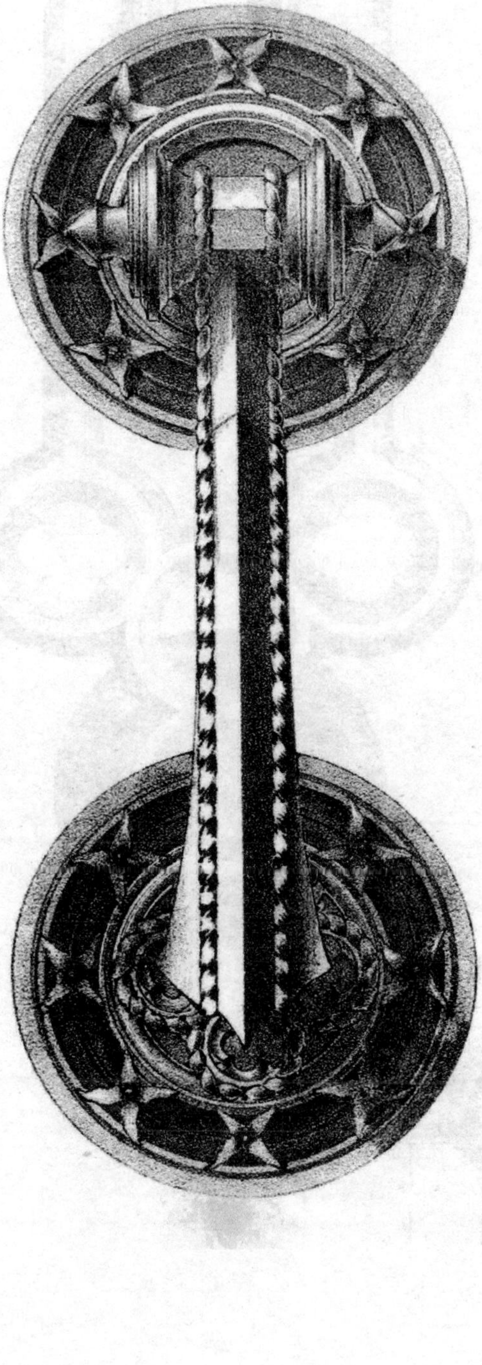

金属装饰

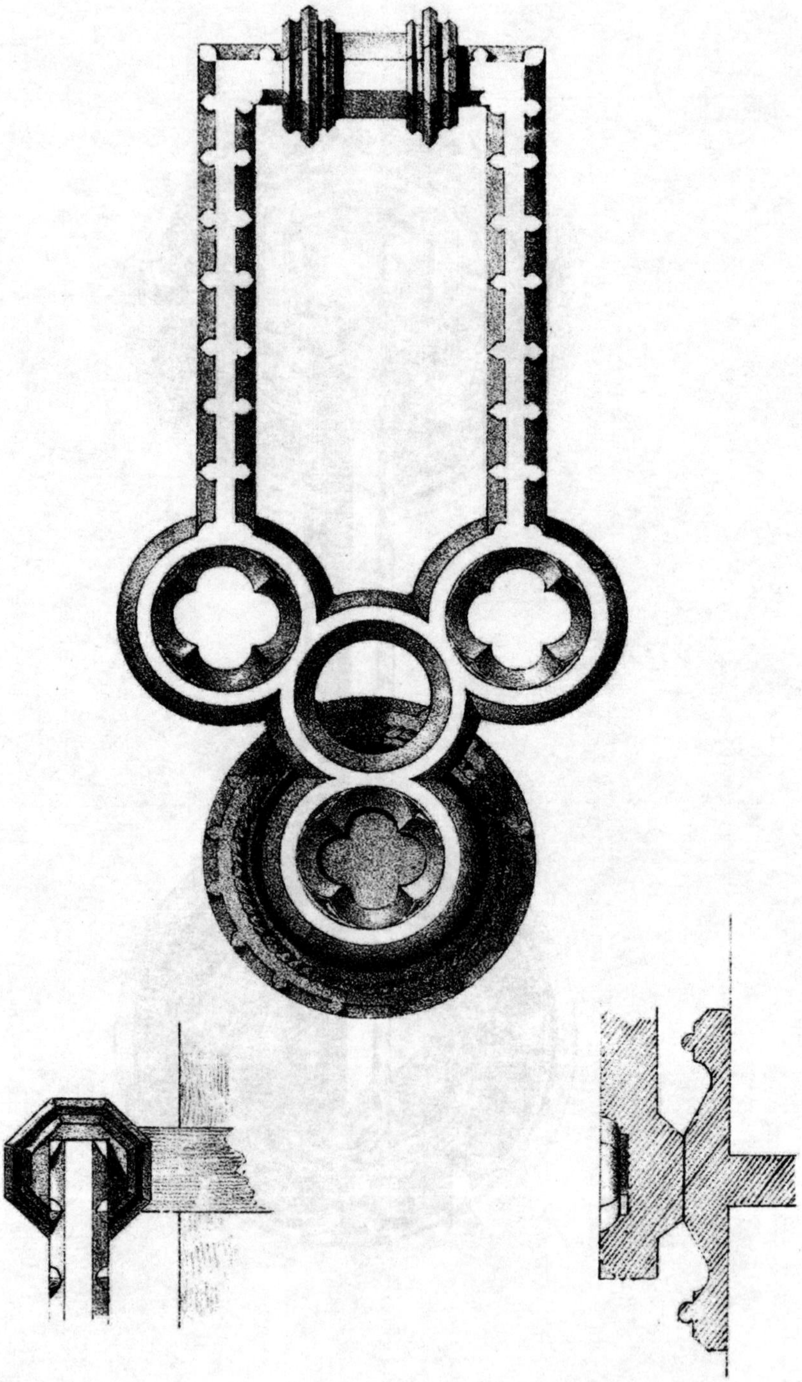

金属装饰

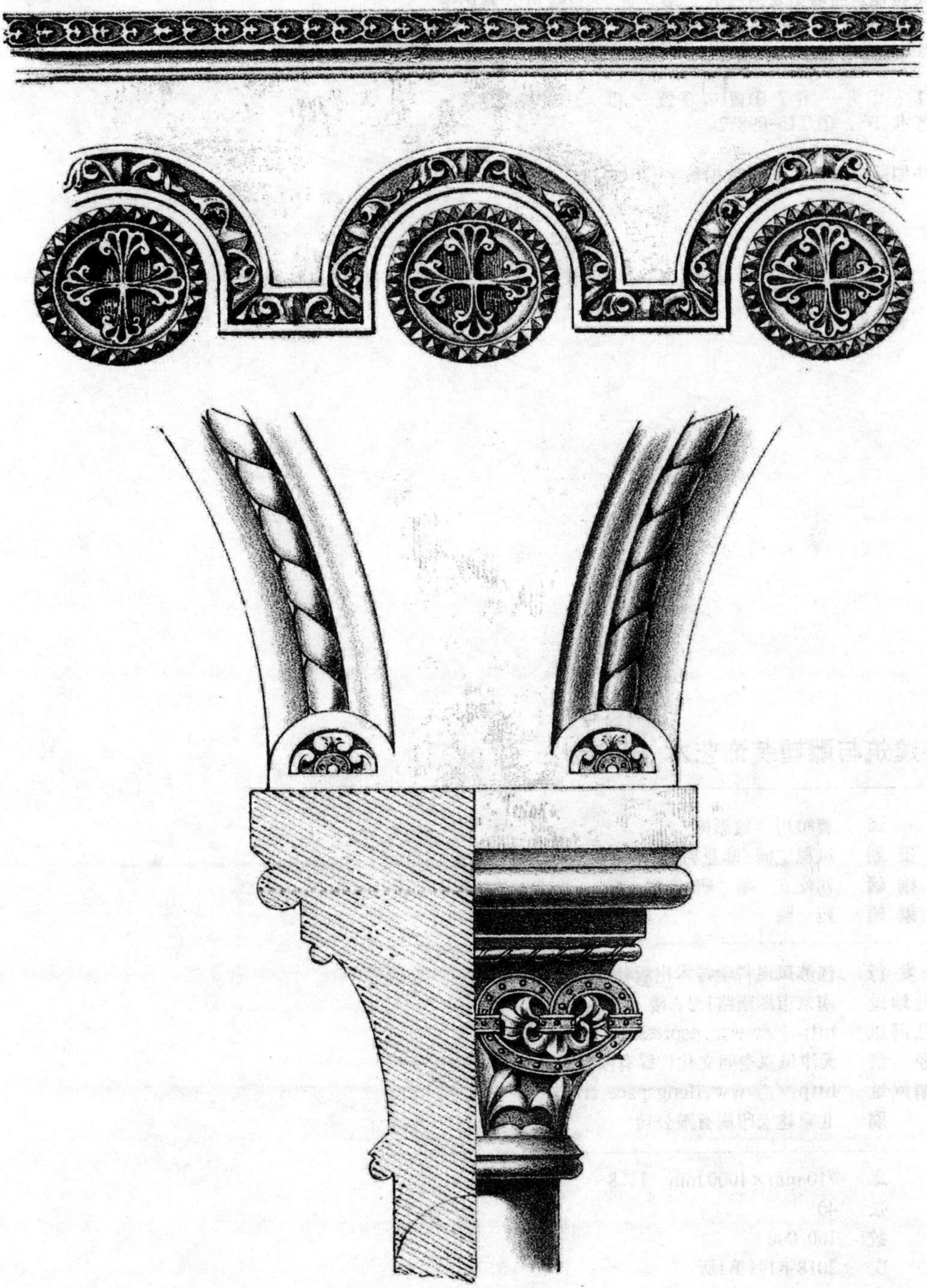

金属装饰

图书在版编目（CIP）数据

哥特建筑与雕塑装饰艺术．第3卷／曹峻川，甄影博编．-- 南京：江苏凤凰科学技术出版社，2018.1
ISBN 978-7-5537-8757-2

Ⅰ．①哥… Ⅱ．①曹… ②甄… Ⅲ．①哥特式建筑－建筑艺术 Ⅳ．①TU-098.2

中国版本图书馆CIP数据核字(2017)第292849号

哥特建筑与雕塑装饰艺术　第3卷

编　　译　曹峻川　甄影博
项目策划　凤凰空间/郑亚男
责任编辑　刘屹立　赵　研
特约编辑　苑　圆

出版发行　江苏凤凰科学技术出版社
出版社地址　南京市湖南路1号A楼　邮编：210009
出版社网址　http://www.pspress.cn
总　经　销　天津凤凰空间文化传媒有限公司
总经销网址　http://www.ifengspace.cn
印　　刷　北京建宏印刷有限公司

开　　本　710 mm×1000 mm　1/8
印　　张　40
字　　数　160 000
版　　次　2018年1月第1版
印　　次　2023年3月第2次印刷

标准书号　ISBN 978-7-5537-8757-2
定　　价　188.00元